The Amygdaloid Nuclear Complex

Vincent Di Marino
Yves Etienne • Maurice Niddam

The Amygdaloid Nuclear Complex

Anatomic Study of the Human Amygdala

Springer

Vincent Di Marino
Faculté de médecine
Marseille
France

Yves Etienne
Unité de médecine légale
Hôpital de la Timone
Marseille
France

Maurice Niddam
Unité SAMU 13, Centre 15
Hôpital de la Timone
Marseille
France

ISBN 978-3-319-79461-7 ISBN 978-3-319-23243-0 (eBook)
DOI 10.1007/978-3-319-23243-0

Springer Cham Heidelberg New York Dordrecht London

Softcover reprint of the hardcover 1st edition 2016

Printed on acid-free paper

Springer International Publishing AG Switzerland is part of Springer Science+Business Media (www.springer.com)

To our beloved wives, our dear children and our adorable grandchildren

To all the neuroscientists and neuroanatomists of the world

Authors

Vincent Di Marino is emeritus professor of anatomy at the Aix Marseille University (AMU), Faculty of Medicine; former director of the Kidney Transplantation Center of the Sainte-Marguerite Hospital; and former director of the Laboratory of Anatomy at the Faculty of Medicine, Marseille, France. Currently, his research topics are focused on the central nervous system and the pelvis.

Faculté de Médecine de Marseille
27 Boulevard Jean Moulin
13385 Marseille Cedex 5
France
(vincent.di-marino@univ-amu.fr).

Yves Etienne is forensic scientist, doctor of medicine, former assistant of anatomy, former house physician in the hospitals of sanitary region and consulting practitioner in the Unit of Forensic Medicine in the Timone Hospital (Marseille, France).

Unité de Médecine Légale
Hôpital de la Timone
264 Rue Saint-Pierre
13385 Marseille Cedex 5

Maurice Niddam is forensic scientist, doctor of medicine, former assistant of anatomy, coordinating doctor of a medical center and regulating doctor of the Emergency Medical Service (SAMU 13, Center 15) in the Timone Hospital (Marseille, France).

Unité SAMU 13, Centre 15
Hôpital de la Timone
264 Rue Saint Pierre
13385 Marseille Cedex 5

Foreword

The amygdala is a small distinct brain structure, located in the anteromedial part of the temporal lobe. Variably designated as a "corpus", a "complex" or a "nucleus", depending on the degree of importance its particular structure has inspired, this, admittedly small quantity of brain tissue, is made up of islets of gray matter and weighs 1 g. It is a basal brain nucleus that came to "glide" along the limbic cortex and appears to have perched on top of the hippocampus. In contrast to other structures comprising the limbic system, the functional identity of the amygdala is clear. It is the nodal point of our emotional life, it controls autonomic functions and their expression and plays a role in the processing of olfactory and auditory stimuli.

The physiology of the amygdala was initially studied in animals, mainly rats. Until the advent of functional imaging, there was no data available allowing to conclude that it had a similar function in humans, all the more since its size varies between species. The role that the amygdala plays in temporal epilepsy was discovered during the first studies with intracerebral electrodes, performed by Talairach and Bancaud on patients with seizures, in the context of operative management. From the outset, the amygdala was proven to be very closely linked to the hippocampus, both during the interictal state, as well as at the outset of epileptic activity. The bidirectional character of these connections has also become a marker of epileptogenic foci in cases of medial temporal epilepsy (studies of Buser and Bancaud in 1983). The spectrum of clinical signs related to the involvement of the amygdala was identified later. The group of Talairach and Bancaud first demonstrated the correlation with oro-alimentary automatisms, then Gloor (this particular neurophysiologist, working at the Neurologic Institute of Montreal, presented his thesis regarding the connections of the amygdala in 1957) demonstrated the association with experiences of "déjà vu" and paroxysmal memory recollections. Autonomic manifestations (heart rhythm acceleration, piloerection) and also emotional ones (anxiety, fear) have been clearly associated with paroxysmal discharges from the amygdala. Electrical stimulation of the amygdala results in reproduction of these same symptoms, especially "déjà vu" experiences. This last fact illustrates its importance in the mechanisms of memory, in combination with the anterior hippocampus and the entorhinal cortex, especially with regard to emotional memory or the influence of emotions in memory encoding.

Functional imaging has majestically demonstrated the role of the amygdala in encoding negative emotions and especially its specificity in deciphering the emotional expression of the human face (Dolan). The team of Joseph Ledoux, having elucidated the functional circuitry of fear and its storage in memory (consolidation and extinction of fearful memories, studies by the Quirk team) in the rat, was also the one to demonstrate in humans, by means of functional MRI, the interactions between the median prefrontal cortex of the cingulate gyrus and the amygdala in regulating the response to a menace. All actual studies in the area of "affective" neurosciences report activation of the amygdala, which has become the cerebral "center of fear". However, just as all sorts of amygdalar stimulations do not result in emotional modifications, all patients with Urbach-Wiethe disease (a genetic disease characterized by calcifications of the amygdala bilaterally) do not manifest with behavioral alterations, nor do they exhibit a much different judgement as regards emotionally charged facial expressions, compared to a control population (Siebert et al. 2003). These observations underline a certain oversimplification characteristic of actual science, at least regarding media announcements resulting, at the end, in affecting scientific approach.

On the opposite side of this oversimplification stands the work of Vincent di Marino; this is why it was absolutely necessary. This work appears to "hit the spot". It opens new horizons to clinical application and research. By means of unique and extraordinary pictures, it perfectly illustrates the anatomic organization of the brain. First evidence of this is the demonstration of the multiplicity of nuclei comprising the amygdala, multiplicity clearly taken into consideration in experimental research but neglected in human imaging studies, due to inadequate spatial resolution. Second evidence is the demonstration of incorporation of the amygdala in the many systems it interconnects. The most apparent is the limbic system, but relations of the amygdala with the neocortex, especially frontal and insular cortex, have just as much significance. Finally, this work is a magnificent demonstration of anatomic evidence provided by the meticulous and rigorous study of macroscopic anatomy. This anatomic evidence transmits an implicit functional meaning. The most striking example is provided by the very clear images supporting the resurrected concept of the "extended amygdala", taking the form of multiple nuclei forming an extension of the central and medial nuclei towards the sublenticular area of the hypothalamus.

Vincent di Marino, whose select sections and original dissections strongly suggest that certain contiguities should be really interpreted as continuities, is also proposing to reconsider the concept of the ventral striato-pallidum. It is clear that the physiology and the pathologic physiology of the relations between the two "tracks" of amygdalar connections are coupled with the obvious anatomic substrate in order to explain the interactions between internal milieu, memory, sensory apparatus and emotion, across this "nuclear cobblestone" between the hypothalamus, the bed nucleus of the stria terminalis, the lenticular nucleus, the septum and the hippocampus.

This immense work of Vincent di Marino and his colleagues has certainly managed to provide us with an inestimable amount of documents of great beauty, but also with a great formal clarity regarding this little heap of brain nuclei, of which so much is talked about these days. Mainly, by means of his intelligent dissection, he has really presented the functional properties of the amygdala, demonstrating its interconnections, specific as they may be. He is proposing to us the study of the circuit of the amygdala, highlighting its unique pivotal place between the diencephalon on the one side, and the paleocortex, the archicortex and the neocortex on the other.

August 9th, 2015 Patrick Chauvel[1]

[1]Patrick Chauvel, Emeritus Professor, Neurophysiologist Clinician, Director of the Institute of Neurosciences of the Systems (Aix-Marseille University), is a specialist in the field of Epilepsy. He practices at present in the USA (Cleveland Clinic, Epilepsy Center).

Preface

De toutes les merveilles de la nature, la plus étonnante, c'est l'homme – Sophocle[1]

Among all of the brain structures that we have re-examined, dissected and photographed during the preparation of our atlas of the central nervous system (Di Marino et al, 2011), it is the cerebral amygdala that has caught our attention the most – not only due to its multiple connections and to the eminent role that it has been acknowledged to play but also and especially due to some of its anatomical characteristics which, we believe, deserve to be studied in further detail. It is on the basis of these observations that the idea of this book project was born.

Very quickly, as from the start of our project, we were challenged by the incredible amount of neuro-scientific publications concerning this formation and by the rarity, not to say the current poverty, of available and relevant neuroanatomical reference material, especially in relation to humans. We were then stunned by the inaccuracy of certain representative diagrams and by the lax nature of the various representations made of its connections. Our surprise was even greater when we observed the discrepancies existing between the descriptions provided by certain authors (a prime example being the connection between the amygdala and tail of the caudate nucleus!). Even more regrettable still were certain articles written about tractography, a remarkable investigation technique for studying the white fibres connected to the amygdala, which appeared to be filled with interpretation errors due to the fact that their authors only accounted for MRI images without making any reference to the actual structures such as they appear when viewed directly, before or after dissection.

Our work was mainly conducted at a macroscopic level with the photographs we took at each dissection stage. All sections were treated with our $FeCl_3$ staining technique (Di Marino et al. 2008) which provided an excellent in situ differentiation of the structures being examined by enhancing their colours. We also compared anatomical MRI images of previously dissected anatomical parts and used MRI studies performed on thick sections that were secondarily subjected to anatomical dissections. We also relied on the microscopic study of certain areas in which the human eye could not identify, without the use of a microscope, the connections of certain amygdaloid areas. We also used a technique for scanning and digitalising microscopic sections and therefore easily progressing from one point to another in amygdala-based preparations, in order to compare, for example, the neurons of the various amygdaloid nuclei.

We believe that the anatomical images that we have collected represent a basis which can be used by neuroscientists in a holistic approach for studying the connections of the amygdaloid complex and the numerous tracts, which can be observed at a macroscopic level and which connect it to other structures of the central nervous system.

We also hope that before extensively investigating an aspect of the amygdala, young researchers will read our book in order to benefit from an overall anatomical view of the area in which they are going to work.

We would also like to reiterate for all students who want to become neurologists, neurosurgeons or neuroradiologists that anatomical dissections or at least their photographs, such as those presented in our book, exceed all diagrams, regardless of their quality, as they offer a direct and irreplaceable vision of the actual formations. It is for this purpose that this book, such as our previous atlas of the central nervous system, has been prepared with great care and with such students in mind.

May 14th, 2015, Marseille, France — V. Di Marino, Y. Etienne, M. Niddam

[1] "Of all the wonders of nature, the most surprising it is the man" *(Sophocles's Antigone).*

Acknowledgements

The authors thank Mrs Sylviane *CONFORT-GOUNY*, research engineer *CNRS* at *CRMBM*[1] and *CEMEREM*,[2] for her precious collaboration, her technical help in obtaining the MR images and her excellent and friendly welcome in her research center.

The authors thank especially, Dr Nadine *GIRARD*, professor chairman of neuroradiology at the Timone Hospital in Marseille, for her expertise, her assistance in the MRI observation of the anatomic slices and parts of dissection and for her remarkable and unforgettable sessions of tracking.

The authors also thank Dr André *MAUES DE PAULA*, University Lecturer (*MCU*), *AMU*,[3] for his advices in the study of the brain slices and the kindness of his reception in the Neuropathology Division of the Timone Hospital.

We greatly appreciate the remarkable expertise of Mr Philippe *MORANDO*, technician specialist in neuropathology who prepared for us, with extreme quality and meticulousness, the most beautiful slices of this book.

We deeply thank Pr Hubert *LEPIDI*, professor of histology at the *AMU* and pathologist in the Pathology Division of the Timone Hospital in Marseille.

We thank wholeheartedly the generous donors who gave their body to the Body Donation Department of AMU (Aix Marseille University) for advancing neuroscience.

[1]CRMBM, Centre de Résonance Magnétique Biologique et Médicale, UMR 7339, Faculté de Médecine, AMU-Université, Marseille, France.

[2]CEMEREM, Centre d'Exploration Métabolique Resonance Magnétique, AP-HM, CHU Timone, Pôle d'Imagerie, Marseille, France (Dir: Pr Maxime GUYE).

[3]AMU: Aix Marseille Université.

Contents

1 Definition and Generalities

1.1 Definition of the "Amygdaloid Body"

The amygdaloid body or amygdaloid nuclear complex is an aggregate of several grey matter nuclei, of diencephalic origin, located in the anterior and medial parts of each of the two temporal lobes, connected not only to the entire limbic system but also to the sensory cortical areas, which receive and process the emotional inputs by triggering, via the thalamus, hypothalamus and brainstem, a series of reactions referred to as emotional expressions, dynamic, neuroendocrine, vegetative and hormonal, of variable intensity according to each individual and to the significance and violence of the triggering emotional factor. This amygdaloid body is controlled by the frontal brain. However, this control can bypassed at a specific moment, in cases of extreme urgency, especially when the survival of an individual is involved.

The name of this formation refers to its shape and comes from the Latin root "amygdala", which itself is derived from a Greek term "amugdalè". As we have observed, it is Burdach (1819) who was the first to mention the existence of this formation and he gave it this name. Many variations have been linked to this morphological root from "amygdalum" or "cerebral amygdala" to "amygdala body", "amygdaloid body" or "corpus amygdaloïdeum" as well as "amygdala nucleus" and "amygdala nuclei" and then "amygdala complex". The term "amygdaloid body" has been chosen by the Human Anatomic Index. Numerous Anglo-Saxons prefer the term "amygdaloid complex" or even better "amygdaloid nuclear complex". This terminology is most certainly not the most simple. However, it appropriately defines the plurinuclear structure of the amygdala and its morphological similarity to an almond.[1]

[1]In this book, two terms will be used invariably: "amygdaloid nuclear complex" (Amn) or "amygdala". However, the last term will be favoured so as to lighten the text.

1.2 General Location of the "Amygdaloid Body"

The amygdaloid body is located in the posterior half of the rostral portion of the temporal lobe, at the dorsal side of its medial part. It caps the first digitations of the head of the hippocampus from which it is separated by the anterior recess of the inferior horn of the lateral ventricle.

Its cortical projection area occupies the top portion of the medial surface at the anterior end of the temporal lobe, directly behind the anterior part of the rhinal sulcus which separates it from the temporal pole, which is even more rostral. This projection area remains clearly above the collateral sulcus (which extends from the rhinal groove).

The gyri corresponding to the amygdaloid projection area (Fig. 1.1) therefore include the gyrus piriformis, the gyrus ambiens of Retzius, the rostral part of the uncinate gyrus (i.e. the part which is at the front of the uncal band of Giacomini), and the top third of the entorhinal area. In order to correctly view the cortical area defined in this manner, the medial surface of a hemisphere is to be observed. It corresponds to an anatomical portion obtained by median sagittal section of the brain (section passing in the longitudinal fissure of the cerebrum). It is also by observing the above-mentioned portion that the extension of the limbic lobe, and even better that of the limbic system to which the amygdala is connected, can be assessed.

V. Di Marino et al., *The Amygdaloid Nuclear Complex: Anatomic Study of the Human Amygdala*,
DOI 10.1007/978-3-319-23243-0_1

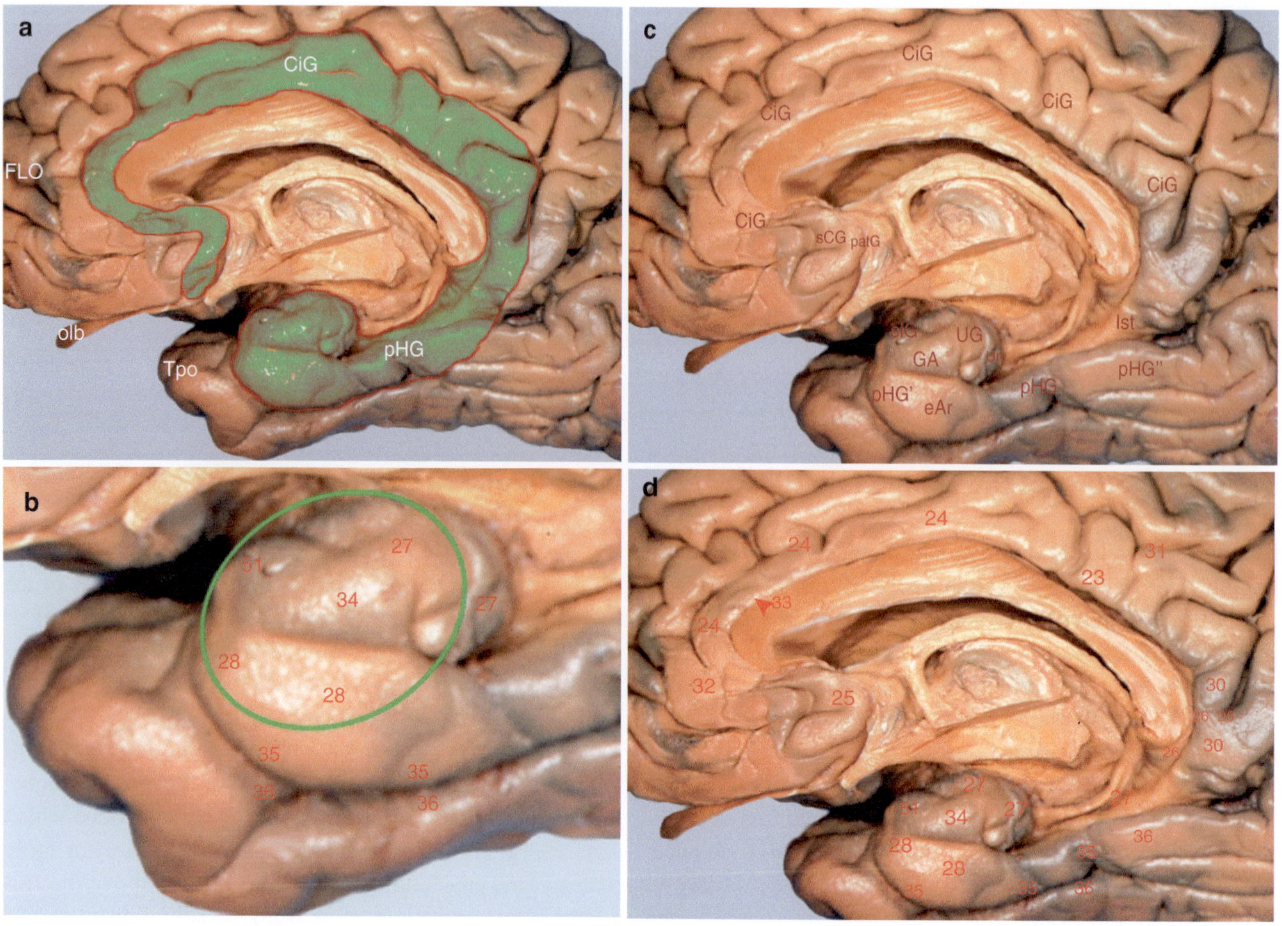

Fig. 1.1 Localisation of the amygdaloid nuclear complex in the limbic lobe. (**a**) Limbic lobe (*green coloured*) on a medial view of the right hemisphere. (**b**) Amygdala's projections (*green circle*) on the temporal areas, according to Brodmann. (**c**) The gyri of the limbic lobe. (**d**) The Brodmann's areas corresponding to the limbic lobe. *CiG* cingulate gyrus, *eAr* entorhinal area, *GA* gyrus ambiens, *FLO* frontal lobe, *Ist* Isthmus, *olb* olfactory bulb, *patG* paraterminal gyrus, *pHG* parahippocampal gyrus, *pHG'* anterior part of pHG (piriform lobe), *pHG"* posterior, part of pHG, *sCG* subcallosal gyrus, *slG* semilunar gyrus, *Tpo* temporal pole, *UG* uncinate gyrus, *Un* uncus

1.3 Amygdala and Limbic Lobe

When Broca described the *large limbic lobe*, he only included, in its composition, the annular structure which surrounds the median inter-hemispheric commissures and the intermediate or diencephalic brain. This structure is essentially formed by the cingulate gyrus (referred to in the past as the pericallosal gyrus) and the parahippocampal gyrus (referred to in the past as the circonvolution of the hippocampus or fifth temporal gyrus).

Following the works performed by PD MacLean (1958a, b), the amygdala (Amn) was included among the components of a "limbic entity" and was therefore joined to the nearby hippocampus (Fig. 1.1). Moreover, the major significance of these two structures was revealed: once was dedicated to the memory and the other to the emotional component of our perceptions. As a result, it was considered that they were practically the sole structures to form the limbic lobe.

In this configuration of the limbic lobe (HM Duvernoy 1988), two concentric circles[2] are observable.

The first circle or large limbic ring is made of the limbic cortex (subiculum, prepiriform cortex, periamygdaloid cortex) and by the associative paralimbic cortex (cingulate gyrus, subcallosal gyrus, parahippocampal gyrus, isthmus). We group from now on, under the word "limbic complex", the amygdala ("basal ganglion") and the "cortex" of this first circle.

The second circle is included in the large limbic circle: it is the intralimbic gyrus. It essentially corresponds to the hippocampus and includes the following: the anterior part

[2]On the basis of the work completed by Papez, some authors consider that there is a third limbic annular structure included in the intralimbic gyrus. It is the annular structure of the fornix which is composed of the fimbria, the fornix, the mammillary body and the mammillothalamic tract (fasciculus of Vicq d'Azyr) to the dorsomedian nucleus of the thalamus.

of the uncus, the band of Giacomini, the apex of the uncus, the Ammon's horn, the dentate gyrus, the gyrus fasciolaris, the fasciola cinerea, the indusium griseum with the lateral and median longitudinal striae, the prehippocampal rudiment or precommissural hippocampus (located between the paraterminal gyrus which covers the septal nuclei and the subcallosal gyrus).

At present, considerable progress has been made in the study of emotions, and it has been shown that the limits of the enlarged limbic lobe (see A Lautin 2001), which go beyond the limits defined by Broca, are too narrow due to the fact that many other structures (cerebral and other types) are involved in our various emotions. We therefore preferred to replace the notion of limbic lobe by the notion of limbic system in which have been included all anatomical regions involved, whether directly or indirectly, in the emotional process, and whose perturbations could have a genetic origin (L Pezawas et al. 2005).

1.4 Amygdala and Limbic System

The *Limbic System* therefore includes:

The cortical structures of the limbic lobe (Fig. 1.2): cingulate gyrus, paraterminal gyrus, subcallosal gyrus, periamygdaloid gyrus, prepiriformis gyrus, subiculum and parahippocampal gyrus and isthmus.

All of the hippocampal structures: uncus and band of the uncus (Giacomini's band), Ammon's horn, dentate gyrus, gyrus fasciolaris and fasciola cinerea, subsplenial gyrus, indusium griseum, median and lateral longitudinal striae and prehippocampal rudiment.

The amygdala (amygdaloid nuclear complex) and the stria terminalis.

The entorhinal, anterior frontal and orbitofrontal cortices, connected to several of the previous structures.

Structures also belonging to the old rhinencephalon: bulb, olfactory tract, olfactory striae, olfactory tubercle and anterior perforated substance. Some authors connect the limen and the insula to these structures:

The septal nuclei, the diagonal band and its nucleus

The habenular nucleus and the stria medullaris

The thalamus and especially the anterior and medial-dorsal nuclei

The hypothalamus and especially the supra-optical hypothalamus and mammillary bodies

Some mesencephalic structures, the interpeduncular nucleus, the ventral region of the tegmentum and the periaqueductal grey matter.

The nuclei of the brainstem and especially the locus coeruleus, the parabrachial nucleus and the raphe nuclei, and this list is not exhaustive.

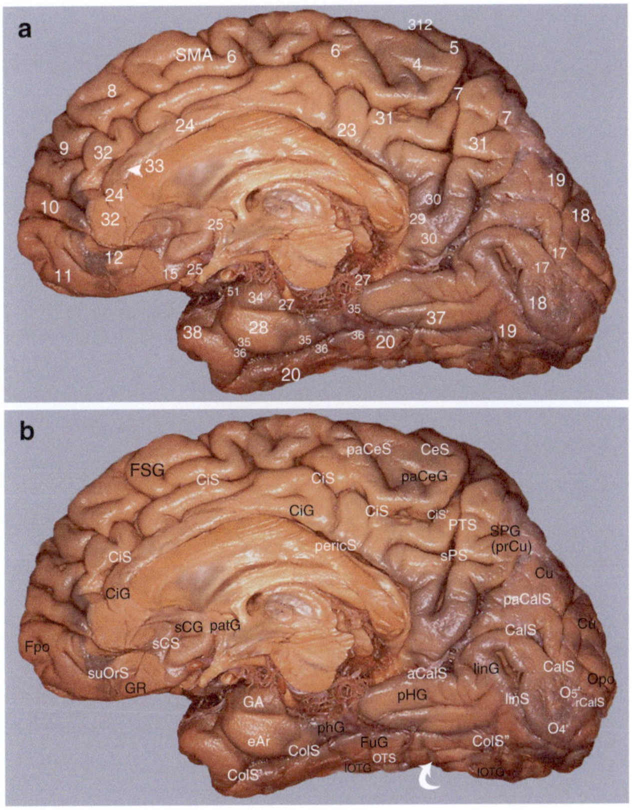

Fig. 1.2 Mapping of the cortical areas according to Brodmann (medial views of the brain): (**a**) numbering of the areas. *White arrowhead* showing the thin area 33. (**b**) correspondences with gyri and sulci. *aCalS* antecalcarinus sulcus, *CalS* calcarine sulcus, *CeS* central sulcus, *CiG* cingulate gyrus, *CiS* cingulate sulcus, *CiS'* marginal part of CiS, *ColS* collateral sulcus, *ColS'* transverse anterior collateral sulcus, *ColS*" transverse posterior collateral sulcus, *eAr* entorhinal area, *Fpo* frontal pole, *FSG* frontal superior gyrus, *FuG* fusiform gyrus, *GA* gyrus ambiens, *GR* gyrus rectus, *linG* lingual gyrus, *linS* lingual sulcus, *lOTG* lateral occipitotemporal gyrus, *O4'* posterior part of FuG, *O5'* posterior part of linG, *Opo* occipital pole, *OTS* occipitotemporal sulcus, *paCalS* paracalcarine sulcus, *paCeS* paracental sulcus, *paCeG* paracentral gyrus, *patG* paraterminal gyrus, *pericS* pericallosal sulcus, *pHG* parahippocampal gyrus, *PTS* parietotemporal sulcus, *rCalS* retro-calcarine sulcus, *sCG* subcallosal gyrus, *sCS* subcallosum sulcus, *SMA* supplementary motor area, *SPG* superior parietal gyrus (*prCu* precuneus), *SPS* superior parietal sulcus, *suOrS* sus orbital sulcus, *white curved arrow* temporo-occipital incisure

We therefore obtain an extremely complex group of structures, which appear as disparate and heteroclite, and whose only common factor is to be involved, at various but always essential degrees, in the expression of our emotions and our behaviour.

This is the reason for which various authors have voiced their opinion and denied the existence of a limbic "system" as such. They prefer to simply refer to each of its components as a "structure involved in the perception of emotions".

Broca's limbic lobe has been greatly discussed (R Olry and X Nicolay 1994) as for some it is only an artificial construction comprised of cortical structures which differ on many levels, especially on an architectonic level (4 or 5 cellular layers for the mesocortex of the cingulate and parahippocampal gyri, 3 layers for the archicortex of the hippocampus and dentate gyrus, 6 layers for the retro-splenial cortex). Luckily, the macroscopic anatomical identity of this lobe remains incontestable and represents an obstacle for all critics.

As for the limbic system, although it may be made, such as mentioned by sad individuals, of pieces and portions, even it has become a sort of patchwork, its unity is obviously functional. All becomes clear if, instead of aligning its various components, we work out from the essential portion, which is the amygdaloid body, central receptor of emotional stimuli, which will be able to trigger, via various systems, extremely rapid reactions in the body, mainly in aim of protecting it.

2 Anatomy of Emotions via the History of Neuroanatomy and Neurosciences

This is the history of successive discoveries which allowed the identification of anatomical organs and systems involved in any emotional process.

Our elders had good knowledge of emotions. They were even capable of arousing them to entertain an audience. This is shown in the theatre and especially Greek tragedies, with expressive masks worn by the actors and scary stories with a preposterous mythological basis.

But, most people believed that the emotion organ was the heart. However, the well-known physicians and philosophers of the Antiquity did not all share the same beliefs. Thus, Hippocrates (460–370 BC) believed that passions, feelings and emotions were controlled by the brain. A few years later, Aristotle (384–322 BC) referred back to the primacy of the heart which governs, according to him, all mental activity, including thoughts. The work conducted by Galien (131–204 AC) on animals was however going to confirm the hippocratic theory by declaring that the brain is "where the directing mind resides".

The Middle Ages suffered from obscurantism and numerous prohibitions. As a result, nearly no progress was made regarding knowledge of the human body and brain, and the rare anatomists of this era, such as Berengario da Carpi (1460–1530), simply based their work on Galien's models.

We had to wait for the Renaissance and its series of discoveries for the scientific study of the brain to finally begin. Anatomical dissections were firstly performed secretly by several audacious scientists. This is supported by the remarkable sketches of the brain by Leonard de Vinci (1452–1519). Such dissections were then authorised. The anatomy era could thus begin, and representations of dissected human brains were going to appear in anatomical books by authors such as Charles Estienne (1504–1564) or André Vesale (1515–1564) in his "de Humani corporis fabrica", published in 1543. With this remarkable book and anatomy lessons based on real dissections, Vesale was going to establish the basis of modern anatomy.

During the seventeenth century, although the anatomy of the nervous system was slowly developing, the role played by the brain was still going to be a major source of philosophical discussions, and the term "heart" was going to be used instead of affective brain.[1] René Descartes (1596–1650) prepared a theory which dissociates the body, whose operation is governed by humours, and the mind, whose mental functions are governed by God and which are joined together by the epiphysis or pineal gland.

Thomas Willis (1621–1675) who was going to discover the vascular polygon at the basis of the brain, rejected Descartes' theory. As a result, he proposed a new theory in his book *de Cerebri Anatome* (1664). According to him, the memory and will are located in the cortex, imagination is located in the corpus callosum and sensations in the corpora striata! It is also Thomas Willis[2] who first introduced the notion of "limb" which is employed to refer to the medial surface of the cerebral hemisphere. This term was to be shortly reused by Thomas Bartholin in 1677. Despite all of the controversy surrounding the Cartesian theories, the brain still played its predominant role. Among Descartes critics, two philosophers are to be mentioned: John Locke (1632–1704) and Etienne Bonnot De Condillac (1715–1780), for whom the brain is the organ which collects "ideas of sensation" and which adjusts "ideas of reflection".

[1] It is in his book entitled *Pensées* (*Thoughts*), whose unfinished manuscript was discovered after his death, that Blaise Pascal (1623–1662) included the famous saying " Le Coeur a ses raisons que la raison ne connaît point" (the heart has its reasons which the reason does not comprehend) by which the author uses a diaphora to demonstrate the dichotomy between the "heart" (which will later become the… emotional brain, whose motivations, "reasons", result from emotions and feelings) and "reason" (future rational brain, cognition brain, reasoning brain, which proposes solutions based on the cold analysis of situations).

[2] Thomas Willis (1621–1675), an English physician, precursor of neuropathology, who discovered the arterial polygon at the base of the brain, wrote his books in Latin, including the following:

"Cerebri anatome, cui accessit nervorum descriptio et usus, studio Thomae Willis…". Londini: typis J Flesher, Impensis J Martyn et JA Allestry, 1664.

V. Di Marino et al., *The Amygdaloid Nuclear Complex: Anatomic Study of the Human Amygdala*,
DOI 10.1007/978-3-319-23243-0_2

During the eighteenth century, the study of the nervous system at a general level and of the brain, in particular, was to be developed via numerous discoveries:

- In the physiology of the nervous system with Felix Vicq d'Azyr (1746–1794), with Albrecht von Haller (1757–1766) and then with Pierre Flourens (1794–1867)
- In nervous electrophysiology with Luigi Galvani (1737–1798)
- In anthropology and evolution, with the remarkable work performed by Jean Baptiste Lamarck (1744–1829) concerning the evolution of species and their respective brains
- In neuroanatomy, with Giovanni Domenico Santorini (1681–1778) and then Samuel Thomas von Sömmering (1755–1830)

However, it was during the nineteenth that the most progress was made in neuroanatomy and histology of the nervous system, in particular with the beginning of the discovery of the brain of emotions:

- Remarkable biologists studied nervous tissue with a microscope. Theodor Schwann (1810–1882) showed that this tissue is an assembly of nervous cells. Then came the specific colouration of nervous cell nuclei by Franz Nissl (1860–1919), the colouration of nervous cells with their extensions by Santiago Ramon y Cajal (1852–1934) and the specific study of cell layers by Vladimir Betz (1834–1894) and Korbinian Brodmann (1868–1910) whose cytoarchitectonic map of the cortex is still used at present!
- Well-known neurologists became the founding fathers of modern neurology: this especially concerns Jean Martin Charcot (1824–1893), who, through is research on hysteria, showed that the brain is where emotions are expressed and that many disruptions in this field are the cause of numerous pathologies.
- In neurophysiology, François Magendie (1783–1855) continued the work of his teacher, Flourens, and studied cerebral locations by observing the results of electrical stimulation of the cortex.
- Extremely well-known anatomists have focused on the brain and nervous system, especially Friedrich Arnold (1803–1890), Sir Charles Bell (1774–1842), Auguste Forel (1848–1931), Achille Louis Foville (1799–1878), Franz Joseph Gall (1758–1828), Friedrich Goll (1829–1903), Louis-Pierre Gratiolet (1815–1865), Johannes Bernhard Aloys von Gudden (1824–1886), François Magendie (1783–1855), Théodore Hermann Meynert (1833–1892), Johann Christian Reil (1759–1813) and Luigi Rolando (1773–1831), Ludwig Türck (1810–1868) and Sir William Turner W. (1832–1916).

Two anatomists, one who is German, Karl Friedrich Burdach (1776–1847)[3], a professor of physiology, and the other who is French, Paul Broca (1824–1880), a physician and anthropologist, made remarkable discoveries which became the foundation of the brain of emotions:

The first anatomist, in his remarkable work (Burdach KF, 1818–1826) covering 3 volumes, entitled "Vom Baue und Leben des Gehirns und Rückenmarks", provides an initial description and the first illustration of the amygdaloid body whose presence is shown on a coronal section.

The second anatomist, who discovered, in 1861, the articulated language area[4], which bears his name (area of Broca)[5], and who revealed the "concept of hemispheric dominance", will be the first to describe what he will call the "limbic lobe" during a conference given at the Society of Anthropology of Paris and published in the "Revue d'Anthropologie" in 1878[6]

At that stage, the developing science of neuroanatomy included the anatomy of emotions in its scope of exploration. Successively William James[7] and Carl Lange[8] developed, in 1884 and 1885, respectively, each working alone, the first theory of emotions which was soon to be unified in scientific circles under the theory of James–Lange: for these authors, emotion is the response to corporal modifications and reactions triggered by an emotional stimulus. If I am sad, it is because I am crying. It is not because when we see a bear, we are afraid and cry; it is because we run when we see a bear that we are afraid. "My theory … is that the bodily changes follow directly the perception of the exciting fact and that our feeling of the same changes as they occur, is the emotion" (Lange).

The twentieth century appears to be promising for research work conducted on the nervous system. During the first part of

[3] Refer to: Meyer A : Karl Friedrich Burdach and his place in the history of neuroanatomy. J Neurol Neurosurg Psychiat, 1970, 33, 553–561.

[4] Carl Wernicke (1848–1905), a neuropsychiatrist, who will describe another centre of language, located in the left temporal lobe. If this centre is damaged, it is responsible for the non-understanding of language.

[5] Broca's area (area of articulated language) is located in the postero-inferior region of the frontal lobe, in front of the premotor area. It is astride on Brodmann's areas 44 and 45.

[6] Broca P – Anatomie comparée des circonvolutions cérébrales: le grand lobe limbique (Compared anatomy of cerebral circonvolutions: the large limbic lobe). Rev Anthropol, 1878, 1, 385–498.

[7] William James (1842–1910), physician, psychologist and American philosopher and author of the "theory of emotion". Théorie de l'émotion – French translation with an introduction written by J. Dumas, 1 vol. in-18, Paris. Alcan, 1902.

[8] Carl Georg Lange (1834–1900), Danish physician, author of "on Emotions: A Psychophysiological Study"

the century, very well-known scholars from diverse and complementary fields emerged. They were to become the protoresearchers in neurosciences. They included:

- Neuroanatomists: Paul Emil Flechsig (1847–1929) and James Papez (1883–1958)
- Neurologists among which Joseph Dejerine (1849–1917) and his wife, Augusta Klumpke (1859–1927); remarkable clinicians as well as researchers; and Henri Gastaut (1915–1995), the French pioneer in electroencephalography
- Psychoanalysts with Sigmund Freud (1856–1939), Carl Gustav Jung (1875–1961) and all of their thought trend which was expressed via the various psychoanalytical societies.
- Embryologists, including Constantin von Monakow (1853–1930) who performed remarkable work on the myelogenesis
- Psychiatrists, such as Christfried Jakob (1866–1956), who is mentioned at the same time as Papez concerning the complementarity of their research
- Neurosurgeons including Harvey Cushing (1869–1939), a pioneer in brain surgery, Otfrid Foerster (1873–1941) and Wilder Penfield (1891–1976), whose experience in peroperative cortical stimulation experiments were to develop knowledge of the cortical areas
- Talented neurophysiologists including Albert von Kölliker (1817–1905), Luigi Luciani (1840–1919), author of a remarkable monograph concerning the cerebellum, or even Hermann Munk (1839–1912), known for his work carried out on the occipital lobes and vision
- Psychologist physicians among which we must mention Pierre Janet (1859–1947), who continued the work of Charcot

All along the twentieth century, the brain and the whole nervous system are going to be better known thanks eminent elders, of which we have only mentioned a few who were especially involved in the development of knowledge.

However, at the same time, serious invalidating pathologies have appeared in the population whose lifetime is extending (Alzheimer[9], etc.). We must therefore accept the challenge, and this is an opportunity to widen our scope of research by providing the many researchers in all newly developing sectors in the field of neurosciences, with diverse and generally converging research possibilities.

And although the treatment of neurodegenerative diseases remains disappointing, the research results that they have generated and which are still being obtained are extremely remarkable:

- Extraordinary and new technologies especially appeared in the field of imaging (electroencephalography, evoked potentials, MRI, magnetoencephalography, functional MRI, stereotaxic exploration) and the engineers and technicians are improving them on a continuous basis.
- Progress also concerns new approaches which have become actual specialities in the large field of neurosciences, neurobiology (within France, Jean-Pierre Changeux and Marc Jeannerod), neuroendocrinology, neuropharmacology, neuropsychology, neuropathology, neuropsychiatry and philosophy of neurosciences.
- As for the older sectors of neurosciences, several have been modified and enriched: thus, neurophysiology, which integrates all neuro-immunology methods and which now benefits from electrophysiology techniques; neuroanatomy which benefits from neural tracing; and neuro-histology which can use the data provided by neuro-cytology, immunohistochemistry, immunostaining and electronic microscopy. Embryology, which is now known as "neurodevelopment", now covers a large sector including neurogenesis, apoptosis, neurotrophins, neural migration, synaptic plasticity and neurosurgery with new techniques of awake brain surgery.[10]

Three fields[11] are targeted by these specialities: cognition (field of the cognitive neurosciences), behaviour (field of the behavioural neurosciences) and emotions (field of affective neurosciences).

During the first half of the twentieth century, cognition was predominant to not say exclusive.

The authors compared the operation of the nervous system to that of a computer and gained a better understanding of the meanders of cognition than of "the dark cloud of subjectivity that hung over the topic of emotion".[12] However, the fields of behavioural neurosciences and affective neurosciences will also be extended. It is relevant to quote the great names which led to this progression and which were used to individualise these sectors of neurosciences.

[9]This disease bears the name of the person who discovered it: Aloysius Alzheimer (1864–1915), a famous German neurologist, psychiatrist and neuropathologist.

[10]The neurosurgeon stimulates the brain of the operated patient (who was woken by the neuro-anaesthetist) to identify the functional areas that must be avoided (intraoperative brain mapping).

[11]Two other sectors of neurosciences, social neurosciences and neurolinguistics, are also under development.

[12]The general influence of cognition is also explained by the fact that it has benefited, without any restriction, from animal experimentation, whereas this was not the case for research concerning emotions as subjective states, whose integration appeared as necessary, were "difficult if not impossible to prove".

- The first milestones have already been laid down by a few pioneers who have attempted to understand its emotional behaviour mechanisms: Charles Scott Sherrington[13] (1857–1952), Nobel Prize winner in 1932, as well as Donald Olding Hebb[14] (1904–1985) and Walter Bradford Cannon (1871–1945) [15] who, with his student Philip Bard (1898–1977), refuted James' theory of emotions by stating that emotional corporal manifestations occur at the same time as the emotion itself and that they do not cause it.
- Then came the era of great theoreticians, James Papez first, followed by Paul Donald MacLean and lastly Walle JH Nauta, who thought that they explained cognition as well as emotion, by each taking, as a basic assumption, their own theory:
  - In 1937, James Papez[16] published in J Neuropsychiatry Clin Neuroscience an original anatomical pathway specifically involving the temporal lobe and which he described as the mechanism of emotions. This pathway could, according to him, be used to collect emotive sensations at the entorhinal area and then subiculum, their transfer to the mamillary body via the alveus and fornix, and then to the anterior nucleus of the thalamus (via the mamillothalamic tract) prior to reaching the anterior cingulate cortex and then the para-hippocampal gyrus with return to the "entorhinal square one". The make-up of this pathway had already been highlighted by Jakob in 1907, but was unknown as it had not published in English. As a fair return, it was the same author who was going to provide evidence of the possibility of a deep connection between the entorhinal area and the subiculum by his work on viral propagation along nerve fibres. The Papez–Jakob pathway, whose anatomical existence was now acquired, and which proved to be the prerogative of mammals[17], was shortly to be rejected outside the field of emotions due to the fact that its role was to preside over the general affective behaviour and modalities of adaptation to the environment.
  - It is in 1960 that Paul D MacLean[18] (1913–2007), an American physician and neurobiologist, presented his interesting model of the "triune brain" with, for man, a reptilian central and primitive brain, which can only serve for performing functions necessary for survival purposes; a limbic brain, involved in emotions, which surrounds this "old reptilian bark" and corresponds to the old description of the limbic lobe by Broca; and lastly, capping the two previous components, a developed brain, which can be used for reasoning, decision-making and abstract thinking, the neocortex.

    The theory appeared to be attractive with the perfect imbrication of three brain layers, deposited and complexified as the organ developed.

    This layered organisation not only satisfied Darwin's supporters but also Freud's and pupils, for whom it became easy to acknowledge the absolute primacy of the neocortex which represses the two others.

    However, MacLean's theory, which had reached its peak in a decade (PD MacLean 1970), was to be swept aside due to the fact that the dichotomy introduced by his author could not withstand the demonstration of the complexity of cerebral functions and the interdependency of all parts of the brain, which new research techniques were starting to reveal. Furthermore, the field of emotions could no longer be dedicated to a single limbic lobe as it now appeared as very complex and involved both the visceral brain and the cognition brain!
- Walle Jetze Harinx Nauta (1916–1994) studies with W Haymaker, in a book on the hypothalamus, published in 1969, the concept that he has already revealed in 1958: within the limbic system, there exists a rostro-caudal continuum including the septal and preoptic regions, the hypothalamus and several mesencephalic structures including, in particular, the periaqueductal grey matter and the dorsal raphe nucleus. Nauta has called this major pathway the septo-hypothalamo-mesencephalic continuum and believes that it plays a major role in the anatomy of emotions. However, it represents only half of the emotional brain as the amygdala is not included. However, this part must be connected to the limbic lobe which is becoming larger! Tenacious, Nauta defended his concept again in 1973. Moreover, other authors will continue his work: Stephan in 1975 and OT Phillipson and AC Griffith in 1985.

On the basis of all of the relevant work, the notion of limbic system has naturally and progressively replaced that of limbic lobe. Several formations were to be connected to it including the hippocampus and the amygdaloid body or complex.

In this system, the hippocampus rapidly appeared as the memory organ playing a secondary role in relation to emotions.

However, the amygdaloid body was to be extremely involved in emotional processes. This eminent role was demonstrated in 1938 by the work completed by Heinrich

[13]Charles Scott Sherrington (1857–1952) is an English physician and scientific and Nobel prize winner in 1932, who discovered the "synapse" and created the related term.

[14]Donald Olding Hebb (1904–1985) is a Canadian neurophysiologist.

[15]Walter Bradford Cannon (1871–1945) is an American neurophysiologist and author of the concept of homeostasis.

[16]James Papez (1883–1958) is an American neuroanatomist.

[17]This pathway is observed in certain reptiles but is then, very basic.

[18]Päul Donald MacLean (1913–2007), an American physician and neurobiologist

Klüver[19] (psychologist) and Paul Bucy[20] (neurosurgeon), who had observed in adult monkeys, following the ablation of their temporal lobes, surprising manifestations of the syndrome which were to their name: among all of the disorders observed: disappearance of reactions generally related to anger and fear, use of the mouth rather than of the hands to discover objects or living beings, insufficient knowledge of objects (psychic blindness), memory disorders, hypersexuality and loss of social behaviour. It appeared later on that only the first symptom in this list was a consequence of the bilateral amygdaloid suppression secondary to temporal resections. Nevertheless, following the war in 1938–1945, publications by Klüver and Bucy were to be set aside and were only to reappear at the end of the conflict.

Another formation was to play a major role in the anatomy of emotions: the frontal brain. This part of the brain could also have been taken into consideration at a much earlier stage, if Doctor John Harlow, physician of Phineas Gage, a patient who suffered from a serious accident in 1848, in which a crow bar cut through his rostral brain from bottom to top, had remembered the observation he had made on his patient since the work accident until he lost his sight!

The observation made by Dr Harlow, which is exhaustive at a clinical level as it described the strange behaviour of the patient following an accident and was documented over the years, was sadly incomplete from an anatomical viewpoint as it did not include a description of the initial cerebral injury. Harlow, who learnt about the death of his patient 5 years later (on May 21, 1861, following an epileptic fit), managed to complete his observation by exhuming his body to recover his skull[21]. He published this observation in 1869[22] without creating any commotion among the scientific community.

Luckily, the study of this exceptional clinical case, reviewed 127 years later by par Antonio Damasio in his remarkable and well-known book entitled *L'erreur de Descartes* (1995) (*Descartes' Error*), based on a medicolegal and neuroanatomical survey conducted by Hanna Damasio (3D reconstruction of MRI images of Ph. Gage's skull), would clearly show the correlations existing between a cerebral injury (bilateral destruction, predominant on the left, of the "ventro-median regions" of the frontal lobes) and deep personality and behaviour disorders of the patient, who became incapable of making any decision.

Over the last decade of the twentieth century, neuroscientific research concerning emotional processes and especially their location at the level of the nervous system· will be significantly developed. This announced a very promising future twenty-first century, with three major strong points:

- At present, neuroanatomists agree on the fact that there is not a single centre of emotions but pathways made of structures or centres connected together such as the fear pathway or the reward pathway. The amygdala plays a key role in these pathways.
- A reconciliation is taking place between "cognition" and "emotion". Thoughts belong to both fields. "Minds are neither cognitive nor emotional; they are both and more". Neuroscientists have understood that all cognitive research not accounting for emotional processes would not correspond to reality.

  Emotions, such as cognitive processes, may be unconscious. "Contrary to popular belief, conscious feelings are not required to produce emotional responses" (LeDoux).
- Psychologists, with Paul Ekman (American psychologist born in 1934), have agreed to classify emotions in three categories:

  Primary emotions (Ekman 1972): joy, sadness, fear, surprise, anger and disgust (which correspond to the main facial expressions)

  Secondary emotions (Ekman 1990): pride, guiltiness, jealousy, shame, embarrassment, amusement and excitement (emotions involving facial expressions are only a part of the individual's global aspect)

  Emotions involving little outward expression but a major internal component for the individual: well-being or ill-being, tiredness or "form", satisfaction, relief and sensory pleasure (P Ekman 1992).

Incredible investigation techniques are available to make major progress in the scientific field: the development of the MRI technique (Paul Lauterbur 1973; Peter Mansfield 1976) who will be together rewarded by the Nobel Prize 2003 in medicine or physiology, allowed the study of the soft tissue of the encephalon, obviously including the amygdaloid body, and then the measurement (1992) of its activity by means of the functional MRI technique (Seiji Ogawa, Pierre Bandetini, John Belliveau). Derived from the diffusion tensor MRI, which was developed by Denis le Bihan in 1985, fibre tractography [23] is now used to visualise the neural pathways that only anatomical dissection could reveal (L Concha et al. 2005).

[19]Heinrich Klüver (1897–1979), an American psychologist from Germany

[20]Paul Bucy (1904–1992), an American neurosurgeon and neuropathologist

[21]The skull and the iron bar are kept in the Warren museum at the Medical Faculty of the University of Harvard, at Boston (USA). The iron bar weighs 6 kg. It is 1.10 m long and had a diameter of 3 cm. Its tapered part measures 18 cm. The terminal point measures 6 mm.

[22]JM Harlow – Recovery from the passage of an iron bar through the head. Publications of the Massachusetts Medical Society, 1868, n°2, 327–347

[23]The fibre tractography is a 3D reconstruction technique using data collected by DTI.

At the beginning of this twenty-first century, most universities in the world have a neuroscience department so that number of studies performed each year amounts to thousands. A considerable quantity of data has been collected, despite the fact that many uncertainties remain because most programmes have only been performed in animals.

Among the numerous and excellent studies dedicated to the amygdala and brain of emotions (which we cannot mention considering their abundance), we want to highlight a few major orientations and several remarkable studies:

- In the field of neuroanatomy, three authors specifically studied the two large limbic structures in humans: Henri Duvernoy for the hippocampus and Lennart Heimer and José S De Olmos for the amygdaloid body.
- In the field of neurophysiology, the studies performed by Hélène Barbas and her team, who showed in primates the existence of connections between the amygdaloid body, the prefrontal lobe and more specifically the orbitofrontal cortex, the insula and other cerebral areas.
- In the field of neuropsychology:
  The studies carried out by Antonio Rosa Damasio[24] and his wife Hanna on the role played by the frontal lobe in the perception of emotions and on the indispensable need of these emotions so that the decision-making can be made
  The work of Joseph Ledoux[25] on the mechanism of emotions and on the connections between the emotions, the brain (especially the amygdala) and the memory
- In the field of neuro-amygdaloid chemistry:
  The study performed by Alexander J MacDonald[26] on the neurochemistry of the amygdala neurons and on their operation under the dependency of monoaminergic neurons of the brainstem and cholinergic neurons of the basal forebrain.
  The studies carried out by MM Mesulam[27] on the amygdala and neural networks, cholinergic pathways and Alzheimer's disease.

It is probable that with such researchers, with the investigation centres that they have developed, with the multiplication of research centres created at the start of this century and with the sophistication of new functional MRIs and new PET scan techniques, the history of the anatomy of the amygdala, in particular, and of the limbic system, in general, should be continued and allow future generations to write new pages while continuing to analyse the large fields of emotions and the emotional brain, etc.

[24] Antonio Damasio, professor of neurology, neurosciences and psychology, born in 1944 at Lisbon, and his wife Hanna, who was also Portuguese, professor of psychology and neurology, co-wrote many articles concerning the role of the frontal lobe. A Damasio is in charge of the institute for the neurological study of emotions and creativity at the University of Southern California.

[25] Joseph Ledoux, as for him, was born in 1949 and is a psychologist and professor in sciences at the University of New York. His work has specifically focused on the mechanism of emotions in his experience concerning conditioned fear. Founder of the Emotional Brain Institute, his theory of emotions is accepted by most scientists who have studied this subject. He is actually the author whose work concerning the amygdala is considered as authoritative on the subject. He demonstrated that the amygdala is coupled with the hippocampus to provide an emotional memory which enriches our experience and reactional capacity regarding event in our lives.

[26] Alexander J MacDonald, American professor, director of the department of pharmacology, physiology and neuroscience at the University of South Carolina, School of Medicine

[27] M Marsel Mesulam, American professor in neurology, director of the CNADC (Cognitive Neurology and Alzheimer's Disease Center), Northwestern University, Feinberg School of Medicine, Chicago.

3 Development of the Human Amygdaloid Complex

There have been few studies of the amygdala's development in humans to date. However, those that have been carried out are sufficiently exhaustive that we have a fairly accurate idea of this development.

Among excellent former works, the works of Johnston in 1923, Elizabeth Caroline Crosby and Tryphena Humphrey in 1941 and George Macchi in 1951 must be cited. But it is the contemporary authors, notably Fabiola Müller and Ronan O' Rahilly, who have brought an unrivalled contribution to the field, thanks to modern computer reconstruction processes of serial sectioning performed on significant numbers of embryos. It is since their remarkable work in particular that we have been able to precisely track the development of the amygdala during embryonic development.

3.1 Development During the Embryonic Phase

It is from 5½ weeks into development (*stage 16*, approx. 6–8 mm) that the primordial, corticomedial and baso-lateral areas of the amygdala develop almost simultaneously (Müller and O'Rahilly 2006).

The first nuclei, however, will not be visible until between *stages 17 and 21*: recall that at stage 17 (F Müller and R O'Rahilly 1989), the embryo is at the beginning of the 6th week of development and that it measures about 11–14 mm. On a transverse section of the forebrain, through the interventricular foramina, it can be clearly seen that the two lateral ventricles (cavities of the two telencephalic vesicles) communicate widely, despite the presence of the choroid plexus, with the median ventricle (the diencephalic cavity) or third ventricle (Fig. 3.1). The roof of these cerebral ventricles (pallium) is rather thin. It continues laterally and thickens ventrally to form the floor of the telencephalon or subpallium.[1] The thickenings of this floor are referred to as the ventricular eminences.[2] The side walls of the third ventricle, which follow those of the lateral telencephalic ventricles, are also the thickened zones that will be at the origin of the diencephalic nuclei. It can thus be observed on the telencephalic wall, and projecting into the lateral ventricular cavity, the lateral ventricular eminence appeared at stage 15 (Müller and O'Rahilly 2006), in the diencephalon, the emergent thalamus (the dorsal thalamus and the ventral thalamus separated by the sulcus medius and the intrathalamic zone), the subthalamus and the hypothalamus which project slightly into the cavity of the third ventricle. At the telencephalo-diencephalic junction, a noticeable protrusion, labelled the medial ventricular eminence, appeared from stage 14, the significance of which a number of embryologists have questioned. At this stage, the two lateral and medial ventricular eminences[2] remain separated by a sulcus. *The medial eminence is particularly important because it is here, as shown by F Müller and R O' Rahilly, that the first nuclei of the future amygdaloid body appear: the cortical, anterior, baso-lateral and medial nuclei.*

It is the cortical nucleus that appears first, followed by the anterior nucleus, then the baso-lateral and the medial nucleus. The baso-lateral complex is larger than the cortico-medial complex (T Humphrey 1968). The medial nucleus is visible in sagittal sections, in contact with the beginning of the dentate gyrus, already showing the amygdalo-hippocampal proximity!

[1]The prosencephalon includes the pallium which forms the vault and primitive envelope of the lateral ventricles (its the primordium of the cerebral cortex); the subpallium or paleopallium, which forms the ventral thickening of the floor of the ventricles (it will form the basal part of the brain); the archeopallium or archipallium, the most primitive part of the pallium, will only occupy a reduced area but will provide a derivative of major importance: the hippocampus.

[2]The ventricular eminences are often wrongly called ganglionic eminences while they not resemble nerve ganglion structures at all!

V. Di Marino et al., *The Amygdaloid Nuclear Complex: Anatomic Study of the Human Amygdala*,
DOI 10.1007/978-3-319-23243-0_3

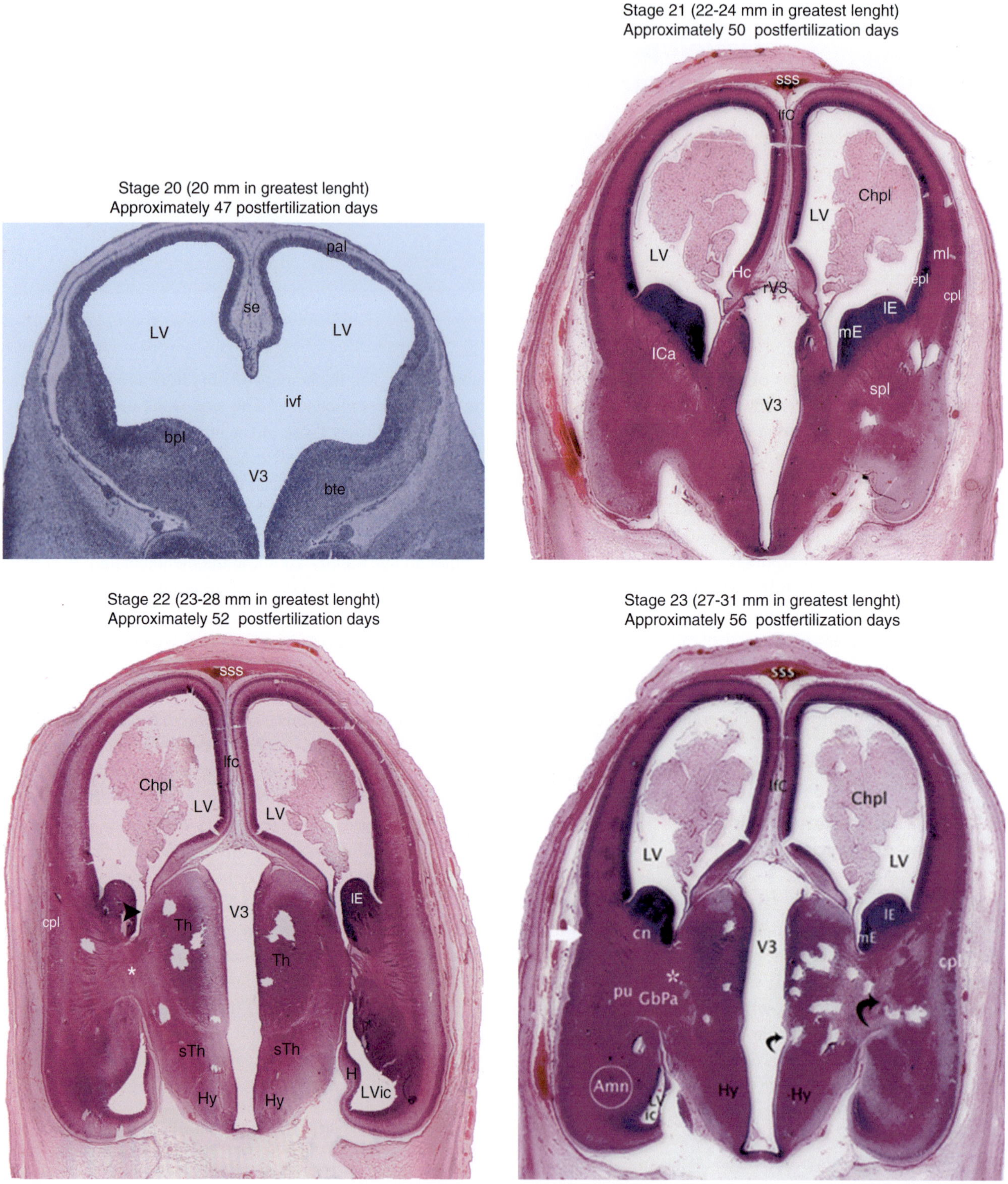

Fig. 3.1 Coronal sections of heads of embryos, showing the evolution of the human brain from stage 20 to stage 23. *Amn* amygdaloid nuclear complex, *bpl* basal plate (subpallium), *bte* basal telencephalon, *Chpl* choroid plexus, *cn* caudate nucleus, *cpl* cortical plate, *epl* ependymal and mantle layers, *GbPa* globus pallidus, *H* hippocampus, *Hy* hypothalamus, *ICa* internal capsule, *ivf* interventricular foramen, *lE* lateral eminence, *lfC* longitudinal fissure of cerebrum, *LV* lateral ventricle, *LVic* lateral ventricle inferior cornu, *mE* medial eminence, *ml* marginal layer, *pal* pallium, *pu* putamen, *rV3* roof of V3, *se* septum, *spl* subpallium, *sss* superior sagittal sinus, *sTh* subthalamus, *Th* thalamus, *V 3* third ventricle *black arrowhead*, sulcus terminalis; *little curved black arrow*, hypothalamic sulcus; *greater curved black arrow*, it shows the diencephalon incorporating itself to the telencephalon; *white asterisks*, they show the lateral migration of the GbPa: *white right arrow*: future insula

The medial eminence was previously thought as of telencephalic origin, a notion that remains, still to this day, in a number of books. But the works of F. Muller and R. O Rahilly have indisputably shown since then, thanks to reconstructions, that this eminence is primitively diencephalic even if it tends to invade the telencephalic area early on, becoming di-telencephalic in this manner.[3] The amygdala arising from this medial eminence is therefore originally a nuclear diencephalic complex.[4] The line of demarcation that separates the telencephalon and the diencephalon is a straight line that links the velum transversum to the preoptic recess: that which lies above this border belongs to the telencephalon. All that is underneath is diencephalic. The medial eminence, by its own expansion, will be quickly divided in two by this line, which explains the discussion on the telencephalic or diencephalic origin of the amygdala. The lateral eminence, originally telencephalic, is smaller. It remains in the subpallium and bulges slightly into the lateral ventricular cavity. As soon as the ventricular eminences cross, the interventricular foramina contract, a phenomenon further accentuated by the expansion of the choroid plexus.

At stage 18 (about 13–17 mm), that is, at 6½weeks of development, the four nuclei of the amygdala are still inside the medial eminence, but the cortical nucleus extends toward the lateral eminence which therefore contributes partially to its formation. This stage 18 marks the start of the rapid development of the hemispheres, which will be further accentuated from stage 20. Muller and O'Rahilly (1990a) have observed at stage 18 the presence of nervous fibres linking the lateral olfactory stria and the olfactory tubercle at the medial amygdala nucleus, as well as linking the olfactory bulb with baso-lateral and anterior nuclei. Other connections are made with the ventral thalamus via the lateral forebrain bundle.

Stage 19 (about 16–18 mm), that is, between 6½ and 7 weeks of development, could be considered to be the stage of striae. Indeed, it is at this stage that the fibres of the stria terminalis appear. It is also the stage that other fibres of the stria linking the baso-lateral nucleus of the amygdala to the lateral nucleus of the habenula take the path of the medullary stria of the thalamus. The lateral eminence crosses rostrally and from there pushes the medial eminence towards the median line, causing it to protrude slightly into the ventricle. The lateral globus pallidus appears inside the diencephalon, in the subthalamic region (Muller and O' Rahilly, 1990a). It is also of note, according to these authors, that the nucleus accumbens is recognisable at this stage.

At stage 20 (about 18–22 mm), the 7th week of development, the medial eminence is entirely in the telencephalic territory. The nuclear complex of the amygdala is from now positioned in the middle part of the subpallium. Other nuclei have appeared: those of the diagonal band and of the septum. Connections are established between the two latter structures, the amygdala nuclei, the hippocampus, the diencephalic structures and the bundle of the prosencephalon (Muller and O'Rahilly 1990a).

At stage 21 (about 22–24 mm), towards the 50th day of development, the future medial globus pallidus starts to differentiate, in the diencephalon. The bundle of the prosencephalon that unites the telencephalon and the diencephalon is at this stage highly visible (Muller and O' Rahilly 1990b).

Stage 22 (about 23–38 mm), towards the 52nd day of development, the hemispheric neopallium is well developed and the internal capsule that leaves it, is visible. The lateral globus pallidus begins its migration towards the telencephalon.

Stage 23 (about 27–31 mm), towards the 56th day, corresponds to the end of the embryonic period. The putamen that arises from the subventricular layer of the lateral eminence is visible from this stage, and due to its rapidly increasing volume, it requires the amygdala, within which the significant central nucleus is formed, to move towards the temporal area. The caudate nucleus itself also becomes morphologically recognisable. The lateral globus pallidus migrates towards the putamen. It is also the stage at which the insula differentiates, beginning to form the lateral or sylvian fossa (compared to the claustrum, which had appeared at stage 22), that the emergent dorsal and ventral thalamic structures appear, that the rhombencephalic nuclei organise and that the anterior commissure is formed (O'Rahilly and Müller 1999). It is at the final stage in which the hippocampus positions itself in proximity to the future temporal pole.

3.2 Development During the Foetal Period

This period is dominated by the completion of the nuclear amygdalar structure, in parallel with the remarkable development of the cortex (development had already commenced at the end of the embryonic period).

The differentiation of the nuclei in the amygdala is completed by the formation of the lateral nucleus. The baso-lateral nucleus becomes the largest in volume. The volume of the baso-medial nucleus increases also, but a little less

[3] Other authors still believe and write that it is the lateral eminence that provides the most significant cellular component for the formation of the amygdala.

[4] For the Swedish authors (Holmgren and Kâllen), the pallium also participates in the formation of the amygdala. It is also the opinion of Puelles and Rubenstein et al., who accept a subpallial origin for the cortical, central and medial nuclei, but who describe a pallial claustro-amygdaloidian primordium at the origin of the basal and lateral nuclei of the amygdala. For T Humphrey it is not until later that the neuroblasts are coming from the "medial striatal ridge" to participate in the baso-medial nucleus formation.

than that of the baso-lateral. The basal complex is the most developed and will remain in this place.

The neocortex of the pallium will progress in two ways:

- First, in a coronal arrangement, laterally and in the caudal direction, to enclose the diencephalon which appears for the first time, in coronal sections, to invaginate between the hemispheres separated by the thin choroidal fissure (see stage 21 in Fig. 3.1). The lateral wall of the diencephalon will soon make contact with the medial walls of the hemispheres, which will lead to the process of the union of neighbouring structures. Of the two classic theories (fusion by expansion of the diencephalon, or fusion by joining), we prefer our vascular theory, evident in the observation of foetal brain sections: it explains the diencephalo-telencephalic coalescence by a considerable lateral development of vessels of the diencephalon which will colonise the neighbouring telencephalon and annexed it (see stages 22 and 23 in Fig. 3.1). Whichever is the case, the diencephalon incorporates itself into the telencephalon, and from this point, it occupies the central part. Some diencephalic elements are therefore facing the structures of the telencephalic subpallium, and it is easy to understand that the lentiform (or lenticular) nucleus is created by single lateral sliding of the medial pallidum that just has to migrate from its origin at the diencephalon towards the neighbouring telencephalon to join the lateral pallidum, which already joined the putamen (see stage 23 in Fig. 3.1).
- Next in a sagittal arrangement, the pallium in full expansion unfurls into a 'C' shape, with an anterior concavity and lentiform centre, developing as much on top and in front (gyrus frontal) as it does from behind (gyrus occipital), as in periphery (cingulate gyrus and parietal gyrus), as underneath and in front (temporal gyrus). This lateral unfurling of each hemisphere clearly participates with the aforementioned coronal growth, in the covering of the diencephalon and of the brainstem that follows it.

Recall that these phenomena of progressive growth in a 'C' shape do not only involve the pallidum. Most of the structures directly underneath are affected, as one can observe in the foetus from the early post-embryonic phase: 9th week of the development, foetus about 30–50 mm (Fig. 3.2).

The caudate nucleus develops mainly at its anterior end (head and body), but its posterior end (tail) that will taper, continue its growth in unfurling around the large thalamus to reach the temporal lobe. During this proceeding, it pushes ahead of it the large amygdaloid body that no longer has space in the area where it came to create the lentiform nucleus.

The ventricular system at the heart of the pallium, follows the unfurling of the C shape of this last structure. It has further developed three expansions, the frontal horn, above and in front; the occipital horn, behind; and the temporal horn, underneath and in front. This last structure is in close relationship with the amygdala and with the head of the hippocampus.

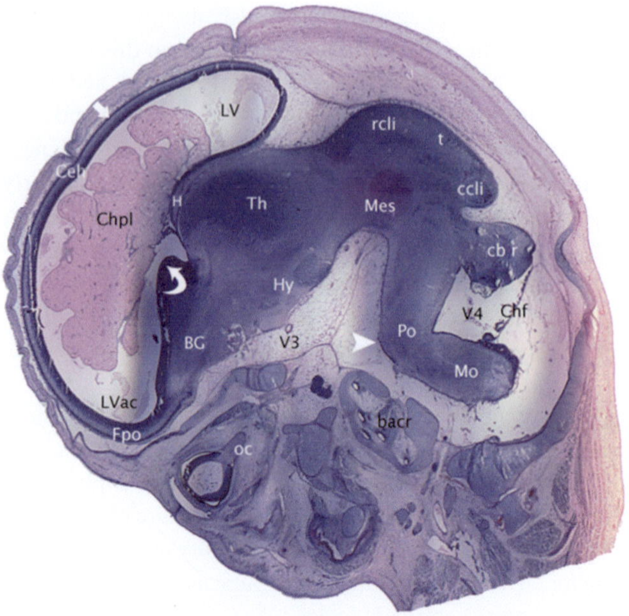

Sagittal section of the head of an human embryo, 65th day (C.R. lenght 40mm) through the right eye and the right orbital cavity, to show the position of the ventricular eminence and of the basal ganglia area

Fig. 3.2 Sagittal section of the head of an human embryo, 65th day (C.R. length 40 mm) through the right eye and the right orbital cavity, to show the position of the ventricular eminence and of the basal ganglia area *bacr* basicranium, BG basal ganglia, *cb r* cerebellar rudiment, *ccli* caudal colliculus, *Ceh* cerebral hemisphere, *Chf* choroid fold of V4, *Chpl* choroid plexus, *Fpo* frontal pole, *H* hippocampus, *Hy* hypothalamus, *LV* lateral ventricle, *LVac* lateral ventricle anterior cornu (or frontal horn), *Mes* mesencephalon (or midbrain), *Mo* medulla oblongata, *oc* orbital cavity, *Po* pons, *rcli* rostral colliculus, *t* tectum of midbrain, *Th* thalamus, *V3* third ventricle, *V4* fourth ventricle *white arrowhead*, pontine flexure; *white right arrow*, it shows the cortical plate (future cortical grey matter); *white curved arrow*, showing the ventricular eminence. *Notice the size of the blooming choroid plexus and the position of the ventricular eminence closely to the dorsal thalamus*

The hippocampus, which has taken part very early, in the embryonic period in the unfurling of the pallium develops at the level of its lower thirds and will conclude in the formation of the gyrus fasciolaris, of the fasciola cinerea, of the Ammon's horn and of the dentate gyrus.[5]

The development of the corpus callosum, incomplete, will be interrupted behind the splenium. It is at the 6th month of development that the corpus callosum takes on its definitive form.

[5]The dentate gyrus, which has formed by migration of the cells from the ventricular wall, only becomes noticeable in the foetus at a stage corresponding to 150 mm (O'Rahilly and Müller 1999).

In contrast, the development of the fornix will be complete, not only the fimbria, the crura, the body and the fornix columns, but also the mammillothalamic bundle.

Among the structures directly connected to the amygdala, we shall mention the stria terminalis whose winding will connect the septal nuclei and the nuclei of the amygdaloid complex.

The formations at the centre of the unfurling, that is to say, the components of the lentiform nucleus, are not or are very little required by the 'C' process. However, their expansion is in no way less important, their growth will take place where they are and they will tend to occupy a large part of the centre of each hemisphere as pushing back the other structures laterally (in the case of the claustrum) or towards the base (which we have already seen occurring for the amygdala).

It is particularly interesting to review more precisely the parallel evolution of the hippocampal head and the amygdala, since these structures are found in the end, in contact, on both sides of the uncal recess (in the anterior cul-de-sac) of the temporal horn of the lateral ventricle.

The amygdala pushed toward the base of the telencephalic floor by the growth of the lentiform nucleus has not had, as we have come to see, the other possibility, to place it against the tail of the caudate nucleus, the only structure occupying little space as it is thin and tapered. However, the caudate nucleus is at a stage of growth and its extremity progresses toward the temporal lobe. The amygdala will be carried by this movement and will end up in an anterior temporal position against the end of the lateral ventricle.

As for the hippocampus, it continues its development to contact with the inferior horn of the lateral ventricle (this later is also in the process of expansion toward the temporal lobe) and will project into this horn, becoming the intraventricular hippocampus.

In this way the amygdala and the hippocampus will have a development linked for, one, to the tail of the caudate nucleus and, for the other, to the inferior horn of the lateral ventricle.

All the aforementioned developments will continue until the second trimester of development. They can be observed between the 9th and 13th weeks of development. (foetus of 50–100 mm).

The second trimester of development (foetus of 100–250 mm) will be marked by a continuation of the development on the inferior part of the hippocampus and dentate gyrus and by the involution of the antero-superior of the hippocampus which occurs after 3½ months and which will leave only residues (prehippocampal rudiment, indusium griseum and median longitudinal and lateral striae), by the termination of the arc of progression by the corpus callosum (which does not continue its growth from the posterior part, the splenium), but most of all by the intense phenomena of the cerebral maturation and the first myelinisation. The amygdala which undergoes a final axial displacement (rotation toward the outside, positioning the lateral nucleus at the base and behind) in relation with the end of the coronal hemispheric unfurling is now in the definitive temporal position that is known in the full-term foetus in intimate contact with the tail of the caudate nucleus which completes its path toward the anterior part of the temporal lobe. It has taken with it the stria terminalis, which links it with some essential structures. The end of the inferior horn of the lateral ventricle has terminated its unwinding movement by flattening itself against the posterior face of the amygdala. As for the head of the hippocampus, pulled by the brutal ventricular rotation, it appears to 'bump' into the wall of the amygdala while it was in the process of extension, so that it has not had other solution that folding back on itself, a phenomenon at the origin of the formation of hippocampal digitations (theory issued by Giacomini in 1884). Recall that there is another theory to explain the existence of the hippocampal digitations: it is the theory of Zuckerkandl (1887) for whom the digitations are the result of the phenomena of cortical atrophy. The observation of certain transverse sections performed on brains of full-term foetus or adults seem rather to support the first mode of explanation (Fig. 3.3).

The third and final trimester of development (foetus of 250–335 mm) will be that of the gyration with a considerable development of sulci and gyri and, in parallel, the very active process of myelinisation. The amygdala is at this stage in contact with the hippocampal head, separated only by the uncal recess. Its nuclei further undergo maturation processes (under hormonal influences) and apoptosis (see Chap. 4) that will make them acquire their structure, shape and final volume (CL Buss et al. 2012).

In this way the influence on the volume of the amygdala of the levels of maternal cortisol at the start of the pregnancy (moment when the amygdala forms) was until this point only considered to be likely the case. Now, it has been demonstrated (CA Sandman et al., 2012) at least in girls. These authors had actually measured the volumes of amygdalae and hippocampi in children with a mean age of 7½ (between 6 and 9 years) of whom the mothers had presented elevated levels of cortisol during pregnancy (measurements were taken at 15 weeks of pregnancy). They had observed in this way that the more the maternal levels of cortisol had been high, the more the right amygdala of girls of 7½ years old had a significant volume. They did not, on the other hand, observe a correlation with the volume of the left amygdala, nor any correlation with the hippocampus, nor any correlation with either the amygdalar or

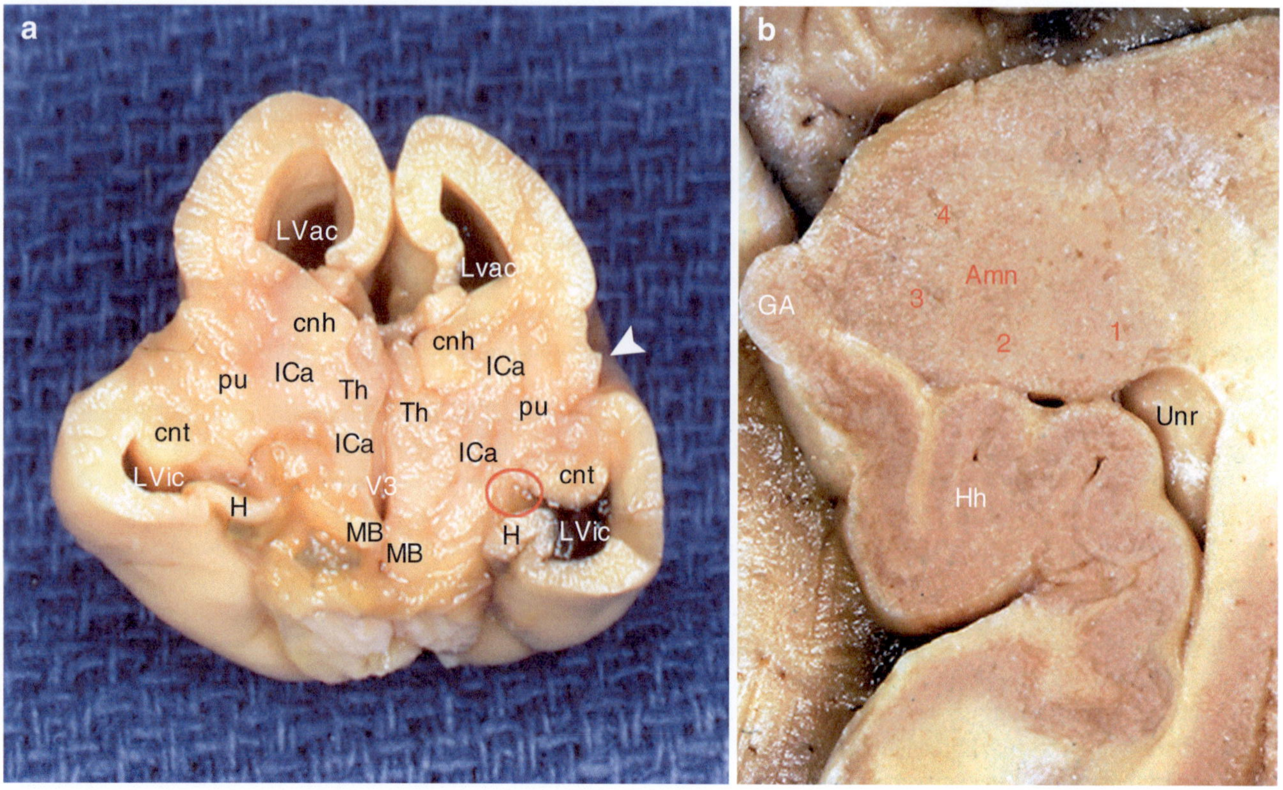

Fig. 3.3 Development of the amygdala and the hippocampal head during the later postembryonic phases (**a**): coronal section of the brain through the interventricular foramina in a foetus of 68 mm (approximately 12 weeks). (**b**): Axial section of the amygdalo-hippocampal area in a foetus of 7 months post-fertilisation. *Amn* (or *red circle*) amygdaloid nuclear complex, 1: lateral nucleus, 2: baso-lateral nucleus, 3: baso-medial nucleus, 4: cortical nucleus, *cnh* caudate nucleus head, *cnt* caudate nucleus tail, *GA* gyrus ambiens, *H* hippocampus, *Hh* hippocampal body, *ICa* internal capsule, *Lvac* lateral ventricle anterior cornu, *Lvic* lateral ventricle inferior cornu, *MB* mammillary body, *pu* putamen, *Th* thalamus, *Unr* uncal recess, *white arrow* it shows the formation of the insular fossa (Sylvius' fossa). *NB: On A, we observe that the amygdala (red circle) pushed forward the base of the telencephalic floor by the growth of putamen and got closer to the hippocampal head. On B, we observe the hippocampal digitations due, according to the Giacomini's theory, to the bumping of the hippocampal head (while it was in extension process) into the wall of the amygdala!*

hippocampal volumes in boys. The authors think that the development of the nervous system could be more rapid in girls and that this could explain why these observations were only made in females. The authors had also noted there was a relationship between elevated cortisol in the mother during the pregnancy and the frequency of emotional and behavioural disorders in girls around 7 years, while this relationship was absent in boys.

4 Morphology of the Human Amygdala

4.1 Macroscopic Aspect

The amygdaloid body or cerebral amygdala appears as a small disc-shaped formation, more or less convex, which caps the top surface of the head of the hippocampus and thus forms thereon a discrete cup-shaped cavity (Fig. 4.1).

The *morphology* of the amygdala is most variable, its surface being dented in a variable manner by the surfacing of its constitutive nuclei. However, we can retain an overall disc-shaped appearance with two slightly convex surfaces. The almond shape described by the ancient authors provides an approach of the morphological reality. However, many amygdalae are hardly dished and have the aspect of a flattened disc, which resembles a flat dragee or a little plate haricot bean. Conversely, some amydalas do not match the above-mentioned comparisons (Fig. 4.2): they are the globular and polylobed amygdalae, clusters of nuclei rather than a unitary organ.

The *colour* of the amygdala is that of the grey substance. When a brain section passing through the amygdala is stained by using our ferric chloride staining process, it is observed that the amygdala is coloured with the central grey nuclei but with different shades: the amygdala stains less intensely than the caudate nucleus or the putamen but more intensively than the pallidum and thus evokes a specific neurochemistry. Some amygdalae are very difficult to colour, and we cannot see why!

The *inclination* of the amygdala is characteristic: it more or less lies on the head of the hippocampus. If we look at the two formations on the medial dissections of hemispheres, it is observed that the major axis of the amygdala forms, with the long axis of the body of the hippocampus, an average angle ranging from 30 (for the most horizontal) to 45° for the most vertical (Fig. 4.1).

The amygdala, in such a position, especially if it is disc shaped, has a dorsal, antero-superior surface directed outwards and a postero-inferior surface facing the head of the neighbouring hippocampus directed slightly inwards (Figs. 4.3, 4.4 and 4.5).

The average *measurements* that we have observed during our dissections clearly show that there are differences between individuals but also slight right–left differences, on the same individual (measurement differences resulting from different morphologies). The length of the major axis varies between 12 and 19 mm; the greatest width varies between 6.5 and 9 mm; the average thickness varies between 1 and 4 mm.[1] The weight is usually less than 1 g, with average values between 0.56 and 0.69 g.

Appreciating the *volume* of the amygdala raises many problems.

It can be easy if it concerns amygdalae obtained by anatomical sampling, and therefore cadaveric amygdalae, keeping in mind that this type of sampling campaign is only conducted on the nuclear complex and therefore eliminates the periamygdaloid cortex.

In our experience, we found with this method an average figure of 1100 mm^3. We observed no significant difference between the volumes of the two amygdalae from the same subject. However, we did observe that male amygdaloid volumes are slightly greater than female amygdaloid volumes, but there again without any major significance due to the fact that men generally have larger brains. J Brabec et al 2010 also carried out anatomical volume measurements based on the flatness of serial sections stained by using the Nissl method. They obtained an average volume of 1.24 cm^3 when not taking into account the circumference of the amygdala and 1.63 cm^3 if this circumference is included. These authors have not observed any intersexual or interhemispheric difference.

The observation of the amygdaloid volume in living subjects is even more interesting. It was made possible by the three-dimensional reconstruction processes using MRI (magnetic resonance imaging). However, once again, difficulties arose due to discrepancies in the landmarks that the

[1] Our measurements are performed on cadavers and thus without any blood flow, and obviously, as a result, they differ from those that can be carried out on living subjects during MRI examinations.

V. Di Marino et al., *The Amygdaloid Nuclear Complex: Anatomic Study of the Human Amygdala*,
DOI 10.1007/978-3-319-23243-0_4

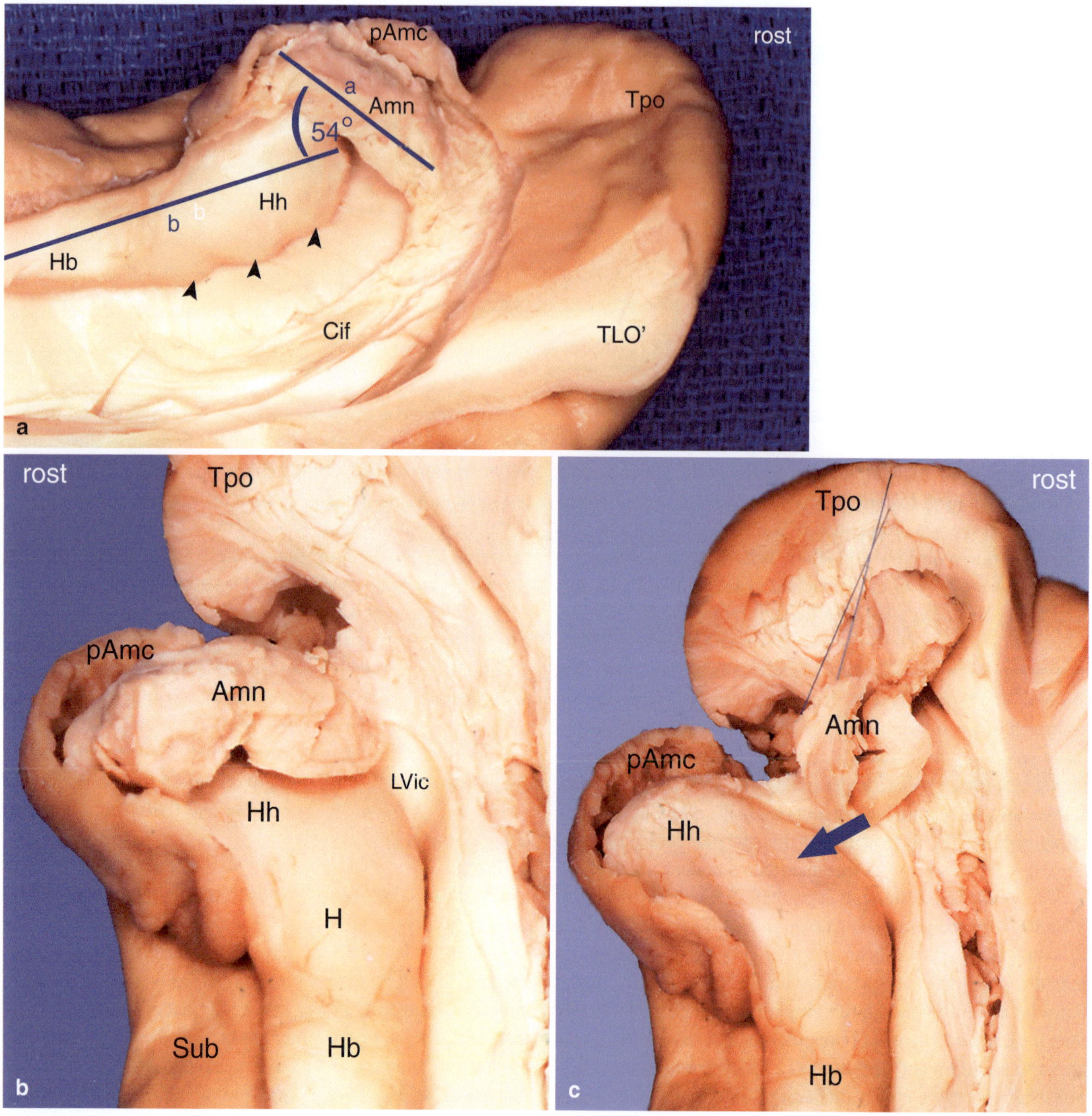

Fig. 4.1 Relations of the amygdaloid nuclear complex with the hippocampal head. *rost: rostral.* (**a**) supero-medial view of the temporal lobe's anterior part. (**b**) superior view of the temporal lobe's anterior part. (**c**) The same view as **b**, with separation of the Amn (lifted). *Amn* amygdala, *Cif* cingulate fasciculus, *H* hippocampus, *Hb* hippocampus body, *Hh* hippocampus head, *pAmc* periamygdalar cortex, *LVic* lateral ventricle, inferior cornu, *sub* subiculum, *TLO'* sagittal section of the temporal lobe, *blue arrow* showing the cupuliform print made by the amygdala's convexity, *black arrowheads* digitationes hippocampi (internal digitations). *Observe on* (**a**): *The great axis of the amygdala and the hippocampal bodies'axis (blue coloured) form an angle of 54°*

writers adopt as contours of the amygdala. This explains that the results vary according to the authors: 1154 mm^3 (right side) to 1160 mm^3 (left side), for JC Pruessner et al. 2000, but 1691.7 mm^3 (right side) to 1726.7 mm^3 (left side), for B Brierley et al. 2002. The divergence in the results is regrettable. That is why, in 1992, C Watson et al. 1992 attempted to standardise the measurements based on anatomical criteria, which are referred to since then as "Watson's criteria". But besides the fact that these criteria were complex to handle and therefore difficult to implement and that they generated numerical results exceeding standard averages, it is the development of new segmentation protocols (JL Hanson et al. 2012; JJ Entis et al. 2012) which should now ensure a more reliable numerical analysis of MRI images and provide

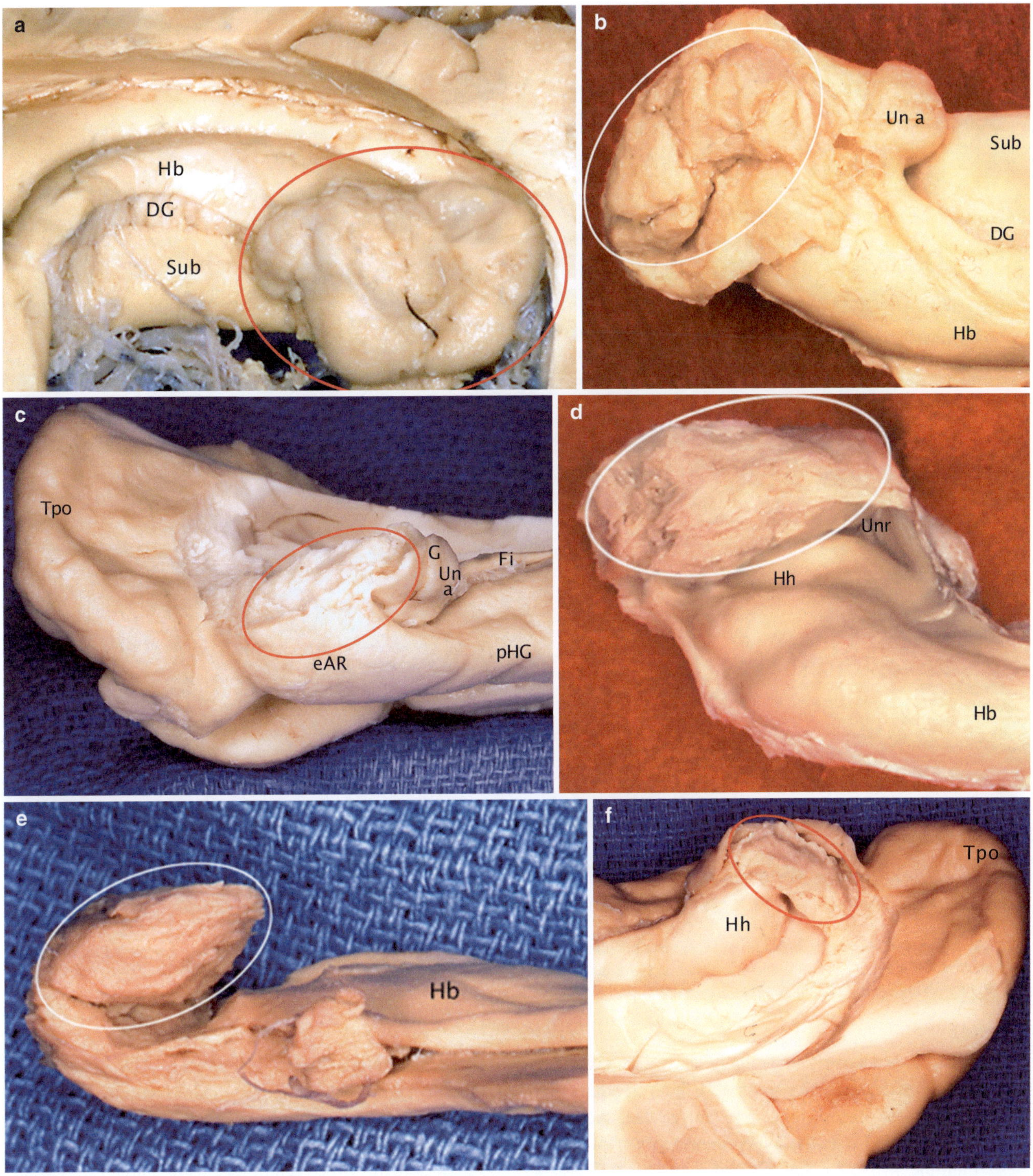

Fig. 4.2 Morphological aspects of the amygdala, in situ, above the hippocampal head. (**a**) Globular with bulging nuclei; (**b**) reniform; (**c**) flattened; (**d**) massive; (**e**) classical, almond shaped; (**f**) in shape of a butter's package. *DG* dentate gyrus, *eAr* entorhinal area, *Fi* fimbria, *G* band of Giacomini, *Hb* hippocampal body, *Hh* hippocampal head, *pHG* parahippocampal gyrus, *Sub* subiculum, *Tpo* temporal pole, *Un a* uncal apex, *Unr* uncal recess. On every picture of the artwork, the amygdala is surrounded by a *circle* so as to be displayed well

comparable volumetric results whatever teams are conducting these measurements. Being able to accurately measure, as a routine process, the volume of the amygdala would be very interesting since some volume variations are due to dysfunctions of the nervous system.

Thus, C Lange and E Irle 2004 observed amygdalae with an increased volume in young women who had recently suffered from a major depression. JP Hamilton et al. 2008 only noted this increase in treated patients (antidepressants seem to secrete neurotrophic factors favouring neurogenesis!).

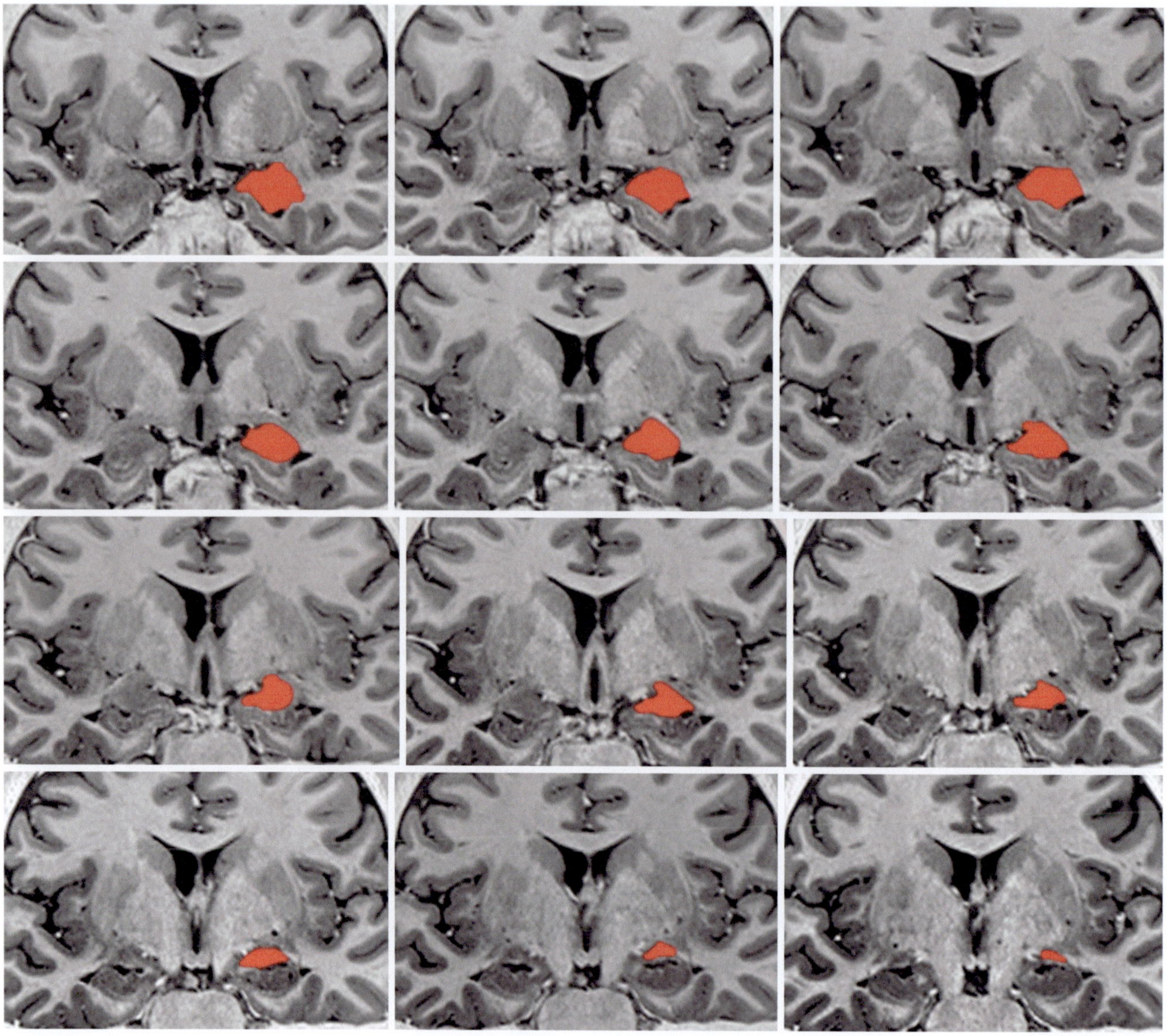

Fig. 4.3 A series of successive MRI coronal sections of the brain with *red coloured* left amygdalae in order to show their coronal main lines. *The slightly oblique main lines at the top of the picture (the more anterior sections) become gradually more or less horizontal on the posterior sections*

In any case, an excessively large amygdala volume is correlated with a greater tendency to anxiety, both in adults and children (in these excessively large amygdalae, there are more synapses, and it is the hyperconnectivity induced by these multiple synapses that generates anxiety). Thus, S Quin 2014 (Stanford University) who examined with fMRI, 76 children aged between 7 and 9 years old, observed an increase of the amygdala volume in anxious children. He also managed to establish that childhood anxiety is predictable based on the increase in volume of the left amygdala or of that of the left and right baso-lateral amygdaloid nuclei.

Equally interesting were the studies conducted on individuals with amygdalae with a volume below average, which are numerous and varied. Thus, correlations with an excessively small amygdala were developed by DA Pardini et al. 2014 who showed that the amygdaloid volume, which is smaller than average, in males, is associated with childhood aggressiveness, future violence and even psychotic features.

Post-traumatic stress syndromes were studied in particular: no amygdaloid changes were observed in adults from various backgrounds (FL Woon and DW Hedges 2008, 2009). The same authors noted no significant change in the volume of the amygdalae in children and adults who suffered from post-traumatic stress syndromes as a result of ill treatment during childhood. However, the study of post-traumatic stress syndromes among veterans revealed a decreased amygdaloid volume. The authors of this study (RA Morey et al. 2012) question whether this decrease in volume does

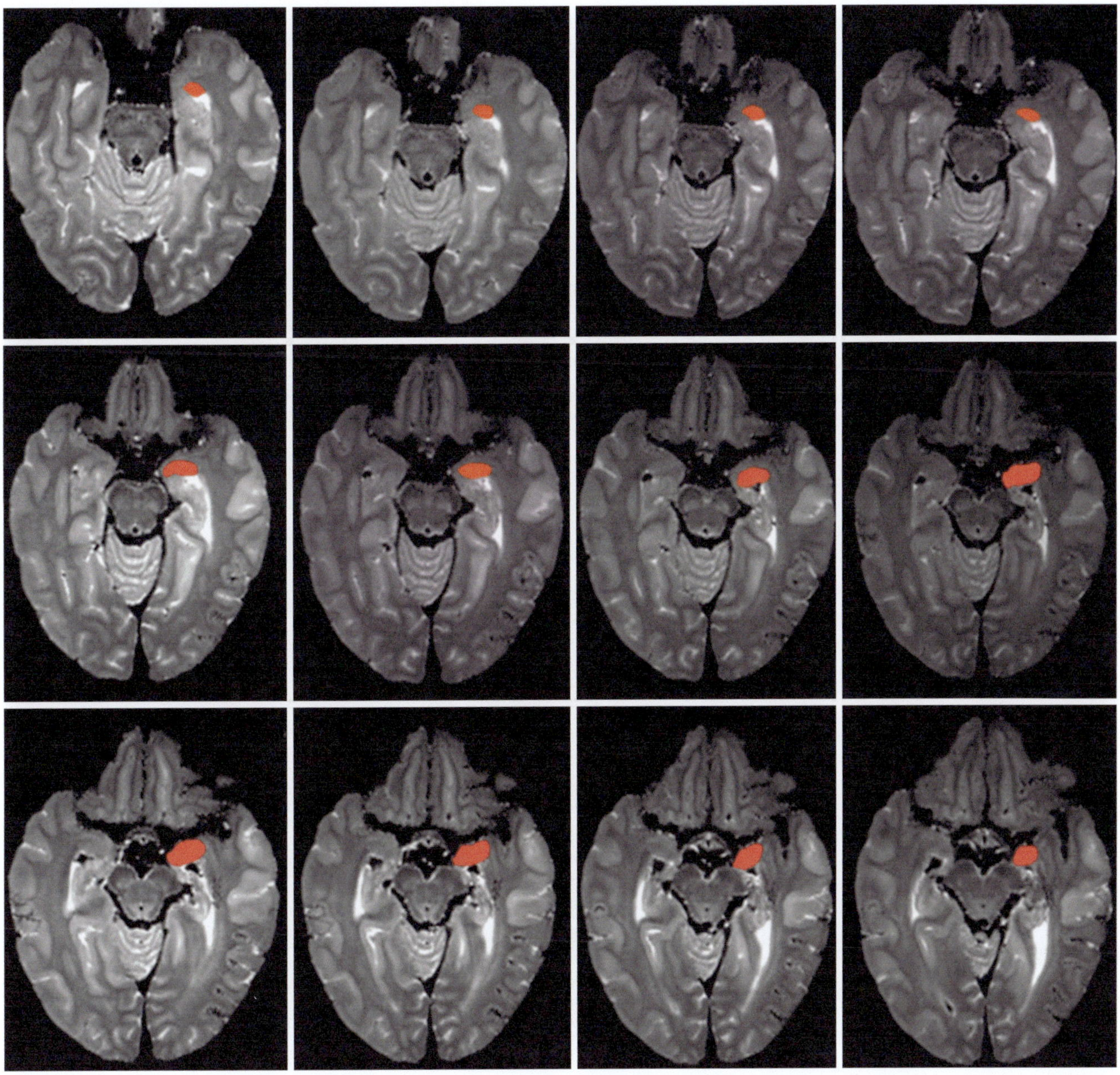

Fig. 4.4 A series of successive MRI axial sections of the brain with *red coloured* left amygdalae in order to show their axial main lines. *The more or less horizontal main lines at the lower level of the sections (top of the picture) become gradually oblique behind and inside at the higher sections*

not reflect the result but rather the vulnerability to this type of syndrome.

In people suffering from schizophrenia, there appears to be no significant reduction in the amygdaloid volume (SA Chance et al. 2002).

However, in bipolar psychotics, studied during their first episode, the volumes of the amygdala and cerebral white substance appear to be significantly reduced (IM Rosso et al. 2007).

SY Hill et al. 2013 also showed that a reduction in the amygdaloid volume is seen in the offspring of families with multiple alcohol-dependent members. Nevertheless, a study conducted by VV Senatorov et al. 2015 showed that alcoholics have an enlarged amygdala and reduced anterior insula.

The daily use of cannabis causes, over the long term, a reduction in volume of the hippocampus and of the amygdala (I Murat Yucel et al. 2008). We also know that a small amygdala increases the risk of addiction (N Makris et al. 2004).

Then there is the work of K Schiltz et al. 2007 who studied the brains of 15 male non-violent paedophiles and observed a size reduction of the right amygdala and a bilateral reduction in the hypothalamic grey substance of the septal region of the innominate substance and of the BST (anomalies unrelated to the age of the subjects being examined).

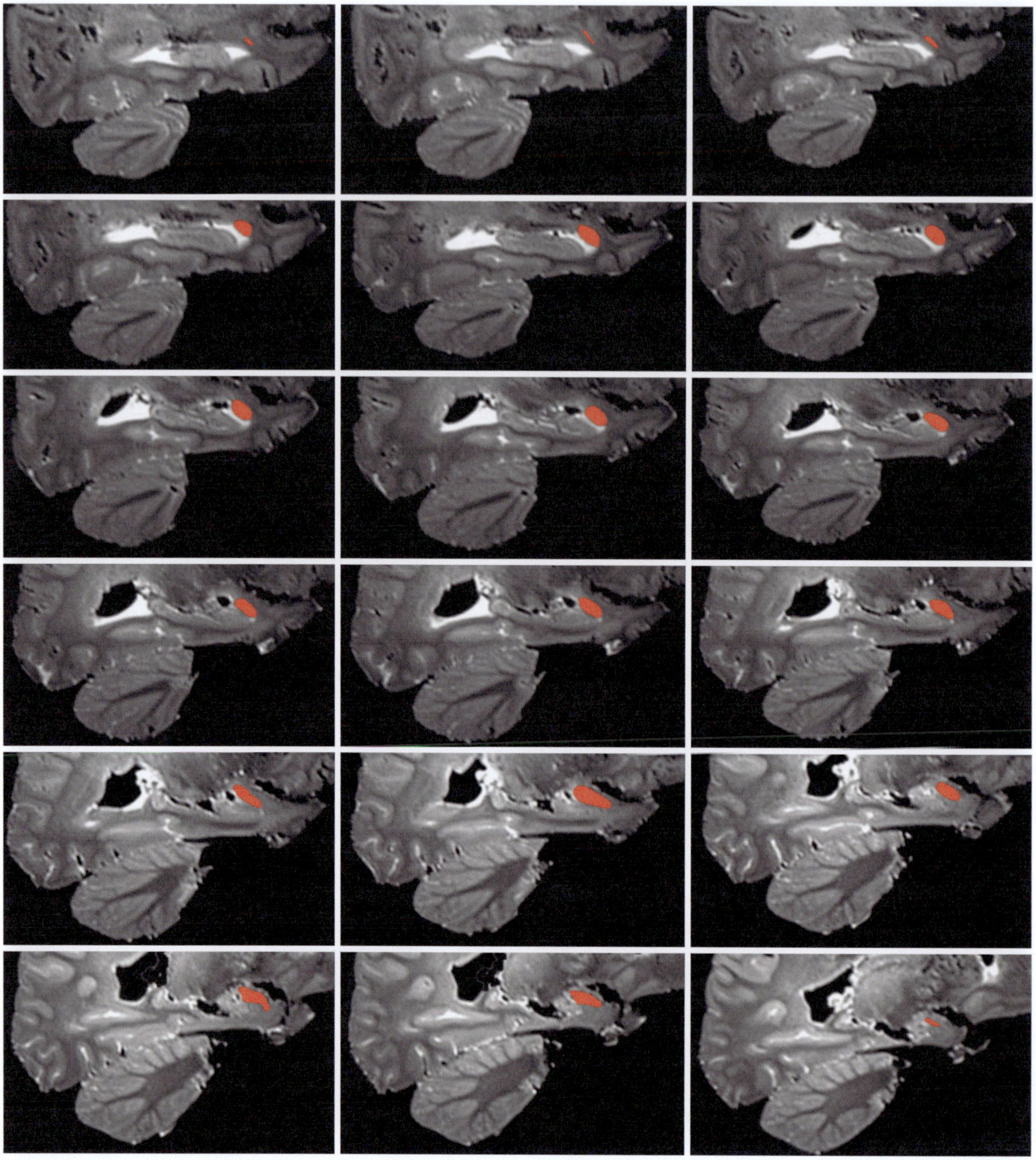

Fig. 4.5 A series of successive MRI sagittal sections of the brain with *red coloured* left amygdalae in order to show their sagittal main lines. *The main line remains oblique below and forwards at all the sagittal sections*

The specific case of autism has greatly interested neuroscientists who have looked for correlations between the size of the amygdalae of children who are affected and the intensity of their disorders (BM Nacewicz et al. 2006). In this puzzling disorder, where we can frequently observe increases in the amygdaloid volume, the most frequently explanation provided is that normal apoptosis phenomena, which normally occur in the amygdalae during the foetal period, do not take place or not sufficiently.

J Munson et al. 2006 focused more specifically on the volume increase of the right amygdala at the ages of 3 and 4 years. The greater the volume increase, the more the communication skills and social life will be disrupted.

CY Schumann et al. 2004 observed no difference in terms of total brain volume among a group of autistic children, aged between 7.5 and 12.5 years, and a control group, while the volumes of the left and right amygdalae of these autistic children were larger than those of the children belonging to the control group. However, in the group from 12.5 to 18.5 years old, the volume differences are no longer observed, which seems to show that the normal volume increase in standard amygdalae during adolescence does not occur in autistic children.

For JE Kim et al. 2010, children with abnormal social behaviours and communication problems have amygdalae whose nuclei of the baso-lateral group have an increased volume.

With the sequential images provided by MRI, we can obtain a relatively accurate idea of the *orientation* of the amygdala within the brain. On the axial sections, the large frontal axis of the amygdala is oriented slightly backwards and inwards. On the sagittal sections, we can observe that the large amygdalar axis is a little less oblique than that of the hippocampal head. On the frontal sections, the direction of the large axis of the amygdala shows that it is the result of the rotational movement undergone by this formation during its evolution. This axis, which was vertical, craniocaudal in primitive mammals, has been titled by 130–140, and in humans, it is directed obliquely downwards and outwards, as shown by the work of T Humphrey.

The special *consistency* of the cadaveric amygdala should also be noted. Its hardness contrasts with the fragility and soft, crumbly consistency of the peripheral cortex. This consistency is actually that of the constituent nuclei that appear dense, compact and crush resistant.

4.1.1 Subdivisions of the Amygdala: The Amygdaloid Nuclei

It is impossible to study these nuclei accurately by only using sections in various spatial planes. However, a well-conducted dissection, extracting the amygdala from its peripheral cortex envelope and then truly sculpting its surface, provides a good idea of the layout of these nuclei. Considering their hardness in anatomical subjects prepared by with formalin injections, such a dissection can generally be achieved in good conditions (Fig. 4.6).

4.1.1.1 Nuclei Forming the Amygdaloid Nuclear Complex

The authors only truly agree to affirm the heterogeneity of the amygdala, which is an assembly of cellular islands referred to as nuclei.

However, the number of constituent nuclei of the human amygdala varies according to the authors, the maximum being 30 nuclei for H Brockhaus 1938. Other authorsl such as EC Crosby and T Humphrey 1962 singularly reduced the segmentation of the human amygdala, not only by acknowledging a smaller number of nuclei but also by grouping these nuclei as superficial nuclei (medial and cortical), which are phylogenetically the oldest, and deep nuclei (lateral, basal and central), which are phylogenetically more recent.

Among the authors who have recently studied the nuclear arrangement of the human amygdala, we should specifically mention L Heimer et al. 1995, 2008, P Sah et al. 2003, JS De Olmos 2004 and CM Schuman and DG Amaral 2006. All of these authors found that each nucleus has specific characteristics not only morphological but also histochemical or cytoarchitectural. Qualitative differences (S Kemppainen et al. 2002) and quantitative differences (N Barger et al. 2012; CY Schumann and DG Amaral 2006) have thus been identified in the neuronal content of some amygdaloid nuclei in humans, and comparisons have been made with the amygdaloid nuclei of great apes. From the various studies, we can conclude that the evolution of the human amygdala did not consist of a simple increase in volume, at the same time as the brain volume increase, but rather of an adaptive reorganisation of the various constituent nuclei (K Semendeferi et al. 2010).

While EC Crosby and T Humphrey 1941, in their aim to summarise, had, as we saw, reduced the nuclei to only 2 groups, the cortico-medial group and the baso-lateral group, most authors acknowledge the current 4 anatomical groups which have different features:

The baso-lateral or deep group (complex) consists of three nuclei: the lateral, the baso-lateral and the baso-medial (or secondary baso-lateral or even secondary basal).

The cortical group or superficial group owes its name to the fact that due to its surface location, it is flush with the cortex, of which it adopts the layered structure. However, we have always found, during each dissection, a real cortical layer which covers these cortical nuclei. This layer is generally thin with respect to other parts of the cortex. Furthermore, several authors include in this cortical group the entire periamygdaloid cortex that they consider as equivalent to a nucleus. The other nuclei belonging to this group are the anterior cortical and posterior cortical nuclei, the nucleus of the lateral olfactory tract and that of the secondary olfactory tract, which amounts to 5 nuclei for the superficial group.

The centro-medial group has 3 nuclei, among which 2 are essential from a functional viewpoint: the central nucleus and the medial nucleus. The third nucleus corresponds to the amygdaloid part of the nucleus of the bed of stria terminalis.

The group of additional nuclei includes the anterior amygdaloid area (visible on axial and sagittal sections), the amygdalo-hippocampal area and the amygdala-piriform area (visible

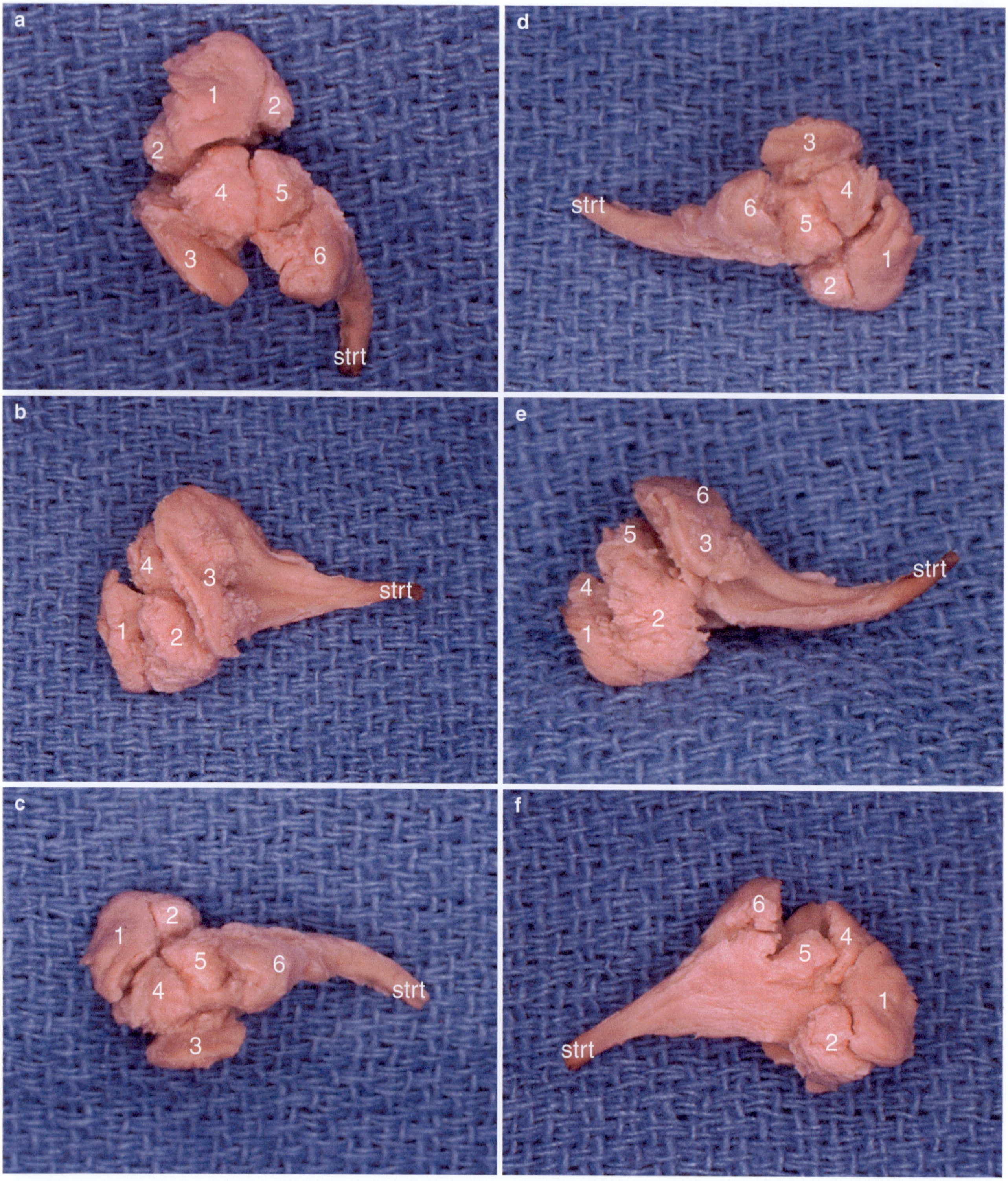

Fig. 4.6 Dissection of the right nuclear amygdaloid complex showing the arrangement of the nuclear subdivisions. *1* lateral nucleus, *2* baso-lateral nucleus, *3* baso-medial nucleus, *4* cortical nucleus, *5* central nucleus, *6* medial nucleus, *strt* stria terminalis. (**a**) antero-superior view; (**b**) antero-medial view; (**c**) superior view; (**d**) lateral view; (**e**) inferior view; (**f**) anterolateral view

on the coronal sections) and smaller clusters of neurons called intercalated nuclei because they are located between the large nuclei that we have just listed and which are as the latter, cellular islands limited by a fibrillar envelope.

4.1.1.2 Nomenclature of the Nuclei

According to us, the nomenclature defined by JS De Olmos 2004 seems to be the most appropriate. De Olmos has the great merit of naming each of the nuclei by abbreviations but also of reviewing and clarifying the amygdaloid nomenclature.

We therefore have: in the latero-basal group, the lateral (La), the baso-lateral (BL) and the baso-medial (BM); in the cortico-medial group, the cortical (Co); and in the centro-medial group, the central (Ce) and the medial (Me).

Each nucleus has subsequently been organised into subdivisions that may confuse novices but which provide a high level of accuracy and a fully functional labelling system (Fig. 4.7).

The *lateral nucleus*, which obviously comprises a lateral portion and a medial portion, was divided into L La and M La divisions. Each lateral and medial division comprises a dorsal portion and a ventral portion, separated by an intermediate division I La. There is also a limiting division labelled Li La.

The *baso-lateral nucleus* has been divided into a dorsal portion D BL and a ventral portion V BL, separated by an intermediate portion I BL.

The *baso-medial nucleus* has been divided into 4 portions, the dorsolateral portion DL BM, the dorsomedial portion DM BM, the ventrolateral portion VL BM and the ventromedial portion VM BM.

The *central nucleus* has been divided into a lateral portion Ce L and a medial portion Ce M, the latter being divided itself into a dorsal portion D Ce M and a ventral portion V Ce M.

The medial nucleus has been divided into an anterior portion Me A and a posterior portion Me P, which itself has been divided into a dorsal portion D Me P and a ventral portion V Me P.

It is the *cortical* nucleus that poses the most problems. Co comprises three parts:

An anterior nucleus ACo with a dorsal division D ACo and a ventral division V ACo

A ventral nucleus VCo with a caudal division C VCo and a rostral division R VCo.

R VCO itself is divided into an inferior portion IF R VCo, a superior portion Sp R VCo and an intermediate portion IT R VCo.

As for the *anterior* nucleus, labelled anterior amygdaloid area AAA, it is divided into two subdivisions, the first one, the deep, Dp AAA, deep division of AAA, and the other, superficial, Sf AAA, superficial division of AAA.

Two junction areas surround the cortical nucleus:

The *amygdalo-piriform area* APir, including an antero-lateral division AL APir and a postero-medial division PM APir

The *amygdalo-hippocampal area* AHi which includes an anterolateral division AL AHi and a postero-medial division PM AHi

We should also mention the *amygdalo-striatal* transitional area AStr and the *amygdalo-claustral* area ACA included in this nomenclature by the author's highly accurate mind.

4.1.1.3 Highlighting the Nuclei

The nuclei are perfectly visible on histological sections after Nissl staining. Using more sophisticated staining techniques, such as silver plating, is not required.

Our surface staining of serial brain sections with FeC13[2] is also a great way of enhancing the visualisation of the amygdaloid nuclei, even if the amygdala is coloured with less intensity than other central grey nuclei. It is above all a simple and inexpensive technique which can be used extemporaneously by a neuroanatomist (Figs 4.8, 4.9, 4.10, 4.11, 4.12, 4.13 and 4.14).

In vivo, the future trends involve the observation of MRI images. While current devices sometimes allow us to assume the outlines of the amygdala, we can hope to achieve such a visualisation as a routine process with the latest generation of devices. The anatomical and clinical research field concerning the amygdala is huge, and one can already imagine that it will be possible to use a superselective MRI mapping technique to establish a correlation between this or that pathology and a specific abnormality of one or more amygdaloid nuclei.

We have also initiated with the CEMEREM,[3] a comparative anatomical study of serial brain sections and their MRI images, and were able to obtain extremely clear images of the amygdaloid nuclei. In vivo attempts are promising: they involved using high-quality diffusion tensor imaging (DTI) at 3 T, which is possible thanks to the diffusion anisotropy of the grey substance of the nuclei. These attempts were performed in living men (Solano-Castiella et al. 2010). Although they only provided segmentation images of the amygdala in lateral and medial areas, this leads us to believe that fast progress will occur in the near future with the new MRI and technical protocols (Entis et al. 2012), which are more efficient for achieving the quality desired for cadaveric sections. It is however to be noted that future imaging must always be confronted with the reality of macro- or microscopic anatomical sections because there are numerous error possibilities which can lead to erroneous interpretations of the so-called nuclei that are in reality artefacts (Fig. 4.15).

---

[2]Di Marino V, Etienne Y, Niddam M, Tacher V (2008) Rehaussement des structures du système nerveux central par traitement des coupes anatomiques avec le chlorure ferrique. Morphologie 90:144–150

[3]CEMEREM: Centre d'Explorations Métaboliques par Résonance Magnétique (APHM, Marseille, France)

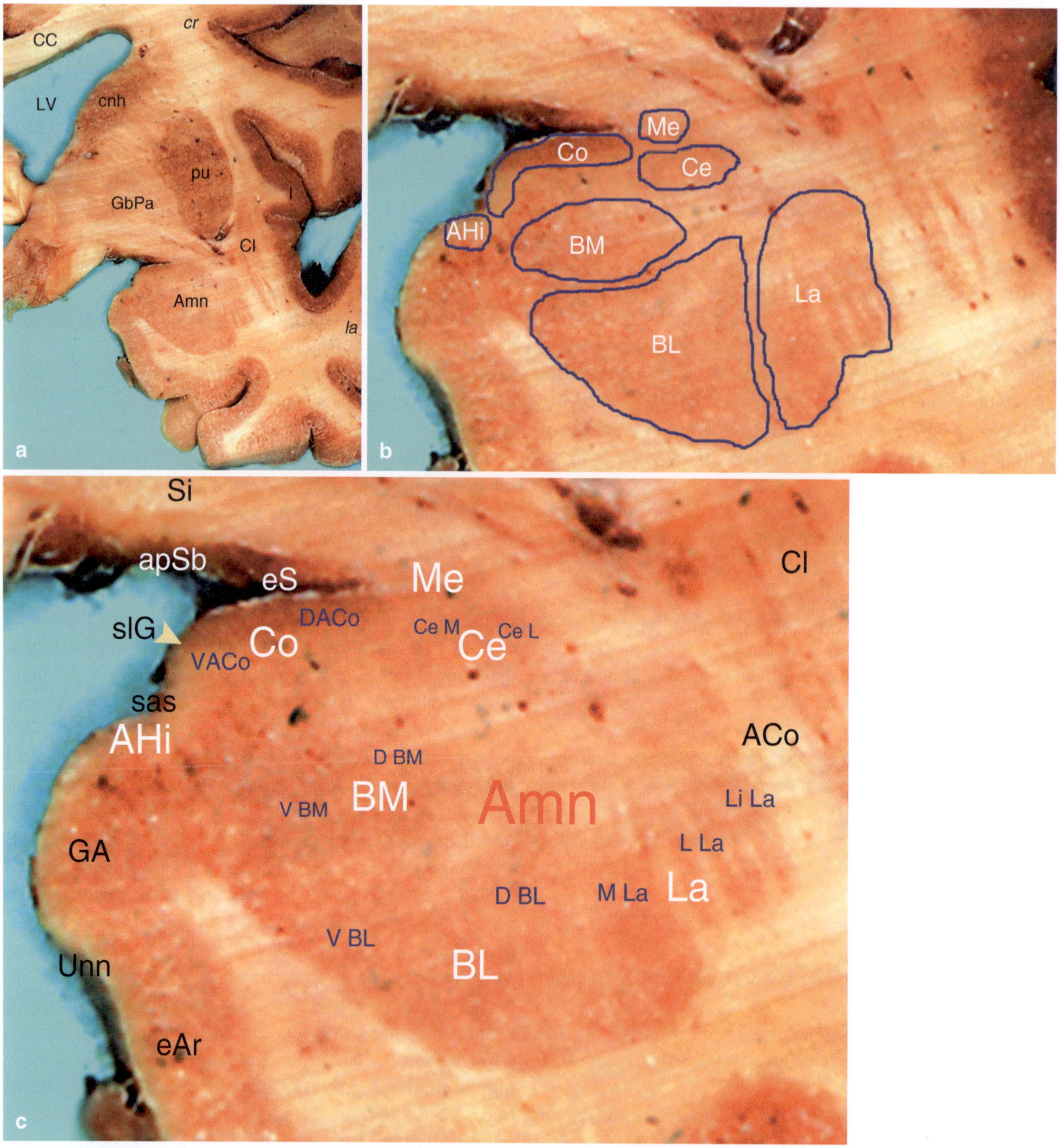

Fig. 4.7 The nuclei of the human amygdala. (**a**) The left amygdala in situ on a coronal section of the brain (*cr,* cranial; *la,* lateral). (**b**) Manual drawing of the amygdala's nuclei coloured by FeCl3. (**c**) An overview of the nuclei according to De Olmos's classification. *ACo* anterior commissure, *Amn* amygdaloid nuclear complex (*AHI* amygdalo-hippocampal area, *BL* baso-lateral nucleus, *D BL* dorsal subdivision sd, of BL, *V BL* ventral sd of BL, *BM* baso-medial n, *D BM* dorsal sd of BM, *V BM* ventral sd of BM, *Ce* central nucleus, *Ce L* lateral sd of Ce, *Ce M* medial sd of BM, *Co* cortical n, *DA Co* dorsal anterior sd of Co, *VA Co* ventral anterior sd of Co, *La* lateral n, *Li La* limitans sd of LA, *L LA* lateral sd of LA, *M La* medial sd of LA, *Me* medial n), *apSb* anterior perforated substance, *CC* corpus callosum, *Cl* claustrum, *cnh* caudate nucleus head, *eAr* entorhinal area, *eS* entorhinal sulcus, *GA* gyrus ambiens, *GbPa* globus pallidus, *I* insula, *LV* lateral ventricle, *pu* putamen, *sas* semi-annular sulcus, *slG* semilunar gyrus, *Si* substantia innominata, *Unn* uncal notch

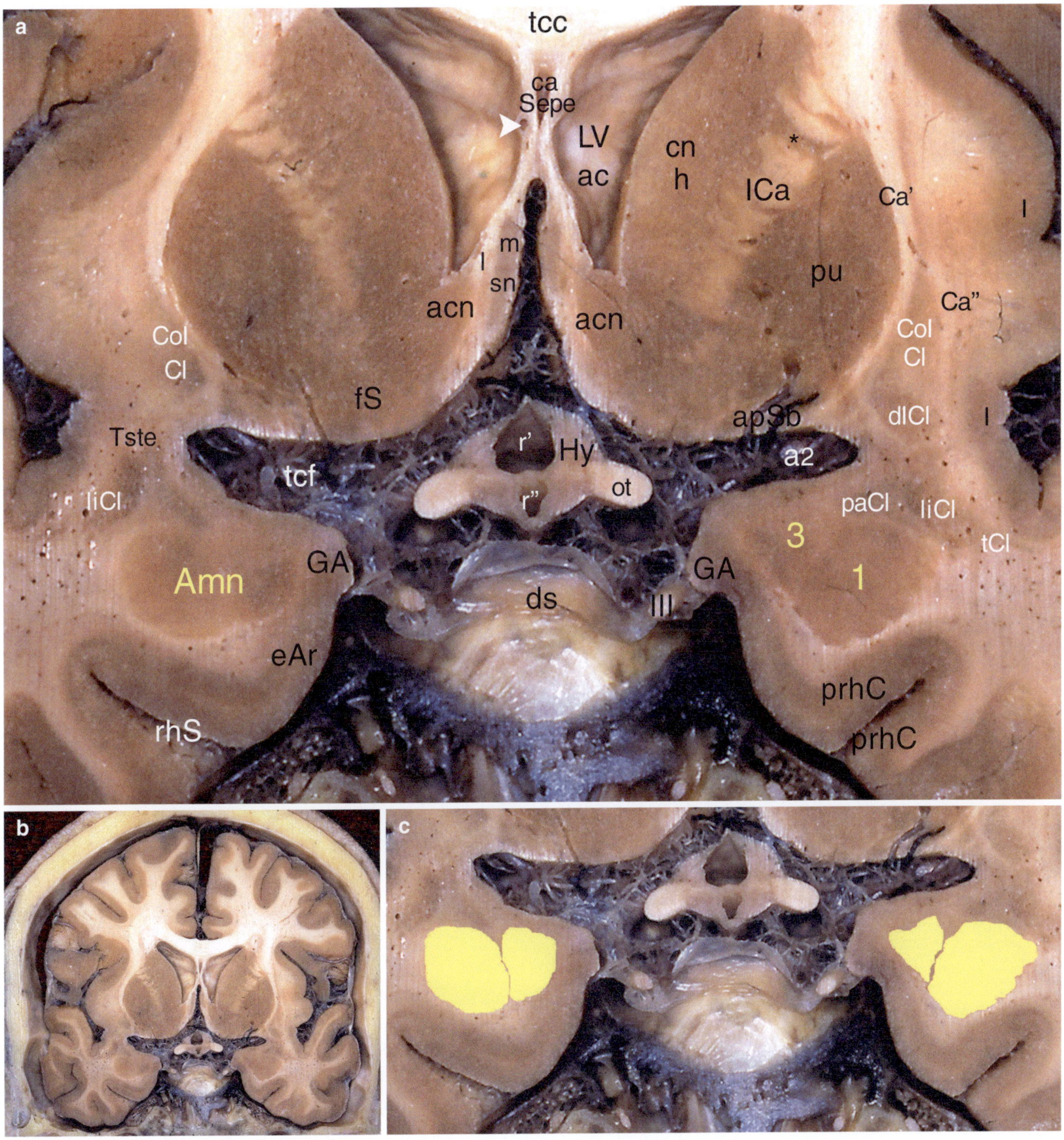

Fig. 4.8 The amygdaloid nuclear complex on a coronal section of the brain through the septal area. (**a**) Details, (**b**) general situation, (**c**) nuclear organisation of Amn. *acn* accumbens nucleus, *Amn* amygdaloid nuclear complex, *1* lateral nucleus, *3* baso-medial nucleus, *apSb* anterior perforated substance, *Ca'* external capsule, *Ca"* extreme capsule, *ca Sepe* cavity of the septum pellucidum, *cn h* caudatus nucleus (head), *CoICl* compact insular claustrum, *dICl* diffuse insular claustrum, *ds* dorsum sellae (sphenoidal bone), *eAr* entorhinal area, *fS* fundus striati, *GA* gyrus ambiens, *Hy* hypothalamus, *I* insula, *ICa* internal capsule, *liCl* limitans claustrum, *LVac* lateral ventricle (anterior cornu), *ot* optic tract, *paCl* pre-amygdalar claustrum, *prhC* perirhinal cortex, *pu* putamen, *r'* optic recess of V3, *r"* infundibular recess of V3, *rhS* rhinal sulcus, *sn* septal nuclei, *m* medial, *l* lateral, *tCC* truncus of the corpus callosum, *tcf* transverse cerebral fissure, *tCl* temporal claustrum, *Tste* temporal stem, *III* oculomotor nerve, *a2* middle cerebral artery, *white arrowhead* venae of the septum pellucidum, *black asterisk* caudato-lenticular grey cell bridge

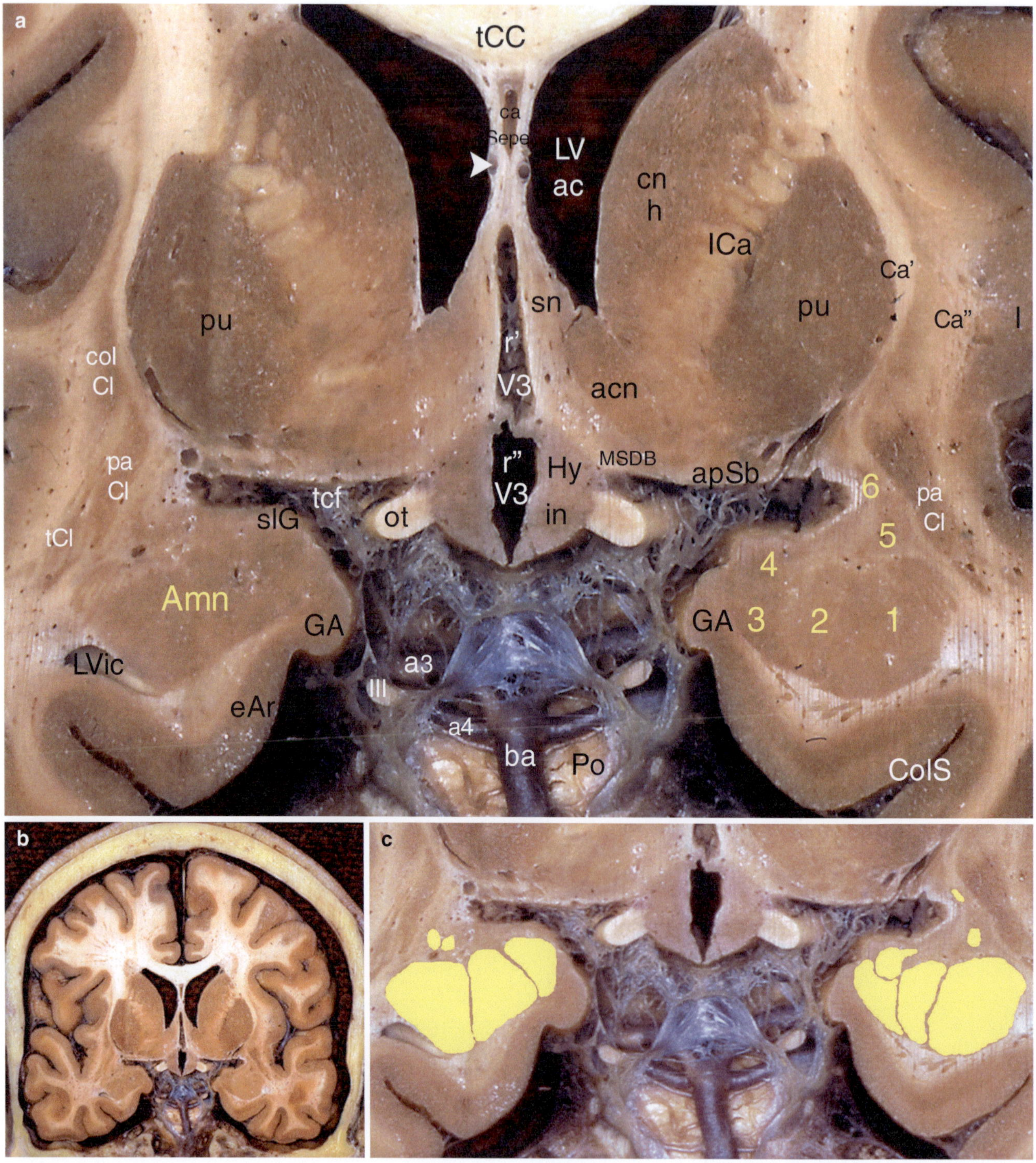

Fig. 4.9 The amygdaloid nuclear complex on a coronal section of the brain through the rostral part of V3. (**a**) Details, (**b**) general situation, (**c**) nuclear organisation of Amn. *acn* accumbens nucleus, *Amn* amygdaloid nuclear complex, *1* lateral nucleus, *2* baso-lateral nucleus, *3* baso-medial nucleus, *4* cortical nucleus, *5* central nucleus, *6* medial nucleus, *apSb* anterior perforated substance, *ba* basilar artery, *Ca'* external capsule, *Ca"* extreme capsule, *ca Sepe* cavity of the septum pellucidum, *cn h* caudatus nucleus (head), *colCl* compact insular claustrum, *ColS* collateral sulcus, *eAr* entorhinal area, *GA* gyrus ambiens, *Hy* hypothalamus, *I* insula, *ICa* internal capsule, *in* infundibular nucleus, *LVac* lateral ventricle (anterior cornu), *LVic* lateral ventricle (inferior cornu), *MSDB* medial septum-diagonal band, *ot* optic tract, *Po* pons, *paCl* pre-amygdalar claustrum, *pu* putamen, *slG* semilunar gyrus, *sn* septal nuclei, *tCC* truncus of the corpus callosum, *tcf* transverse cerebral fissure, *tCl* temporal claustrum, *V3* third ventricle, *r'* optic recess of *V3*, *r"* infundibular recess of *V3*, *III* oculomotor nerve, *a3* posterior cerebral artery, *a4* superior cerebellar artery, *white arrowhead* venae of the septum pellucidum

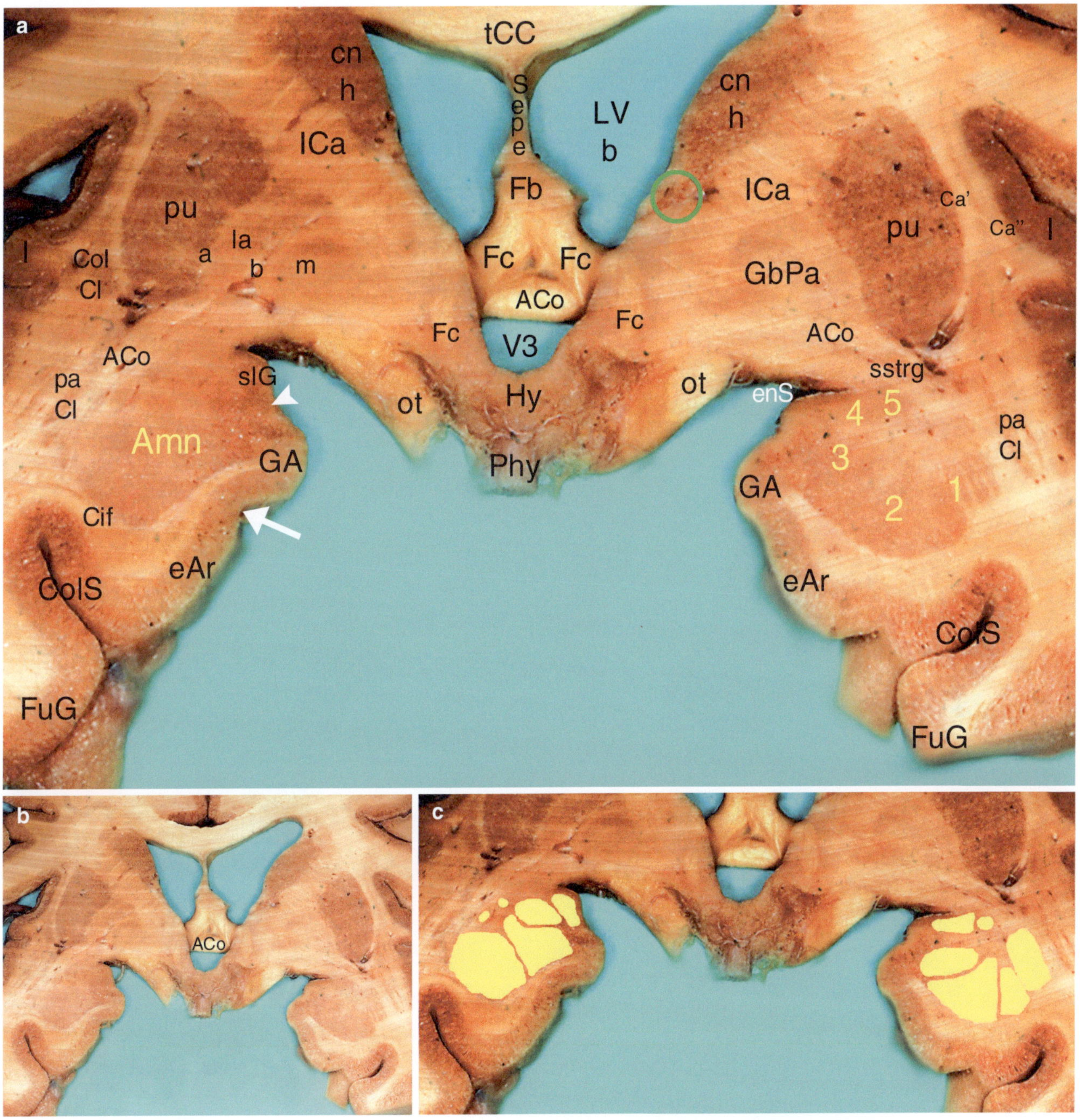

Fig. 4.10 The Amygdaloid nuclear complex on a coronal section of the brain through the anterior commissure. (**a**) details (**b**) general situation (**c**) nuclear organisation of Amn. *ACo* anterior commissure, *Amn* amygdaloid nuclear complex, *1* lateral nucleus, *2* baso-latéral nucleus, *3* baso-medial nucleus, *4* cortical nucleus, *5* central nucleus, *Ca'* external capsule, *Ca"* extreme capsule, *Cif* cingulate fasciculus, *cn h* caudatus nucleus head, *ColCl* compact insular claustrum, *ColS* collateral sulcus, *eAr* entorhinal area, *enS* endorhinal sulcus, *Fb* body of the fornix, *Fc* column of the fornix, *FuG* fusiform gyrus, *GA* gyrus ambiens, *GbPa* globus pallidus, *la* pars lateralis, *m* pars medialis, *a* external medullary lamina, *b* internal medullary lamina, *Hy* hypothalamus, *I* insula, *ICa* internal capsule, *LVb* lateral ventricle body, *ot* optic tract, *paCl* pre-amygdalar claustrum, *Phy* peduncle of the hypophysis, *pu* putamen, *Sepe* septum pellucidum, *slG* semilunar gyrus, *sstrg* substriatal grey, *tCC* truncus of the corpus callosum, *V3* third ventricle, *green circle* BNST, *white arrowhead* semi-annular sulcus, *white arrow* uncal notch

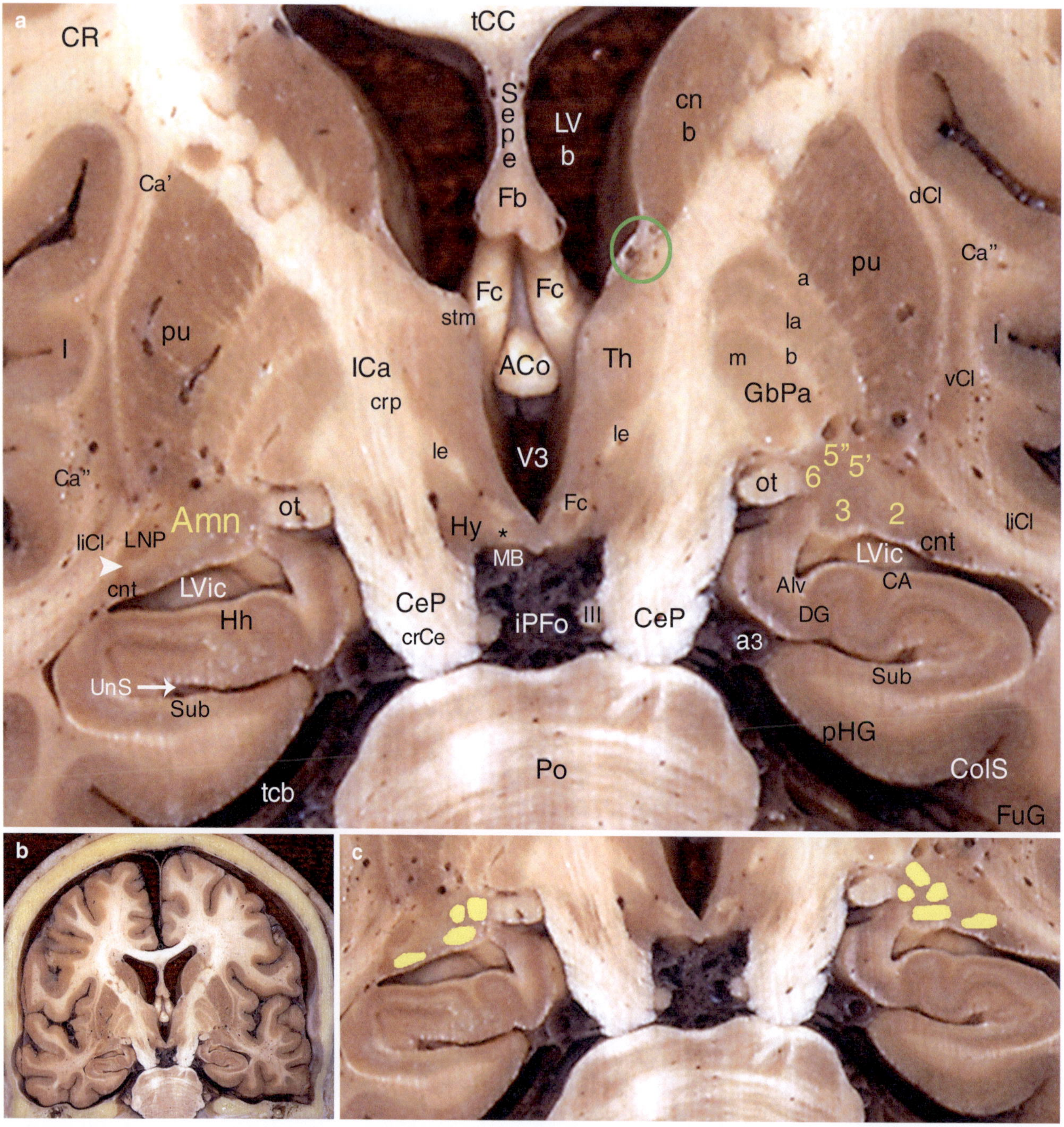

Fig. 4.11 The amygdaloid nuclear complex on a coronal section of the brain through the cerebral peduncles and the pons behind the columns of the fornix. (**a**) Details, (**b**) general situation, (**c**) nuclear organisation of Amn. *ACo* anterior commissure, *Alv* alveus, *Amn* amygdaloid nuclear complex, *2* baso-lateral nucleus, *3* baso-medial nucleus, *5'* central nucleus (lateral part), *5"* central nucleus (medial part), *6* medial nucleus, *CA* cornu ammonis, *Ca'* external capsule, *Ca"* extreme capsule, *CeP (crCe)* cerebral peduncle (crus cerebri), *cnb* body of the caudatus nucleus, *cnt* tail of the caudatus nucleus, *CR* corona radiata, *ColS* collateral sulcus, *dCl* dorsal claustrum, *DG* dentate gyrus, *Fb* body of the fornix, *Fc* column of the fornix, *FuG* fusiform gyrus, *GbPa* globus pallidus, *la* pars lateralis, *m* pars medialis, *a* external medullary lamina, *b* internal medullary lamina, *Hh* hippocampus head (digitationes hippocampi, *Hy* hypothalamus, *I* insula, *ICa (crp)*: internal capsule (crus posterior), *iPFo* interpeduncular fossa, *le* fasciculus lenticularis (field H2 of Forel), *liCl* limitans claustrum, *LNP* lentiform nucleus peduncle, *LVb* body of the lateral ventricle, *LVic* inferior cornu of the lateral ventricle, *MB* mamillary body, *ot* optic tract, *pHG* parahippocampal gyrus, *Po* pons, *pu* putamen, *Sepe* septum pellucidum, *stm* stria medullaris thalami, *Sub* subiculum, *tcb* tentorium cerebelli, *tCC* truncus of the corpus callosum, *Th* thalamus, *UnS (white arrow)* uncal sulcus V3: third ventricle, *vCl* ventral claustrum, *III* oculomotor nerve, *a3* posterior cerebral artery, *black asterisk* mamillothalamic fasciculus, *green circle* NBST, *NB* notice the grey bridge cell (*white arrowhead*) between the tail of the caudatus nucleus and LNP

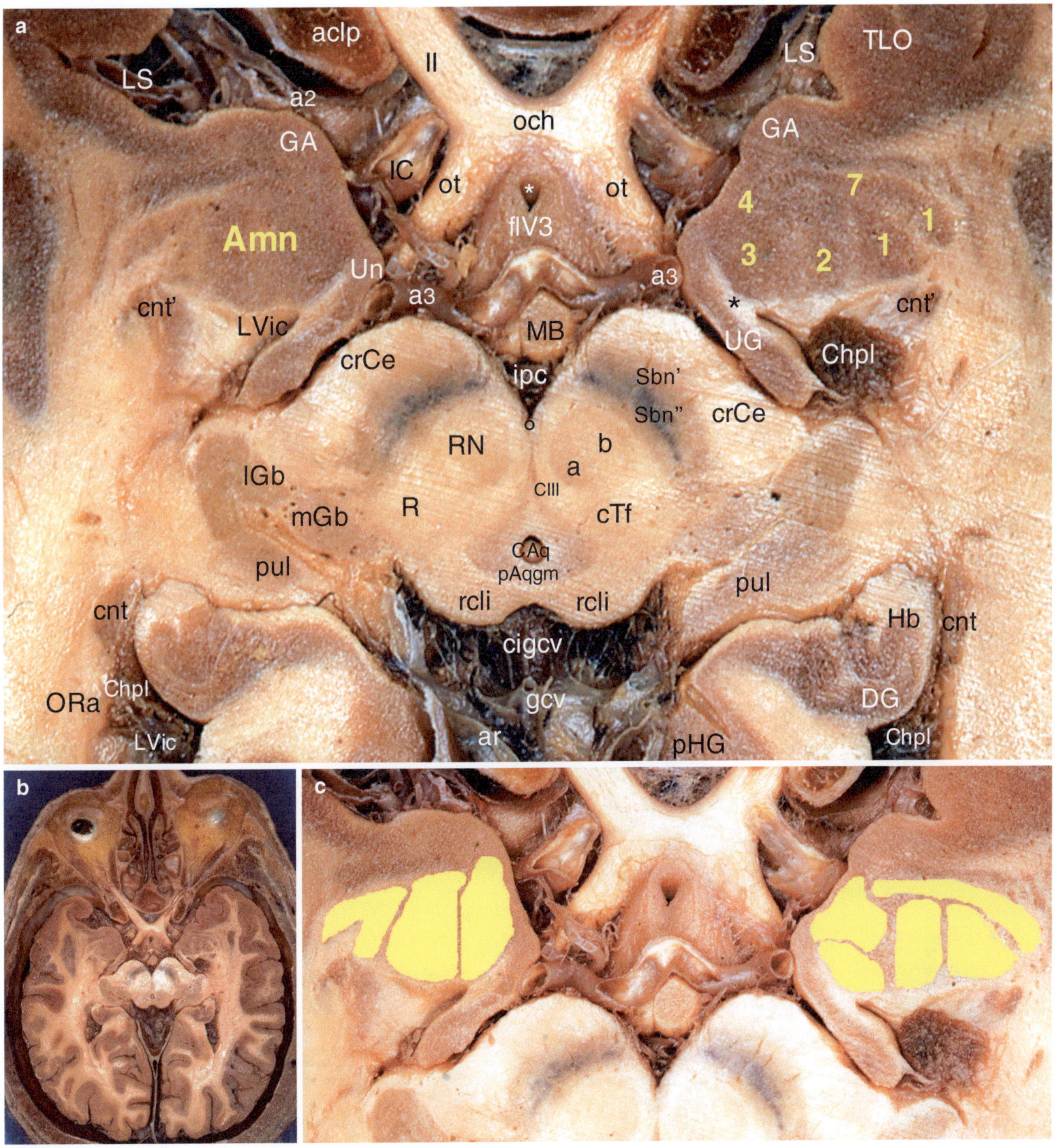

Fig. 4.12 The amygdaloid nuclear complex on an axial section of the brain through the mesencephalon (midbrain) and the optic chiasm. (**a**) Details, (**b**) general situation, (**c**) nuclear organisation of Amn. *Amn* amygdaloid nuclear complex, *1* lateral nucleus, *2* baso-lateral nucleus, *3* baso-medial nucleus, *4* cortical nucleus, *7* anterior nucleus, *aclp* anterior clinoid process, *ar* arachnoid mater, *CIII* oculomotor nerve's nuclear complex, *Caq* cerebral aqueduct, *Chpl* choroid plexus, *cigcv* cisterna of the great cerebral vein, *cnt* caudate nucleus tail, *cnt'* para-amygdalar part of the caudate nucleus tail, *crCe* crus cerebri, *cTf* central tegmental fasciculus, *DG* dentate gyrus, *flV3* floor of V3, *GA* gyrus ambiens, *gcv* great cerebral vein, *Hb* hippocampus body, *IC* internal carotid, *ipc* interpeduncular cisterna, *lGb* lateral geniculate body, *LS* lateral sulcus, *LVic* lateral ventricle, inferior cornu (temporal horn), *MB* mamillary body, *mGb* medial geniculate body, *och* optic chiasm, *ORa* optic radiations, *ot* optic tract, *pAqgm* periaqueductal grey matter, *pHG* parahippocampal gyrus, *pul* pulvinar, *R* reticular formation, *rcli* rostral colliculus, *RN* red nucleus, *a* magnocellular part, *b* parvocellular part, *Sbn'* substantia nigra, reticular part, *Sbn"* substantia nigra, compact part, *TLO* temporal lobe, *UG* uncinate gyrus, *Un* uncus, *a2* middle cerebral artery, *a3* posterior cerebral artery, *II* optic nerve, *black asterisk* cingulate fasciculus(Cif), *white asterisk* infundibular recess (inr), *black circle* interpeduncular nucleus

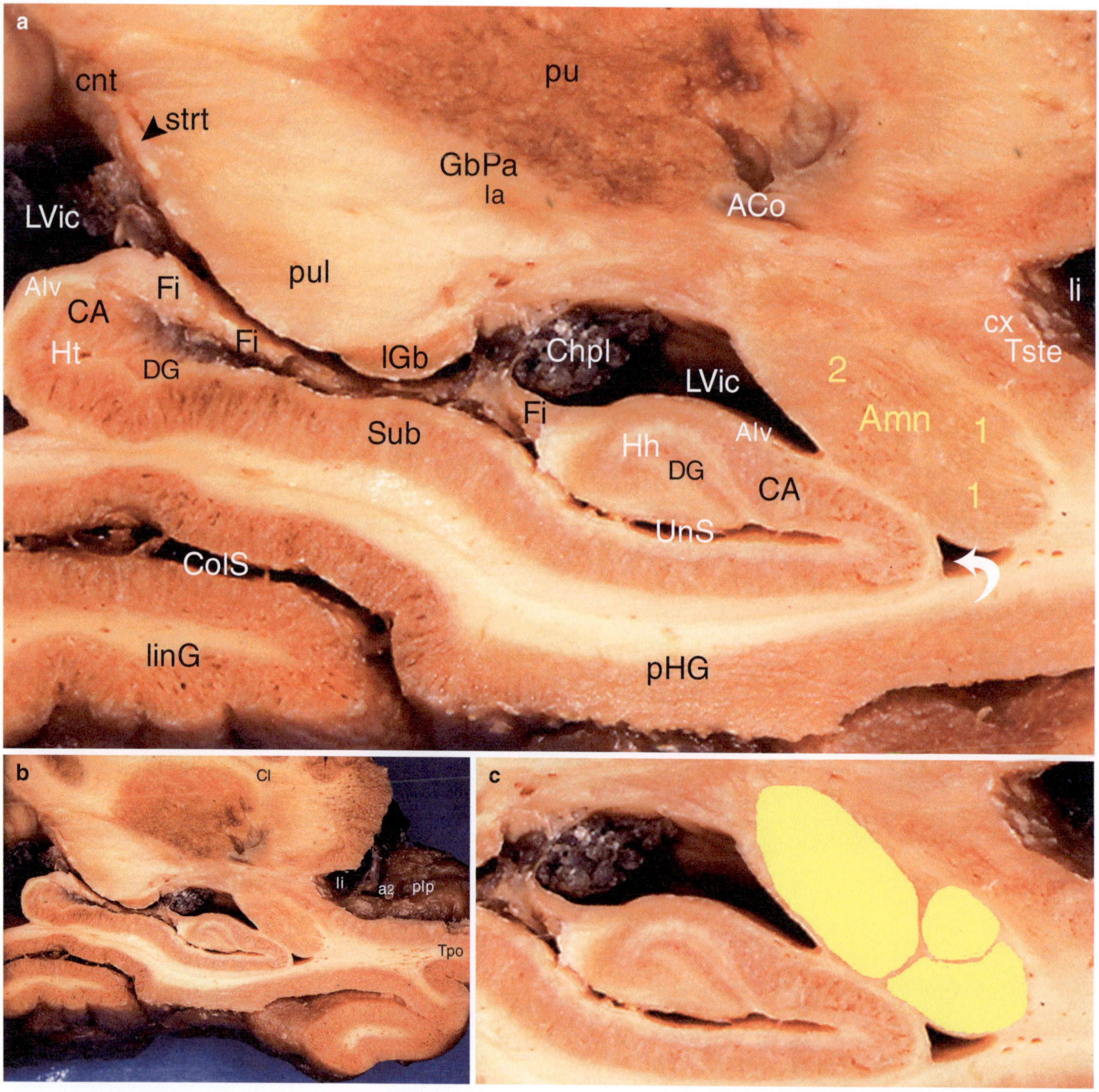

Fig. 4.13 The amygdaloid nuclear complex on a sagittal section of the brain through the lateral part of the lentiform nucleus. (**a**) Details, (**b**) general situation, (**c**) nuclear organisation of Amn. *ACo* anterior commissure, *Alv* alveus, *Amn* amygdaloid nuclear complex, *1* lateral nucleus, *2* baso-lateral nucleus, *CA* cornu ammonis, *Chpl* choroid plexus, *Cl* claustrum, *cnt* caudate nucleus tail, *ColS* collateral sulcus, *DG* dentate gyrus, *Fi* fimbria of the hippocampus, *GbPa* globus pallidus, *la* pars lateralis, *Hh* hippocampus head, *Ht* hippocampus tail, *lGb* lateral geniculate body, *li* limen insulae, *linG* lingual gyrus, *LVic* lateral ventricle, inferior cornu, *pHG* para hippocampal gyrus, *plp* planum polare, *pu* putamen, *pul* pulvinar, *strt* stria terminalis (*black arrowhead*), *Sub* subiculum, *Tpo* temporal pole, *Tste cx* cortex of the temporal stem, *UnS* uncal sulcus, *a2* middle cerebral artery, *white curved arrow* uncal recess of the VL's inferior cornu

4.1.1.4 Descriptive and Topographic Anatomy of the Main Nuclei

- The *lateral nucleus*, lateral amygdaleum profundum (H Brockhaus 1938, 1942a), is the most voluminous nucleus. This volume, which corresponds to a high neuronal density, is explained by the functional importance of this structure. The lateral nucleus is the main receptor of the sensory and sensitive impulses reaching the amygdala, what explains its eminent role in the emotional learning (DM Hilmazer-Hancke, 2012). This nucleus is positioned against the medial part of the temporal stem. More or less cubic on coronal sections, it comprises a lateral portion and a medial portion, separated by a thin intermediate

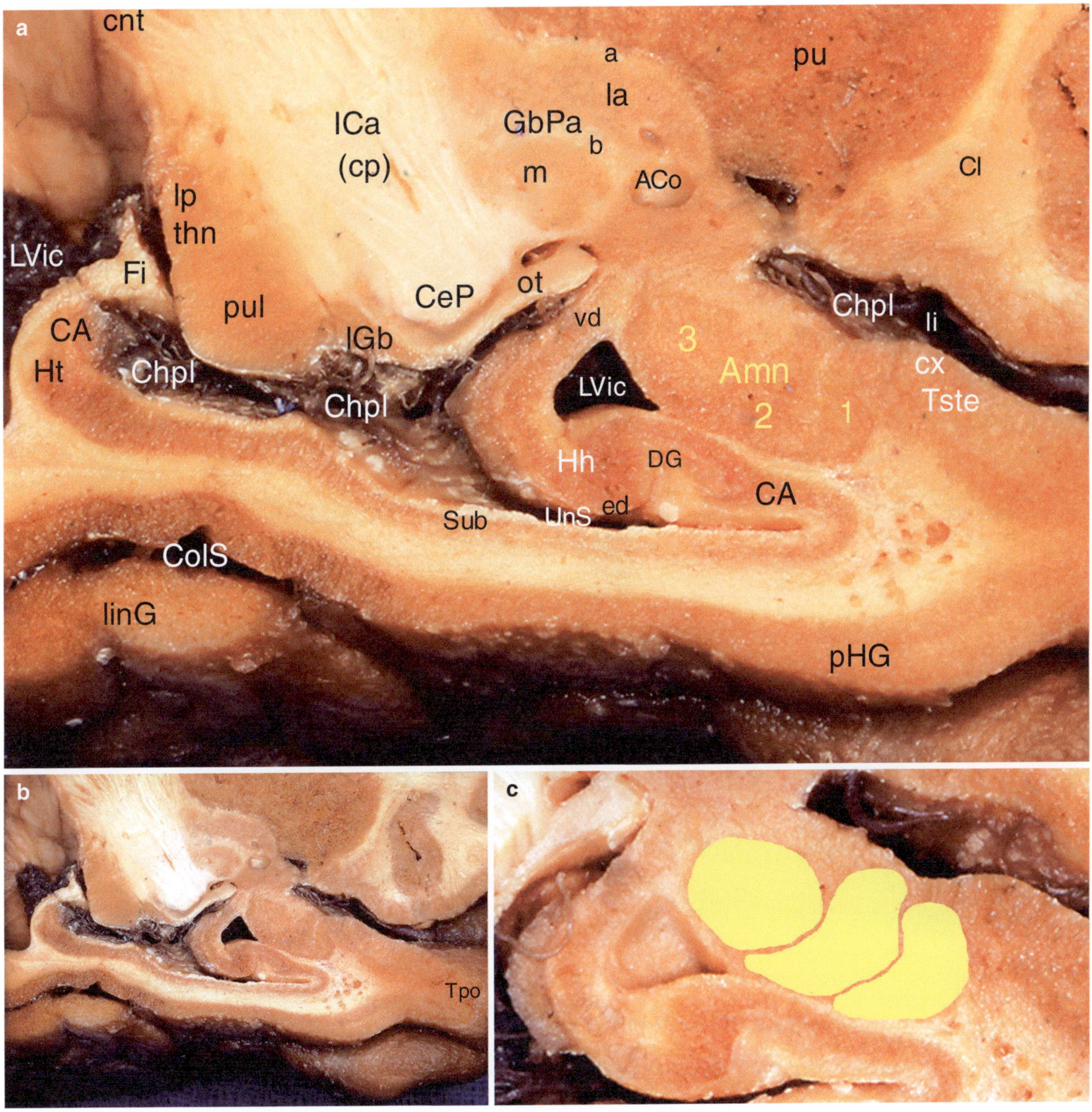

Fig. 4.14 The amygdaloid nuclear complex on a sagittal section of the brain through the medial part of the lentiform nucleus. (**a**) Details, (**b**) general situation, (**c**) nuclear organisation of Amn. *ACo* anterior commissure, *Amn* amygdaloid nuclear complex, *1* lateral nucleus, *2* baso-lateral nucleus, *3* baso-medial nucleus, *CA* cornu ammonis, *CeP* cerebral peduncle, *Chpl* choroid plexus, *Cl* claustrum, *cnt* caudate nucleus tail, *ColS* collateral sulcus, *DG* dentate gyrus, *ed* external digitation (of the hippocampus), *Fi* fimbria of the hippocampus, *GbPa* globus pallidus, *la* pars lateralis, *m* pars medialis, *a* external medullary lamina, *b* medial medullary lamina, *Hh* hippocampus head, *Ht* hippocampus tail, *ICa* internal capsule, *cp* crus posterior, *lGb* lateral geniculate body, *li* limen insulae, *linG* lingual gyrus, *lp thn* lateral posterior thalamic nucleus, *LVic* lateral ventricle, inferior cornu, *ot* optic tract, *pHG* parahippocampal gyrus, *pu* putamen, *pul* pulvinar, *Sub* subiculum, *Tpo* temporal pole, *Tste cx* cortex of the temporal stem, *UnS* uncal sulcus, *vd* vertical digitation (of the hippocampus)

separation. The most lateral area, isolated and referred to by De Olmos as the limiting division, has on the sections, a typical striated appearance. This striation is often interpreted as being produced by the passage of the medial, vertical fibres of the uncinate fasciculus, rather than by the more horizontal fibres of the inferior occipito-frontal pathway. We fully agree with this option because we have always found that the lateral surface of the amygdala was pushed against the powerful uncinate fasciculus in which are routed many amygdaloid axons that provide an excel-

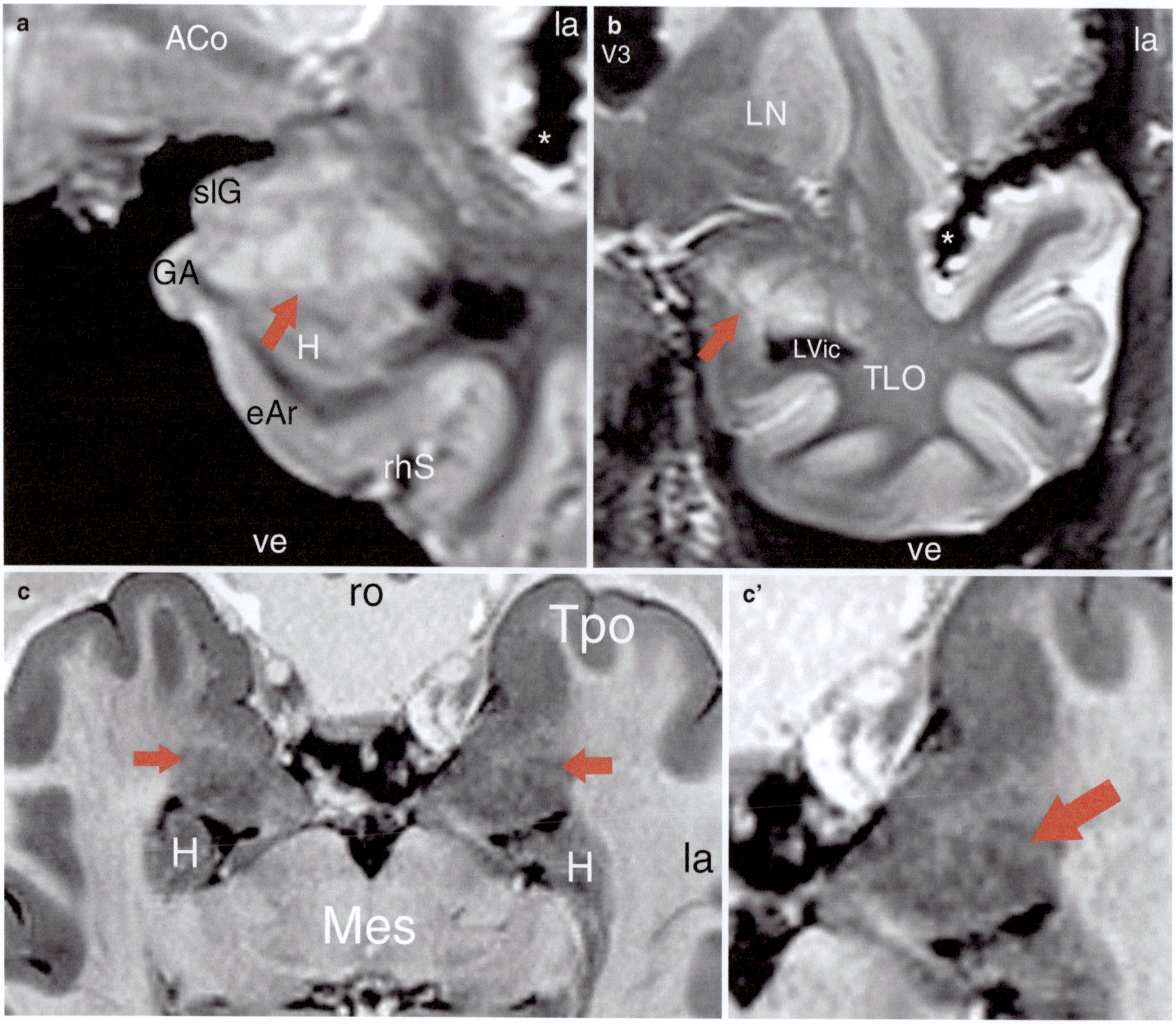

Fig. 4.15 Visualisation of the amygdala's nuclei on MRI images. (**a**, **b**) Good visibility of the segmentation on two anatomical coronal sections. (**c**, **c**') (enlargement of **c**): Images of a clinical MRI (axial section). The segmentation observed on the images **c** and **c**' is probably an artefact. *la lateral; ve ventral; ro rostral. ACo* anterior commissure, *eAr* entorhinal area, *GA* gyrus ambiens, *H* hippocampus, *LN* lentiform nucleus, *LVic* lateral ventricle, inferior cornu, *Mes* mesencephalon, *rhS* rhinal sulcus, *slG* semilunar gyrus, *TLO* temporal lobe, *Tpo* temporal pole, *V3* third ventricle, * asterisk, *ilcSI* inferior limiting circular sulcus of the insula; *red arrows* they show the amygdalae

lent amygdalo-frontal communication route. According to another interpretation, the striated appearance of the limiting portion of the lateral nucleus is defined by the lower extension of fibres from the external capsule. As for the anterior commissure, it is directly related to the lateral surface of the lateral nucleus of the amygdala, and although it distributes branches to the amygdala, it does not contribute to the striation of the lateral nucleus because its fibres are still, at this level, grouped in a single bundle which passes behind the top of the convexity of the uncinate bundle to develop in the centre of the temporal lobe.

Anyway, the lateral surface of the lateral nucleus is flat, more or less compressed by the numerous bundles of nerve fibres, which pass through the temporal stem and which separate it from the bottom of the claustrum.

It should also be recalled that the lateral nucleus is particularly rich in VIP (vasoactive intestinal peptide).

- *The baso-lateral nucleus* (Johnston's basal nucleus) is slightly smaller than the lateral nucleus which it is adjacent to (its lateral surface is separated by a thin fibrillary partition, the lateral medullary lamina, from the medial aspect of the lateral nucleus). This nucleus is more or less pyramidal on axial sections. On coronal sections, it has a triangular shape with an inferior medial apex, a narrow lateral and dorsal portion and a spread out ventral medial portion, whose inferior base corresponds to the inferior edge of the amygdala. Its constituent cells are similar to those of the lateral nucleus. Their average diameter is, however, greater than that of the neurons of the lateral nucleus (P Sah et al. 2003). A study conducted by AJ McDonald et al. in 1987 showed that the baso-lateral nucleus comprises two types of projection neurons to the MD-PFC system (to the medio-dorsal nucleus of the thalamus and to the prefrontal cortex): class I neurons of Golgi (large pyramidal or piriform cells), the most numerous, which project to the prefrontal cortex; and class II neurons of Golgi (small neurons with thin star-shaped dendrites) project to the medio-dorsal nucleus of the thalamus. More recent work, which was performed on the neurons of the baso-lateral nucleus (McDonald 1992) and based on morphological and immunochemical similarities (presence of GABA, neuropeptide Y, vasoactive intestinal peptide, or VIP and cholecystokinin, or CCK), supports the idea that there is a similarity with cortical neurons (although there is no layered organisation, as is the rule at the cortical level). Some authors (GD Petrovich and LW Swanson 1997, LW Swanson and GD Petrovich, 1998) believe that, in fact, this nucleus, as well as the lateral nucleus, has the same embryological origin as the claustrum (itself being derived, according to these authors, from the deepest part of the insular cortex), of which they would appear to be a ventromedial extension. However, this theory is not consistent with the embryological observations made by F Muller and R O'Rahilly (see Chap. 3) who showed that this nucleus is one of the first 4 amygdaloid nuclei which appear at the medial eminence, at stage 17.

H Brockaus, in 1938, referred to the baso-lateral nucleus as the "amygdaleum profundum intermedium" and showed that we can divide it into a lateral magnocellular portion, an intermediate medio-cellular portion and a medial parvocellular portion.

- The *baso-medial nucleus* (nucleus basalis amygdalae medialis) is located within and slightly above the most medial part of the baso-lateral nucleus, from which it is separated by the medial medullary lamina. On coronal sections, it has an oval appearance (see Atlas of JK Mai et al. 2008) and is inserted under the cortical nuclei, beyond and below the amygdalo-hippocampal area. It is sometimes called the secondary basal nucleus, in which case the baso-lateral nucleus becomes the main basal nucleus. In fact, it is not secondary and shares the functions of the baso-lateral nucleus, with which it receives sensory information from the lateral nucleus. It processes this information locally, such as evidenced by the existence of inhibitory synapses and then transmits the processed data to the medial and central nuclei. H Brockaus has called it "amygdaleum profundum medial" and believes it has a magnocellular subdivision and a parvocellular subdivision, separated by an intermediate subdivision. Its neurons resemble those of neighbouring nuclei (baso-lateral nucleus and amygdalo-hippocampal area) with a large irregular cell body to which are connected a few rare primary dendrites.
- The *cortical nucleus* (nucleus corticalis amygdalae) is, as its name implies, superficial. It is the first nucleus to appear during the formation of the amygdala and is visible as from stage 17. It is flush with the cortex but is not, as is often seen written, in a strictly cortical position. Such as we have observed in our dissections, this nucleus is covered by a cortex layer, which is sometimes thin but always present. We can conclude that there is always, on the peripheral surface of the amygdaloid nuclei, a layer of periamygdaloid cortex, even at the periphery of the cortical nuclei. It seems normal to consider this periamygdaloid cortex as part of the amygdala (which is the case in the classification of De Olmos). The cortical nucleus is sometimes shaped as a band but more often it has the aspect of a dished, more or less spherical structure, which lifts up the semilunar gyrus. The cortical nucleus, is in its rostral portion, close to the medial nucleus to which, as we have seen, it is often associated: this formation is referred to as the cortico-medial group. The 2 nuclei both have a layered structure such as the cortex. The cortical nucleus is directly connected to the olfactory centres and pathways, which it is closed to and belongs to what anatomists have referred to as the rhinencephalon. The olfactory inputs from the bulbs pass through the lateral olfactory stria to reach the cortical nucleus (see Fig. 7.1), which modulates the olfactory information before sending it back to the olfactory areas. Sometimes the cortical nucleus is split into an anterior cortical nucleus and a posterior cortical nucleus. The anterior cortical nucleus is then in a ventral position relative to the medial amygdaloid nucleus. As for the posterior cortical nucleus, positioned even more ventrally, it is adjacent to the amygdalo-hippocampal area.
- The *medial nucleus* (nucleus medialis amygdalae), superficiale supraamygdaleum of Brockhaus, is either studied, as we have seen, with the cortical nucleus with which it shares a laminar structure or with the central nucleus, with which it shares functionalities. It occupies the dorsomedial portion of the amygdaloid body. It has the appearance of a small oval nucleus (often bilobed), from which the stria terminalis seems to develop (see Fig. 4.6). It is located in the area where the anterior perforated substance continues by reflecting off the periamygdaloid cortex, in the angle formed by the frontal portion and the temporal portion of the cortex of the piriform lobe. It was initially integrated in the group of the cortical nuclei because, like them, it consists of 3 neuronal layers and because, like them, it

receives projections from the medio-dorsal nucleus of the thalamus. It was removed from this group following the work conducted by McDonald et al. Many arguments support the idea that it has a striatal origin, which is virtually incontestable according to LW Swanson et al and yet disagrees with the formation of this nucleus from the medial eminence (see Chap. 3).

- The *central nucleus* (nucleus centralis amygdalae), supraamygdaleum profundum of Brockhaus, probably is, with the lateral nucleus, one of the two main nuclei in the amygdaloid nuclear complex. It is the "output nucleus", counterpart of the "input nucleus" represented by the lateral nucleus. Small sized and more or less spherical, this nucleus is positioned outwards, in a slightly ventral configuration relative to the medial nucleus. It is dorsal with respect to the nuclei of the baso-lateral group. It has a lateral portion and a medial portion. Due to many arguments, it is now considered, with the medial nucleus, as a differentiation of the striatum. These arguments have been recalled by LW Swanson et al: (1) both nuclei express the same peptides as the ventral striatum (nucleus accumbens and olfactory tubercle), as the medial striatum (lateral nuclei of the septum) and as the dorsal striatum (caudate nucleus and putamen); (2) both nuclei have extrinsic GABAergic projections, such as the cells of the neighbouring striatum; and (3) the neurons of both nuclei have the same characteristics (including immunohistochemical characteristics) as the striatal neurons. In fact, this striatal origin is not necessarily in contradiction with the embryological data because, as we have seen, during the embryonic development phase, the extension of the medial eminence in the telencephalic territory generates overlapping areas likely to give rise to structures resembling the neighbouring striatum. Finally, it is to be recalled that the medial nucleus and the central nucleus both target the hypothalamus and, thereby, the vegetative centres. The central nucleus also directly targets the brainstem and spinal cord, thus acting as an eminent visceral and somatic effector of the amygdala.
- The *other nuclei*: these nuclei include the anterior amygdaloid area, the amygdalo-piriform area and the amygdalo-hippocampal area (the most caudal area of the amygdaloid body). Only the first area, which is clearly visible on axial and sagittal sections, as a prebasal narrow band, can be considered as an actual nucleus. The others are transition zones.

  As for the intercalated cell masses of EC Crosby and T Humphrey, they correspond to small groups of neurons which occupy internuclear fibrillar areas. They play a very important role despite their small size and their dispersion. They receive direct projections from the posterior orbitofrontal cortex (see chapter inputs) and thus are major players in the amygdaloid physiology.

4.2 Microscopic Aspects

4.2.1 Histological Study

Histological studies with a low magnification show that each amygdala consists of a homogeneous assembly of medium-sized neurons (Fig. 4.16). At a higher magnification, we can see that each neuron has a triangular or polygonal aspect and has a weakly basophilic cytoplasm with a large light nucleus and a chromatin consisting of clearly visible hyperbasophilic clods. The smallest neurons are also the lightest.

We can differentiate, such as we have just observed, these neurons depending on their location (Fig. 4.17):

- The neurons belonging to the baso-lateral group resemble the neocortical neurons (OE Milhouse and J De Olmos 1983). They are either pyramidal neurons with spiny dendrites (class I) or smaller neurons with a nonpyramidal appearance (class II), provided with dendrites comprising few spiny extensions. It was believed that only the first type of neurons had the characteristics of projection neurons and that the latter were local circuit neurons. However, we now know, thanks to the work of AJ McDonald (in 1987), that they are also projection neurons, with different targets than class I neurons.
- The neurons belonging to the cortical group "tend to be divided into 3 layers" (De Olmos 2004), molecular, superficial, paucicellular, medium, with a high cell density, and deep, slightly less rich in cells that the middle layer.
- The neurons belonging to the medial and central groups (especially those of the lateral portion of this group) are medium-sized spiny neurons that resemble those of the neighbouring striatum (MacDonald 1992).

For our part, histological studies have been conducted at a modest level due to the fact that our study material came from cadavers injected with formalin and in which the cellular material within the nuclei was frequently lysed or damaged. However, on specimen preserved satisfactorily, we were able to use our island cell enhancement technique with $FeCl_3$, which provided a good and immediate view of the nuclei and thus of the architecture of the amygdala. These specimen were entrusted to histologists and neuropathologists who stained them with luxol fast blue in order to study cell bodies (in pink), myelinated fibres (blue) and haematoxylin with iron (Loyez staining) for displaying the myelin sheaths in black (Fig. 4.16).

The amygdaloid nuclei are surrounded and separated from each other by means of envelopes and partitions formed by axonal projections. We were able to verify, such as

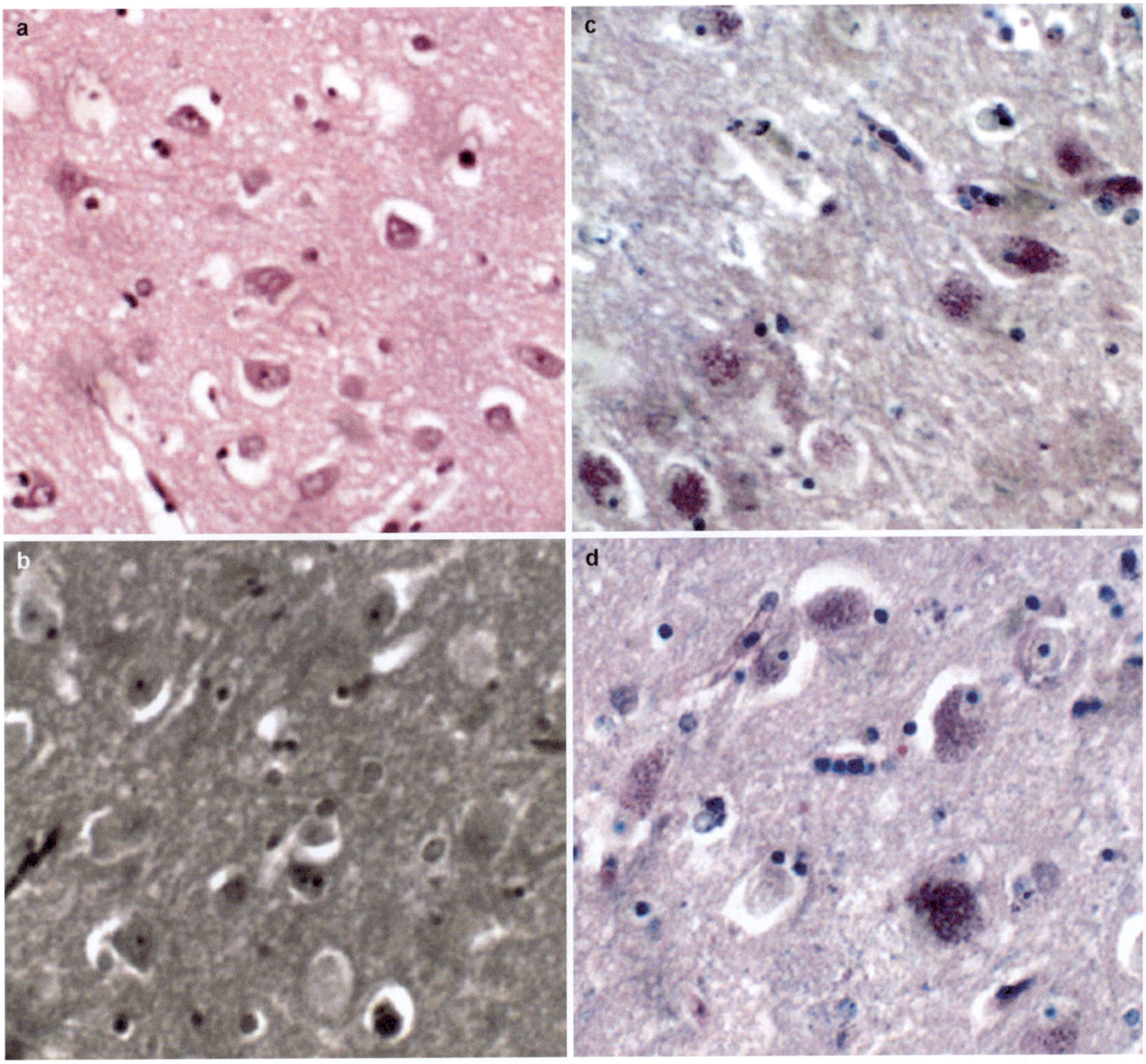

Fig. 4.16 Morphological aspects of the amygdala's neurons. (**a**) Classical colouration (haematoxylin-eosin). (**b**) Loyez's myelin staining. (**c, d**) Luxol fast blue–Pas staining. *On C and D, the colouration reveals globular-shaped neurons (X400) with clear nuclei and conspicuous nucleoli. Their cytoplasms have PAS-positive granular material due to their lipofuscin content*

already noted by Ch Foix and J Nicolesco 1925, that these boundaries are due to the development of the homolateral stria terminalis (if we consider the afferent axons) or that they converge to form the stria (if we consider efferent axons).

4.2.2 Specialised Study (Immunohistochemistry Study of Receptors)

Considering the importance of the amygdala in the behaviour of individuals, the neurons of the amygdala have been, for the last few years, the subject of extensive research using optical and electron microscopy techniques combined with neurochemical and immunohistochemical methods (AJ McDonald 1987, 1992). This research[4] is intended not only to study the morphology and physiology of the neurons in the different nuclei (including the study of the various receptors) but also to clarify the role of the neurotransmitters and function of specific molecules.

[4]Among the top teams conducting such research, we may especially include notably that of AJ McDonald et al (University of South Carolina).

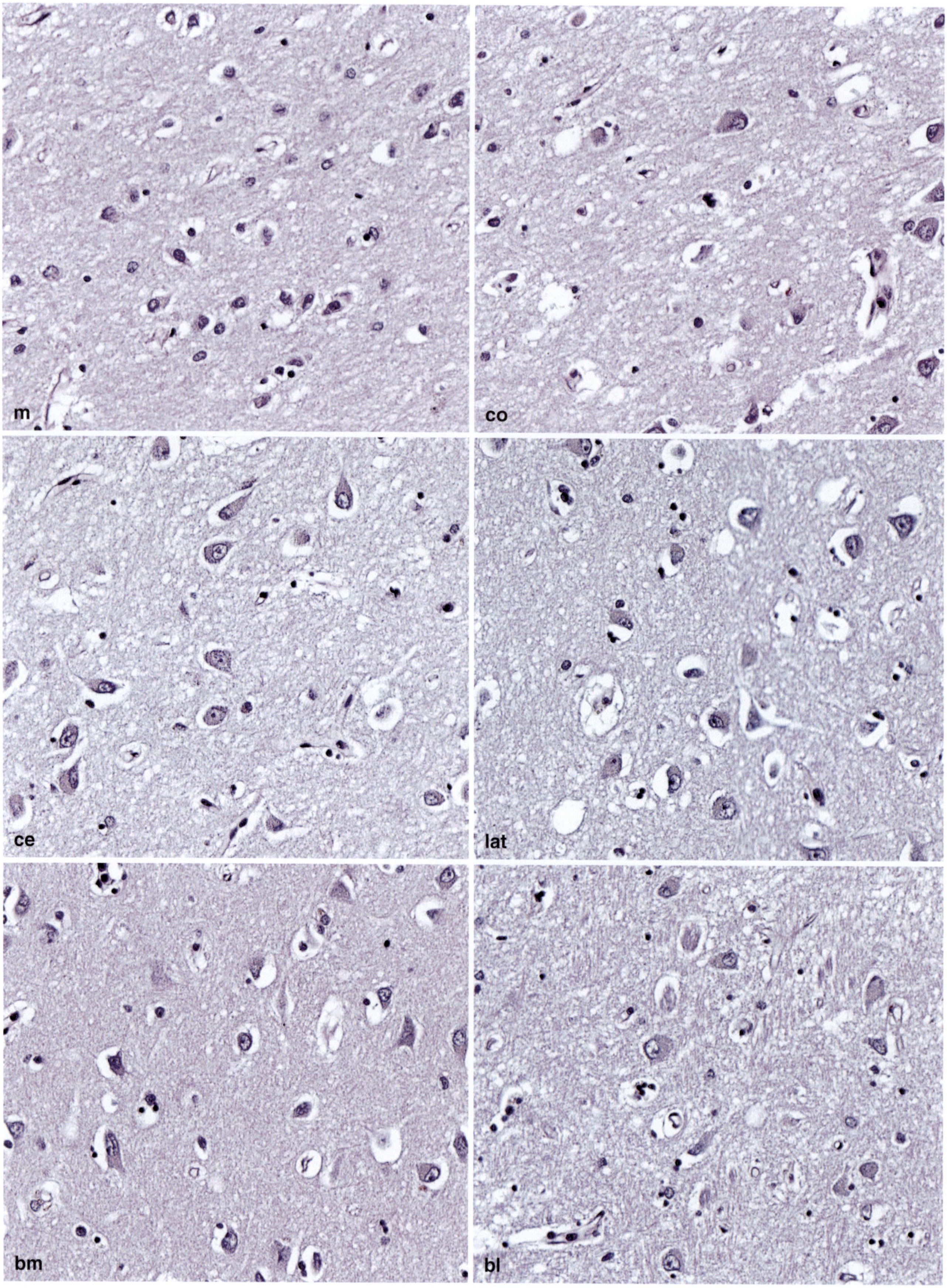

Fig. 4.17 Morphological aspects of neurons in the nuclei of the amygdala (× 200). *m* medial nucleus, *co* cortical nucleus, *ce* central nucleus, *lat* lateral nucleus, *bm* baso-medial nucleus, *bl* baso-lateral nucleus. NB: Although all these nuclei seem quite similar, we observed that the medial nucleus hosts smaller neurons that the other nuclei of the amygdala

Thus it is shown that:

- Adrenaline is used to activate the human amygdala during encoding of emotional material (AH van Stegeren et al. 2005), processes disturbed by the beta-blockers.
- Emotional stimuli processed by the amygdala are modulated by the serotonergic axons afferent to the nuclei of the basal complex, which act on a subpopulation of interneurons secreting neuropeptide Y and which have major anxiolytic properties (M Bonn et al. 2013).
- The baso-lateral complex and the central nucleus of the amygdala contain a wide variety of GABAergic and glutamatergic neurons (M Capogna 2014).
- Serotonergic neurons of the raphe nucleus are connected by their axons to the neurons of the baso-lateral complex of the amygdala, on which are located many 5HT receptors allowing the transmission of their inhibitory effects (RY Wang and GK Aghajanian 1977).
- The human amygdala has a very rich activating cholinergic innervation: this is the region of the telencephalon (except for the striatum), which has the highest density of cholinergic axons. The immunohistochemical studies performed with a ChAT (choline acetyltransferase) antibody have clarified that the nuclei which the highest content in acetylcholine were the baso-lateral nucleus and the lateral portion of the central nucleus (MM Mesulam and Geula 1992; M Emre et al. 1993).
- The administration of dopamine activates the amygdala (fMRI study of subjects viewing pictures representing terrorising or scary faces) (A Tessitore et al. 2002).

All of these examples have been extracted from a particularly abundant literature and show, if required, that the neurons of the amygdala are exposed, at various degrees, depending on the stimuli, on the nuclei being considered, on the subdivisions of each nucleus and on the receptors present within each of them, to the activating or inhibiting influences of the various neurotransmitter systems. This recalls the extreme complexity of the amygdala's operating process and the extraordinary capacities it has for processing stimuli.

4.3 Specificities and Intra-amygdaloid Connections of the Nuclei

Such as described above, the amygdaloid complex has 2 levels, the inferior or deep level, consisting of the lateral nucleus and the 2 basal nuclei, and the superior level, above the previously described level, consisting of the cortical, central and medial nuclei. More or less dense and more or less complete fibrillar walls separate the different nuclei and 2 levels and envelope the peripheral surfaces. They thus define this formation as a single anatomical structure: the cerebral amygdala or amygdaloid complex.

In this complex, numerous scientific studies, above all those conducted during the last decade, especially the works of JE LeDoux (1998, 2008), have helped to assign a specific physiological role to each of the constituent nuclei:

Somatic sensory inputs (whether olfactory, gustatory, visual or auditory) penetrate the amygdala via the lateral nucleus, the main reception nucleus.

- Inputs from the entorhinal cortex or the hippocampus can penetrate not only via the lateral nucleus but also via the basal nuclei.
- The visceral sensory inputs, including painful inputs, directly reach the central nucleus.

  The basal nuclei receive more inputs from the polymodal cortex and prefrontal cortex.

  The intercalated nuclei also receive inputs from the prefrontal cortex.

  The nuclei of the cortico-medial group receive olfactory inputs.
- The outputs of the entire amygdaloid complex leave, for the major part, the central nucleus, the main output nucleus, and travel towards the hypothalamus and brainstem, effectors for the bodily expression of emotions.
- The inputs transferred to the striatum leave the basal nuclei.

  It is also JE LeDoux who demonstrated that emotional stimuli could reach the amygdala either via a short circuit (e.g. directly from the thalamus to the amygdala) or via a long circuit (thalamus, cortex, amygdala) or via a succession of both circuits; the action of the cortex modulates or even interrupts the primary reaction.

We also have a better understanding of how the inputs exert their action on the neurons of each nucleus. If we consider a nucleus of the amygdala, the afferent neurons (therefore the inputs), which belong to circuits using the various neurotransmitters and which are, for example, dopaminergic or glutamatergic[5], will exert on the neurons of the said nucleus, inside which they terminate, a neuromodulatory action. To do so, their axons bind to the neuron receptors of

[5]In the nervous system, there are two types of neurotransmitters or neuromodulators that act at the synapses and especially those concerning the neurons of the amygdaloid nuclei:

- Neurotransmitters distributed within the nervous system by large specific systems (dopaminergic, cholinergic, adrenergic, serotonergic, GABAergic, histaminergic systems). They are either amino acids, such as glutamic acid, or molecules synthesised from an amino acid such as catecholamines (dopamine, epinephrine or adrenaline, norepinephrine and noradrenaline, serotonin, GABA, histamine) or specific molecules such as acetylcholine.
- The neuropeptides consisting of a short chain of amino acids, synthesised such as the previous, in the neuronal soma and released at the synapses: these include endorphins, enkephalins, VIP, neuropeptide Y, somatostatin, cholcystokinin, substance P, etc.

this nucleus and release the neurotransmitter, which will activate the said neurons. We now know that GABAergic interneurons are intercalated between certain dopaminergic or glutamatergic ends and the receptors[6] present on the neurons of the amygdaloid nucleus under study and can thus exert an inhibitory action, particularly at the level of the lateral nucleus and baso-lateral nucleus (Spampanato et al. 2011). At the level of the central nucleus, certain axon terminations of intercalated nuclei or of the BST can play the same inhibitory role as the GABAergic interneurons.

Another research field is opened: That of the intra-amygdaloid connections, in order to know how the stimulus progresses inside the amygdaloid complex. Several studies have been devoted to this subject, since JE Krettek and JL Price in 1978, A Pitkanen et al. in 1991, JE LeDoux in 1998, P Sah et al. in 2003, etc.

Inside the amygdala, it is as if the stimuli collected laterally were returned to the medial portion of the amygdala. "Thus the major flow of information through the amygdala is unidirectional and follows a lateral to medial direction" (R Nieuwenhuys et al. 2008).

The lateral nucleus projects directly to all the other nuclei, including the intercalated nuclei and, for P Sah et al, the periamygdaloid cortex. Only the direct connection to the central nucleus is subject to discussion (Nieuwenhuys et al). Reciprocal connections between these nuclei and the lateral nucleus exist, but they are very modest (Sah et al. 2003).

The baso-lateral nucleus projects to all nuclei which are more medial: the baso-medial nucleus, with which it exchanges internuclear connections (Sah et al. 2003), the medial nucleus and the medial portion of the central nucleus (JP Aggleton 1985).

The baso-medial nucleus projects to the medial and cortical nuclei and towards the medial portion of the central nucleus (Sah et al. 2003).

The medial nucleus projects to the cortical nucleus and the medial portion of the central nucleus.

The central nucleus projects to the cortical nucleus. Its lateral portion, that receives no amygdaloid projections but which does receive cortical and subcortical inputs (Sah et al. 2003), has significant projections to the medial portion. As for the medial portion, it only sends a small amount of return projections to the other amygdaloid nuclei (Sah et al. 2003). However, it is the major source of amygdaloid outputs, which will specifically target the hypothalamus and brainstem (JE LeDoux et al. 1998; P Sah et al. 2003).

All previous data were collected in animals (especially monkeys) and were extrapolated to the human amygdala, which may be a false approach, and in any case it involves a probable error coefficient. However, we are beginning to see opportunities for performing studies with humans (A Pitkänen and DG Amaral 1998) based on data provided by the MRI study of temporal epilepsies, which we know can damage one or even 2 amygdalae or even alter the local inhibitory circuits and cause hyperexcitability of the amygdala.

The selective destruction of one or more nuclei observed on MRI images and the clinical confrontations studying the behaviour problems of patients suffering from such lesions should provide further insights into the physiological role played by each amygdaloid nucleus in humans. Similarly, the study, following the post-operative culture of resection specimens obtained during surgery performed on epileptic subjects, should provide valuable information on human amygdaloid nuclei.

4.4 An Example of Functional Duality Within a Single Nucleus

This example is a very interesting study conducted by researchers at the University of Lausanne (D Huber et al. 2005) who analysed the action of two hypothalamic neuropeptides, oxytocin and vasopressin, on the central nucleus of a rat amygdala. The researchers had previously studied the 2 substances[7] and had observed that they have opposite effects on anxiety: oxytocin[8] is reassuring and favours attachment, sociability and calmness, and agonising vasopressin promotes stress and aggressiveness.

They then found that each of the subdivisions of the central nucleus of the amygdala (output nucleus via which occurs the cascade of bodily reactions to emotions), the lateral and capsular portion, CeL/C, and the medial portion, CeM, have specific receptors (oxytocin receptors for CeL/C and vasopressin receptors for CeM) whom the other portion lacks.

[6] Such as all neurons, those of the amygdala have receptors on which the neuronal terminations of the above-mentioned systems will synapse. These receptors, key components of the interneuronal dialogue and of the action of the neurotransmitters, are numerous and are starting to be counted. Thus, in the amygdala, we found NMDAR NR1, NR1a and NR2c receptors; mGluRS, mGluR1a, R2 and R7 receptors; oestrogen–parvalbumin receptors; receptors 1 for the corticotropin-releasing hormone; receptors 2 for somatostatin; VIP receptors; glucagon peptide Y receptors; cannabinoid 1 receptors; and glucagon peptide Y receptors; and the list is constantly growing.

[7] Oxytocin and vasopressin are synthesised at the level of the magnocellular secretory neurons of the supraoptic and paraventricular nuclei of the hypothalamus and transported in the posterior pituitary as neurosecreta which travels along the pituitary stem. They are then released into the systemic circulation. They are therefore neurohormones acting as neurotransmitters acting remotely.

[8] The injection of oxytocin in rats (work of Stoop 2012) significantly reduces the reaction of immobility (freezing) induced by fear, without removing the fear itself (persistently altered heart rhythm).

They further noted that CeL/C neurones send projections to the CeM portion which is not the case for CeM neurons.

By means of an electrophysiological recording, the researchers demonstrated that oxytocin increases the activity of CeL/C neurons, which are GABAergic inhibitory neurons whose projections will inhibit the activity of CeM neurons, that were for the major part excited by vasopressin and that were going to produce outputs generating bodily effects of emotion.

In other words, oxytocin will offset the excitatory response of vasopressin, a remarkable example of both the modulation of amygdala activity and of the dichotomy of the two portions of the same nucleus! The authors of the study also suggest therapeutic possibilities that may result from the use of molecules acting at the level of the two types of receptors to "regulate responses to fear and control physical manifestations of anxiety".

4.5 Characteristics of the Nuclear Neurons

Such as observed above, each nucleus has its own morphological and functional characteristics. And yet we have only glimpsed at the problem because as it is not the subject of this book. Let's just say, for example, that neurons not only differ from one amygdaloid nucleus to another, from one nucleus portion to another, but all the neuronal structures differ; thus, certain receptors present in a nucleus will be different in a neighbouring nucleus. Even more surprising still, the different categories of neurons, within the amygdala, are involved in different aspects of the same function. Thus, R Adolphs has been able to identify, in humans, neurons that recognise an entire face and two other specific categories, one which recognises the eye area and the other which recognises the area of the mouth (the latter appears to be overactive in autistic subjects and, in any case, certainly much more active than the eye recognition category!). Similarly, they have identified a category of amygdaloid neurons located within the right human amygdala, which specifically recognises animals (F Mormann et al. 2011).

More extraordinary, U Rutishauser et al. 2013, conducted the first recording of an isolated neuron in a rare category of patients (suffering from autism and epilepsy and in the surgical preparatory phase for the resection of the epileptic source and thus wearing intracerebral electrodes implanted in the amygdalo-hippocampal region to detect from which point the seizures start). Such patients were subjected to face detection experiments and uni-neuronal activity recordings were carried out at the same time. Continuing to work with such recordings should contribute to better understanding the causes of autism and the functioning of the amygdala.

4.6 What Are the Roles and Functions of the Amygdala?

We will only mention a few, which we consider as essential because they provide an idea of the significance of the amygdala in our daily lives:

Major role in fear conditioning (M Davis 1997; JE LeDoux 2000; GJ Quirk et al. 1995, 1997)
Vital site for the expression of fear and anxiety (M Davis 1992; R Adolphs et al. 1995; EA Phelps and JE LeDoux 2005; JS Feinstein et al. 2011)
Important structure regulating anxiety (B Myers and B Greenwood-Van Meerveld 2009)
Structure regulating emotional behaviour (JE LeDoux 2000)
Regulatory role in visceral pain (B Myers and B Greenwood-Van Meerveld 2009)
Role in the regulation of altered visceral or somatic sensations (P Sah et al. 2003; MP Jones et al. 2006)
Role in the perception of conscious or unconscious facial expressions (R Adolphs 2008) and in facial evaluation (AL Hariri 2002; A Todorov 2012)
Role in neuromodulation (HC Pape and D Pare 2010)
Role in the memorisation of emotions and in learning new cues (S Pirot 1999; JS Morris et al. 1998)
Role in decision making (role particularly highlighted by the work of A Bechara et al. 1999)
Role in social behaviour (S Pirot 1999; CJ Machado et al. 2008) and 'social cognition' (R Adolphs 2010)
Role in attention (PC Holland and M Gallagher 1999)

And this list is far from exhaustive! (See for more, PJ Whalen and EA Phelps 2009, JP Aggleton 2000) However, the amygdala is the major source of amygdaloid outputs which will especially project to the hypothalamus and brainstem (JE LeDoux 1998; P Sah et al. 2003) and which will allow the management of numerous body functions.

4.7 Single Structure, Assembly of Nuclei? and Others

This is obviously the question that arises at the end of such a chapter, and other questions were asked before us: "What is the amygdala?" (Swanson and Petrovich 1998). There is no lack of arguments according to which the amygdala is a basal nucleus[9], such as the others. Firstly, from an anatomical viewpoint: its morphological aspect is that of a single struc-

[9]We use the collective term "basal nuclei" or "basal ganglia" to refer to the different structures (caudate nucleus, putamen, globus pallidus and amygdala) that are formed during development of the embryo from the thickened base portion of the wall of the telencephalon.

ture even if the frequent polylobed aspects lead us to believe that there are components that lift the cortical shell. And even more components are observed in the sections; the example of the lenticular nucleus with its subdivisions is here to show us that a complex internal structure is no longer considered as a real entity. From a functional viewpoint, the amygdala can be considered holistically, as a warning system for informing the body of anything that might disrupt the bodily homeostasis, each of its nuclei contributing to the same goal.

But there are just as many reasons to consider the amygdala as an assembly of nuclei. Firstly, from an anatomical viewpoint, these nuclei appear during the dissection phase or are highlighted by ultra-dissection. Secondly, there is the fact that these nuclei have very different characteristics and connectivities, such as recalled by many authors (Holland and Gallagher 1999). For Swanson and Petrovich 1998, the central nucleus is only an outpost of the autonomic nervous system, while the lateral and basal nuclei are merely expansions of the ventromedial claustrum and the cortical and medial nuclei are only parts of the olfactory system, thus very disparate functional units. Embryology itself shows that the amygdala is formed by the junction of nuclei which do not appear all at the same time.

However, this argument is not absolutely rigorous because why should we consider as holistic formations other neighbouring grey nuclei such as the thalamus and the hypothalamus also consisting of nuclei? How can we explain how the neurons of most nuclei project to the same central nucleus? How can we explain that the fibres of most nuclei converge to a single terminal stria? How can we explain that most neuroscience experiments involving the amygdala caused, in fMRI, a visible activation of the entire amygdala and often both formations?

It is therefore difficult to conclude, and we believe that the truth often appears as a compromise. Yes, the amygdala is a whole structure with an overall unit operation. Yes, it consists of nuclei, each with specific functions but which all collaborate to perform the same function for vigilance purposes and maintaining body homeostasis.

This shows how adapted we believe the following terminology to be *amygdaloid nuclear complex*. It summarises the unitary aspect, the assembly of the nuclei, the almond-shaped morphology and the extreme complexity of the interconnections and interrelations of the nuclei with each other and all brain structures.

5 Technique for Dissecting the Amygdaloid Body and Its Close Connections

This dissection is performed on a half-brain, preferably prepared by a post-mortem injection of embalming fluid with formaldehyde 10 %[1] (which is the only product capable of ensuring an effective preservation and consistency of the nerve tissues, therefore allowing dissection without any decaying).

It is at the level of the medial surface of the temporal lobe that the amygdalo-hippocampal area is identified. It is hidden in the front region by multiple vessels located at the terminal arborisation of the internal carotid artery and whose small branches seem to intermingle before penetrating the anterior perforated substance and in the rear region by half of the brainstem that, in general, has been preserved with the hemisphere to be dissected and which is difficult to retract. Therefore, the first part of the dissection involves removing the brainstem by making an incision in the mesencephalon with a scalpel and eliminating the fine vascular network and the arachnoid's remains that hinder the vision of the area of interest. On the other hand, we will retain, as far as possible, the vessels of the arterial circle of the base which are important references.

We therefore obtain the view shown in picture a (Fig. 5.1), where the most important characteristics appear: the sectioned internal carotid, the uncus of the hippocampus, the temporal pole and in front of the anterior perforated substance (apSB), the orbital surface of the frontal lobe, under which is routed the olfactory tract.

At the level of the uncus, it is possible to observe the underlying uncal sulcus and the band of Giacomini, which borders, in the front region, the apex of the uncus. Further down, another major sulcus borders, in a downward direction, the area to be dissected: this is the perirhinal sulcus, above which the cortex forms a large and almost flat area, the entorhinal area. It is above this area that we can observe the gyri belonging to the periamygdaloid cortex: from back to front, the semilunar gyrus and the gyrus ambiens, separated by the sulcus semi-annular or amygdaloid sulcus, and then at the level of the uncus (hence its name), the uncinate gyrus.

The next phase involves creating a cortical window in order to view the above area. We will also make sure that a large cortical surface is not immediately resected in order to retain the overall architecture. Cortical resection will remove, on the one hand, all of the medial surface of the periamygdaloid cortex (with the exception of the cortical band which borders the deep endorhinal sulcus) but also the upper three quarters of the entorhinal cortex and impact the uncinate gyrus while remaining at ¼ cm from the anterior edge of the band of Giacomini. This band marks the border area between the upper digitation of the head of the hippocampus and the superior pole of the amygdala. Once the cortical window is created, picture b (Fig. 5.1), it is observed that the amygdala is still not visible: a fibrous wall is present and appears as a thick white rectangular band, which can be easily separated from the cortex which covers it. This band is the inferior end of a large and remarkable white substance pathway: the cingulate fasciculus (see Chap. 7). It is noted that the pathway bends when it reaches the amygdala, such as a protective shield (see Fig. 7.15).

The second part of the dissection, which seems to have reached a sort of blocking point, obviously requires enlarging the cortical window by performing an additional resection of the cortex of the anterior half of the parahippocampal gyrus (Fig. 5.2). But once this procedure has been performed, we once again encounter the cingulate fasciculus that continues backwards. It therefore becomes necessary to perform a resection of this fasciculus facing the entire cortical window. We can then observe that the cingulate pathway is very adherent to the hippocampus against which it is pushed and is present as two layers, the medial layer being the thickest. By proceeding carefully, we can resect the particularly resistant fibres of this pathway up to the antero-superior surface of the amygdala, without harming the body and head of the hippocampus. Towards the front area, we can also note that the cingulate pathway has an important extension towards

[1]The injection site is also important due to the fact that the best results are obtained by injecting embalming fluid in one of the primitive carotid arteries.

V. Di Marino et al., *The Amygdaloid Nuclear Complex: Anatomic Study of the Human Amygdala*,
DOI 10.1007/978-3-319-23243-0_5

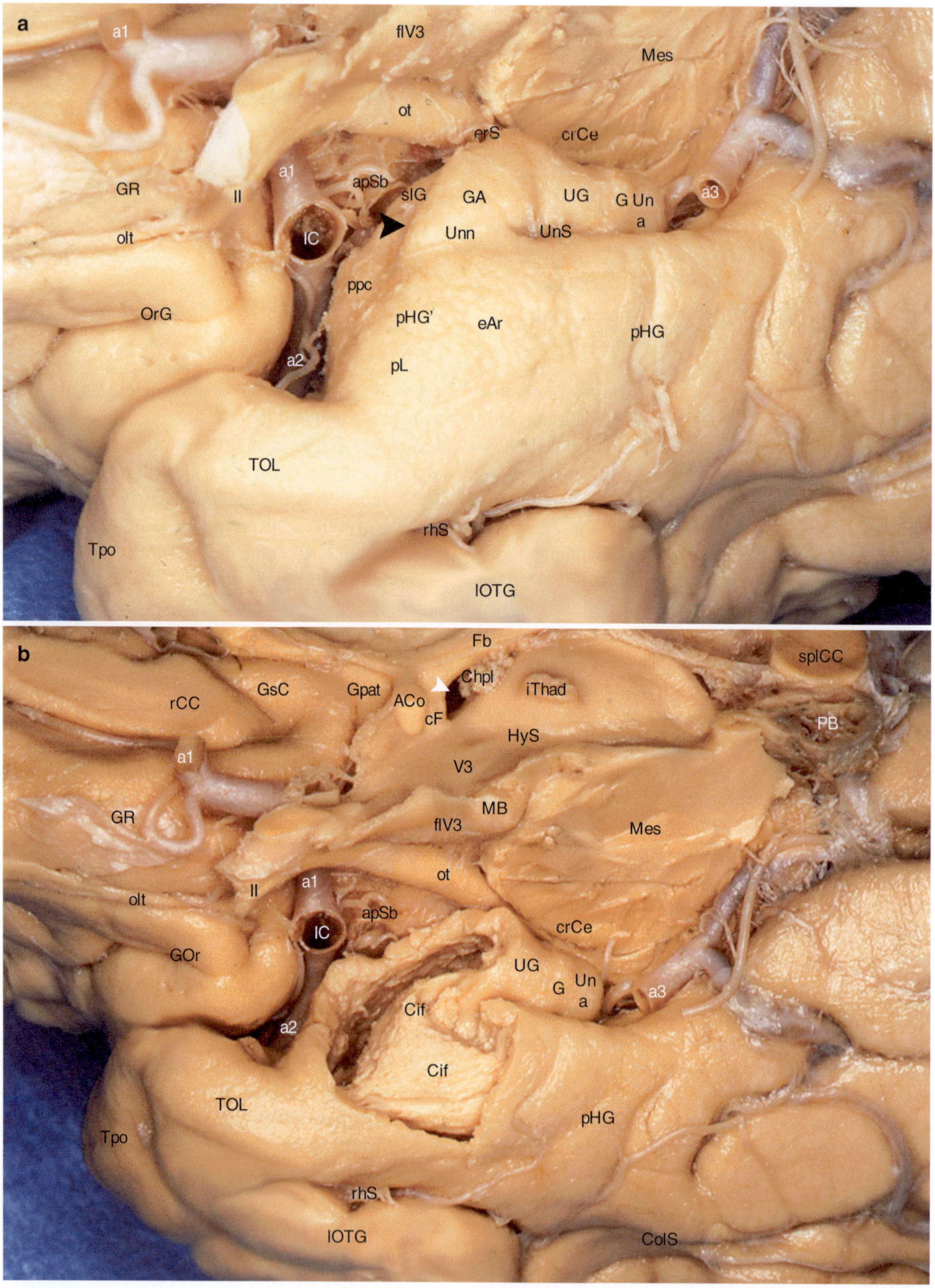

Fig. 5.1 Dissection of the amygdaloid complex (successive stages). (**a**) Initial aspect (medial view of a right half-brain). (**b**) Cortical temporal window *Aco* anterior commissure, *apSb* anterior perforated substance, *Chpl* choroid plexus, *Cif* cingulate fasciculus, *ColS* collateral sulcus, *crCe* crus cerebri, *eAr* entorhinal area, *erS* endorhinal sulcus, *Fb* fornix body, *Fc* column of the fornix, *flV3* floor of the third ventricle, *G* band of Giacomini, *GA* gyrus ambiens, *GR* gyrus rectus, *HyS* hypothalamic sulcus, *IC* internal carotid, *iThad* interthalamic adhesion, *lOTG* lateral occipitotemporal gyrus, *MB* mammilliary body, *Mes* mesencephalon, *olt* olfactory tract, *OrG* orbital gyri, *ot* optic tract, *patG* paraterminal gyrus, *PB* pineal body, *pHG* parahippocampal gyrus, *pHG' (pL)* anterior part of the pHG (piriform lobe), *ppc* pre-piriform cortex, *rhS* rhinal sulcus, *sCG* subcallosal gyrus, *slG* semi-lunaris gyrus, *splCC* splenium of the corpus callosum, *TOL* temporal lobe, *Tpo* temporal pole, *UG* uncinate gyrus, *Un a* uncus apex, *Unn* uncal notch, *UnS* uncal sulcus, *V3* third ventricle, *a1* anterior cerebral artery, *a2* middle cerebral artery, *a3* posterior cerebral artery, *II* optic nerve, *white arrowhead* interventricular foramen (*ivf*), *black arrowhead* semi-annular sulcus (*sas*)

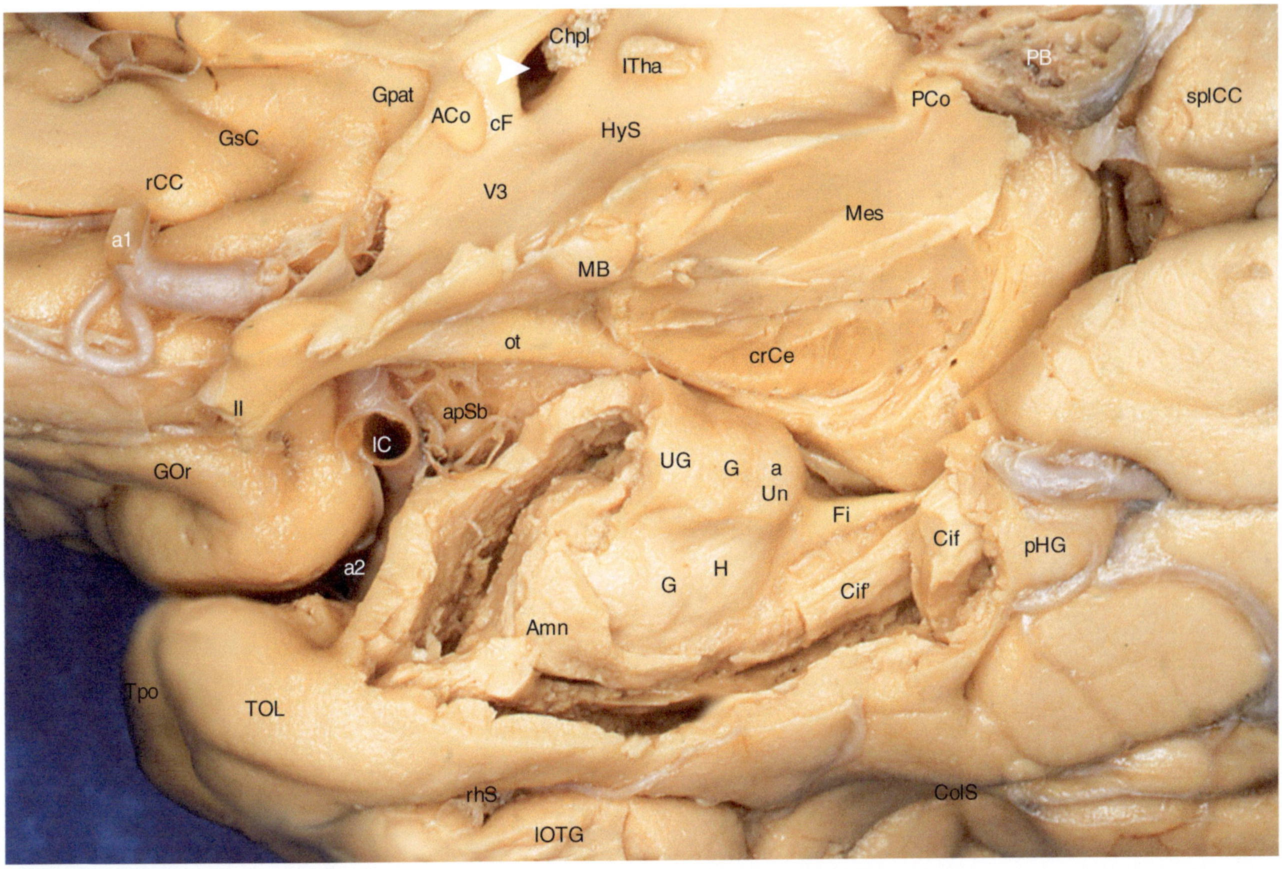

Fig. 5.2 Dissection of the amygdaloid complex (successive stages). Enlargement of the cortical temporal window *ACo* anterior commissure, *Amn* amygdaloid nuclear complex, *apSb* anterior perforated substance, *Chpl* choroid plexus, *Cif* cingulate fasciculus, *Cif'* juxta-hippocampic part of the cingulate fasciculus, *ColS* collateral sulcus, *crCe* crus cerebri, *Fc* column of the fornix, *Fi* fimbria, *G* band of Giacomini, *H* hippocampus, *HyS* hypothalamic sulcus, *IC* internal carotid, *iThad* interthalamic adhesion, *lOTG* lateral occipitotemporal gyrus, *MB* mamilliary body, *Mes* mesencephalon, *OrG* orbital gyri, *ot* optic tract, *patG* paraterminal gyrus, *PB* pineal body, *PCo* posterior commissure, *pHG* parahippocampal gyrus, *rhS* rhinal sulcus, *sCG* subcallosal gyrus, *splCC* splenium of the corpus callosum, *TOL* temporal lobe, *Tpo* temporal pole, *UG* uncinate gyrus, *Un* a uncus apex, *V3* third ventricle, *a1* anterior cerebral artery, *a2* middle cerebral artery, *II* optic nerve, *white arrowhead* interventricular foramen (*ivf*)

the temporal pole (see Chap. 7). Once the cingulate pathway has been resected, it is finally possible to see the amygdala which rests against the head of the hippocampus.

The next phase of the dissection involved further widening the cortical window backwards, up to the level of the splenium of the corpus callosum, and to continue resecting the lateral layer of the cingulate pathway, which adheres to the hippocampus (Fig. 5.3). We therefore discovered the entire region in more detail with, from front to back, the amygdala, the head then the body and lastly the tail of the hippocampus as well as the dentate gyrus and the fimbria. It was also possible to observe connections with the neighbouring mesencephalon and more remotely, with structures of the diencephalon, the hypothalamus and the thalamus.

Many investigations can be undertaken from this dissection in order to highlight the proximity connections and relationships of the amygdala. The most interesting option is to visualise the stria terminalis in its lower juxta-amygdaloid portion. To achieve this (Fig. 5.4), we will perform a large resection of the posterior portion of the thalamus and metathalamus (pulvinar, geniculate bodies and colliculi). The stria terminalis will then be perfectly free, and we can observe its overlying path, parallel to the fimbria. By reflecting the fimbria, we can also see the tail of the caudate nucleus, which is rarely visible in situ other than on diagrams! (See Chap. 7.)

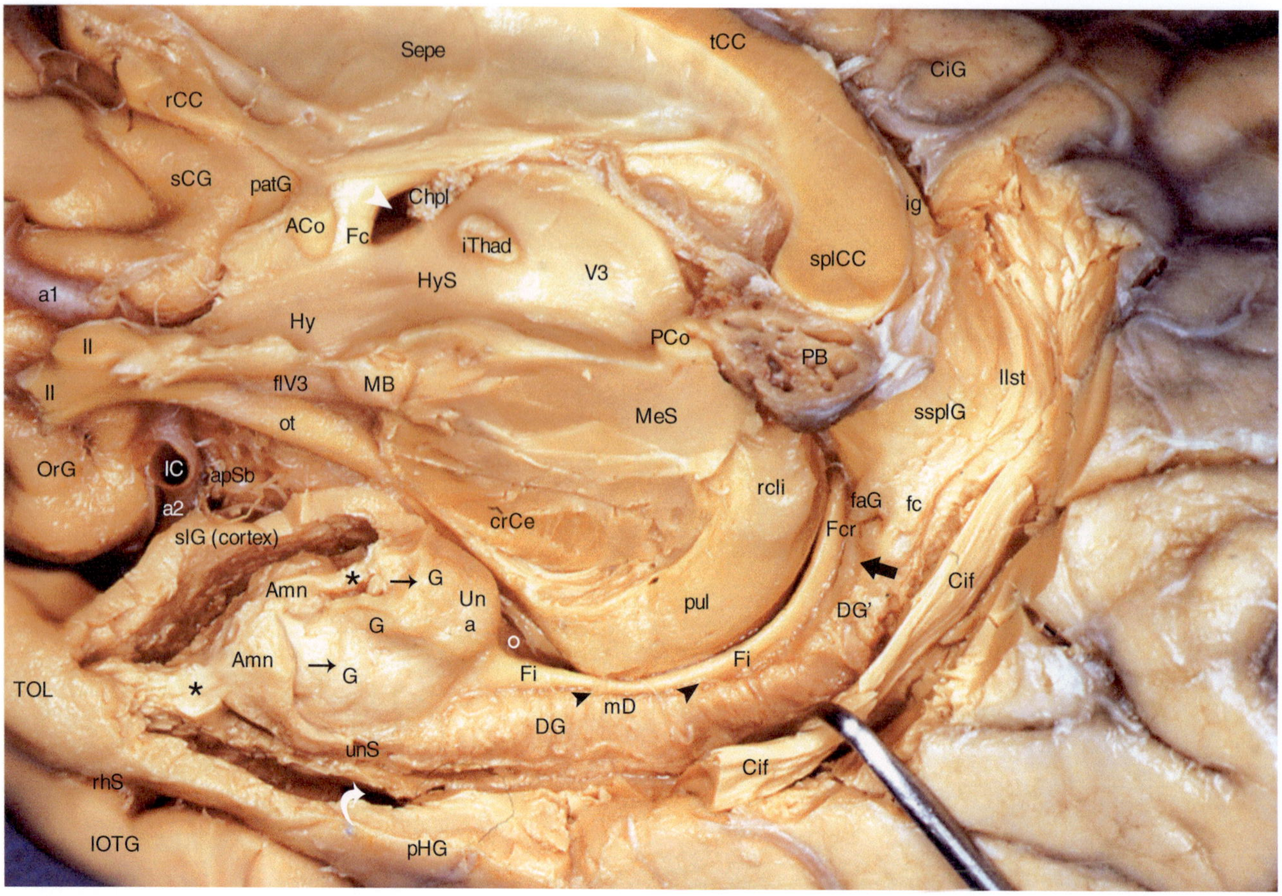

Fig. 5.3 Dissection of the amygdaloid complex (successive stages). Large opening of the temporal cortical window *ACo* anterior commissure, *Amn* amygdaloid nuclear complex, *apSb* anterior perforated substance, *Chpl* choroid plexus, *Cif* cingulate fasciculus, *CiG* cingulate gyrus, *crCe* crus cerebri, *DG* dentate gyrus, *DG'* dentate gyrus in the hippocampal tail, *faG* fasciolaris gyrus, *Fc* column of fornix, *fc* fasciola cinerea, *Fcr* crus of fornix, *Fi* fimbria, *flV3* floor of the third ventricle, *G* band of Giacomini, *Hy* hypothalamus, *HyS* hypothalamic sulcus, *IC* internal carotid, ig indusium griseum, *iThad* interthalamic adhesion, *lOTG* lateral occipitotemporal gyrus, *llst* lateral longitudinal stria, *MB* mamilliary body, *mD* margo denticulatus, *Mes* mesencephalon, *OrG* orbital gyri, *ot* optic tract, *patG* paraterminal gyrus, *PB* pineal body, *PCo* posterior commissure, *pHG* parahippocampal gyrus, *pul* pulvinar, *rCC* rostrum of the corpus callosum, *rcli* rostral colliculus, *rhS* rhinal sulcus, *Sepe* septum pellucidum, *splCC* splenium of the corpus callosum, *sCG* subcallosal gyrus, *slG* semi-lunaris gyrus, *ssplG* subsplenial gyrus, *tCC* trunk of the corpus callosum, *TOL* temporal lobe, *Un* a uncus apex, *unS* uncal sulcus, *V3* third ventricle, *a1* anterior cerebral artery, *a2* middle cerebral artery, *II* optic nerve, *white arrowhead* interventricular foramen (*ivf*), *black arrowheads* fimbrio-dentate sulcus, *large black arrow* dentate-fasciolaris sulcus, *thin black arrows* hippocampal sulcus, curved *white arrow* inferior horn of the lateral ventricle, *white circle* velum terminale of Aeby, little *black asterisk* fasciculus connecting the amygdaloid nuclear complex to the hippocampal head, great black asterisk fasciculus connecting the amygdaloid nuclear complex to the rostralis part of pHG, NB presence of a metallic hook pushing down the cingulate fasciculus

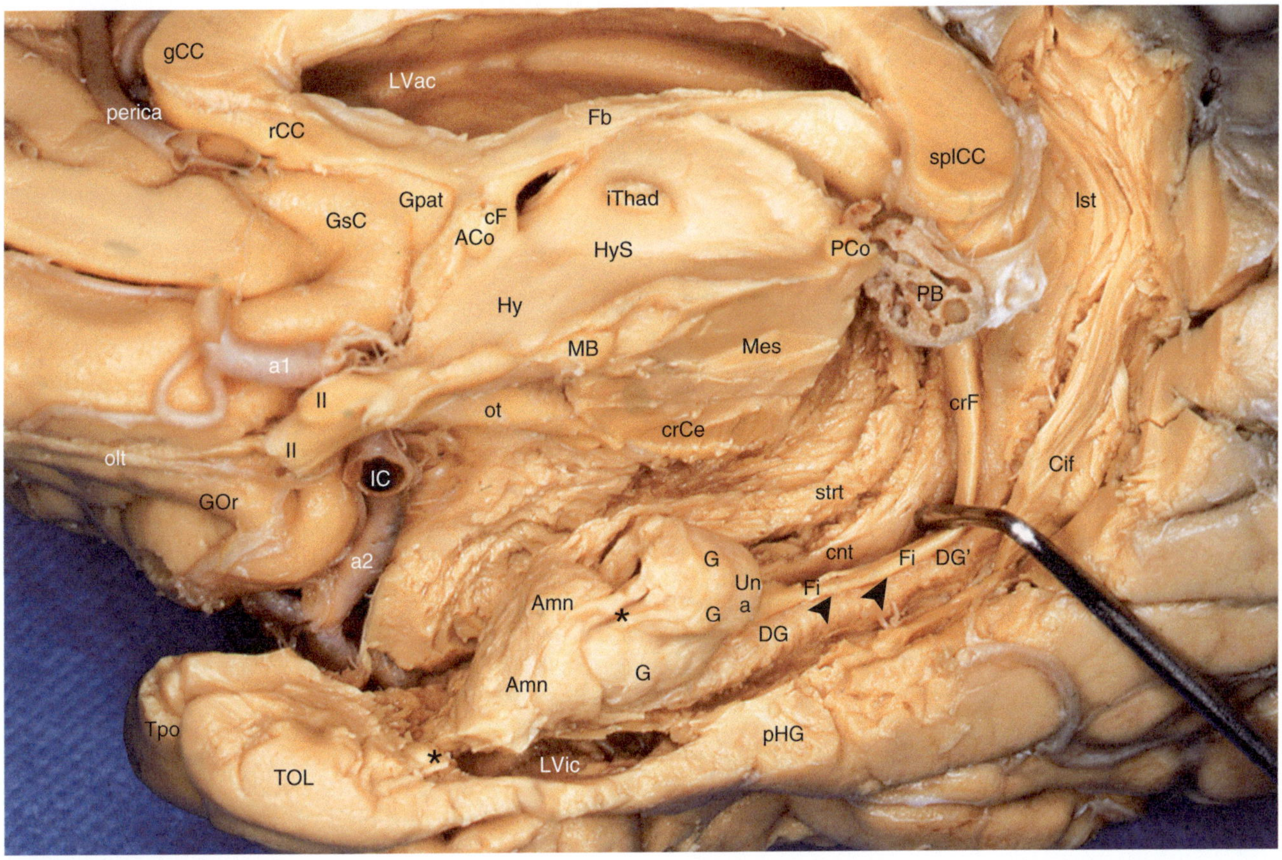

Fig. 5.4 Dissection of the amygdaloid complex (successive stages). After resection of the pulvinar and the colliculi *ACo* anterior commissure, *Amn* amygdaloid nuclear complex, *Cif* cingulate fasciculus, cnt tail of the caudatus nucleus, *crCe* crus cerebri, *DG* dentate gyrus, *DG'* dentate gyrus in the hippocampal tail, *Fb* body of fornix, *Fc* column of the fornix, *Fcr* crus of fornix, *Fi* fimbria, *G* band of Giacomini, *gCC* gyrus of corpus callosum, *Hy* hypothalamus, *HyS* hypothalamic sulcus, *IC* internal carotid, *iThad* interthalamic adhesion, *LVac* lateral ventricle (anterior cornu), *LVic* lateral ventricle (inferior cornu), *llst* lateral longitudinal stria, *MB* mamilliary body, *Mes* mesencephalon, *olt* olfactory tract, *ot* optic tract, *OrG* orbital gyri, *PB* pineal body, *PCo* posterior commissure, perica pericallosal artery, *pHG* parahippocampal gyrus, *rCC* rostrum of the corpus callosum, *sCG* subcallosal gyrus, *splCC* splenium of the corpus callosum, *strt* stria terminalis, *TOL* temporal lobe, *Tpo* temporal pole, *Una* uncus apex, *a1* anterior cerebral artery, *a2* middle cerebral artery, *II* optic nerve, *black arrowheads* fimbrio-dentate sulcus, *little black asterisk* fasciculus connecting the amygdaloid nuclear complex with the hippocampal head, great *black asterisk* nuclear complex to the rostralis part of pHG, NB presence of a metallic hook pushing down the fimbria

Connection Pathways of the Cerebral Amygdala

6

The cerebral amygdala is connected to the septum, the diencephalon and brainstem via two pathways which are a dorsal pathway, the stria terminalis, and a ventral pathway called the ventral amygdalofugal pathway[1]. These pathways include both centripetal fibres to the amygdala (amygdalopetal fibres) and centrifugal fibres (amygdalofugal fibres), arising from the amygdala's neurons. Their point of origin can therefore be described either from the targets of the amygdala fibres or from the amygdala itself. It is this second option that we have chosen. However, an inter-amygdaloid pathway of great importance should be added, the anterior commissure, which is included in this chapter.

6.1 The Dorsal Connection Pathway or Stria Terminalis

The stria terminalis is the current name of the former pathway of Foville[2] or "taenia of Tarin"[3] and became over time the semi-circular tenia (tenia semi-circulaire) according to French authors. It has been studied in depth as early as 1895 by J Déjerine (1901) and in parallel by A Koelliker (1896). It is a band of white substance, which appears as a flattened, very thin, 2–3 mm-wide band, with a fairly homogeneous diameter before it divides. It is very fragile and therefore difficult to dissect. Its length (6.6 cm, on average) slightly varies from one individual to another. It connects the amygdala to the septal region and to the hypothalamus. It is formed of a pathway of long fibres among which several are myelinated and whose origins are at the level of the nuclei of the amygdala, such as has already been demonstrated in animals, by CA Fox in 1943, T Ban and F Omukai in 1959, E Hall in 1960 and WJH Nauta in 1961. According to WM Cowan et al. 1965, only cortical and medial nuclei appear to be involved in the stria terminalis. Our observations made with humans have shown that, in fact, all of the nuclei are involved (Fig. 6.1) with a predominance for the medial nucleus. Yet, the stria terminalis appears to laterally leave the amygdala, which is paradoxical at first glance, since it emerges from the lateral slope of the superior pole (where is located the medial nucleus!). Such as Ch Foix and J. Nicolesco (1925) had already noted, the partitions between the different nuclei of the amygdala or surrounding them, as well as the peripheral capsule from which depart these partitions, are white substance lamina, formed by the neurofibres of neurons forming the nuclei. The stria thus results from the assembly of all the axons of the amygdaloid neurons, which are joined by the axons of the neurons from the centres, which it connects to the amygdala. The stria terminalis then follows the trajectory of the concave edge of the caudate nucleus. It initially runs along the medial surface of the tail of this formation. It will firstly progress to the rear, pass in the depth of the endorhinal sulcus, at the level of the uncus of the hippocampus, to reach the roof of the temporal lorn of the lateral ventricle, parallel to the fimbria from which it is only separated by a few millimetres. At the junction of the body and head of the hippocampus, the tenia of the fimbria and that of the stria terminalis unite (H.M. Duvernoy, in 1988), thus forming a small crescent moon-shaped lamella: the velum terminale of Aeby[4]

[1]Sometimes, we add to the dorsal and ventral connections a previous connection which connects the amygdala to the lateral olfactory stria. Due to its characteristics, this connection will be reviewed in Chap. 7.

[2]Foville Louis Achille (1799–1878) was not only an excellent anatomist and physiologist but also a renowned psychiatrist. Professor of Physiology at the Rouen School of Medicine and Chief Physician at St. Yvon asylum in Rouen, he wrote a "Traité complet de l'anatomie, de la physiologie et de la pathologie du système nerveux cérébro-spinal" (Comprehensive treaty on the anatomy, physiology and pathology of the cerebrospinal nervous system) (LA Foville 1844).

[3]Pierre Tarin (1725–1761), a simple physician, proved to be a talented anatomist, who wrote numerous articles in the first encyclopaedia (including a high-quality article on anatomy) and several books, the most notable was "Anthropotomie, ou l'art de disséquer" (Anthropotomy or the Art of dissection) (P Tarin 1750).

[4]Christopher Theodor Aeby (1835–1885), professor of Anatomy in Switzerland. He taught at Berne (CT Aeby 1871).

V. Di Marino et al., *The Amygdaloid Nuclear Complex: Anatomic Study of the Human Amygdala*,
DOI 10.1007/978-3-319-23243-0_6

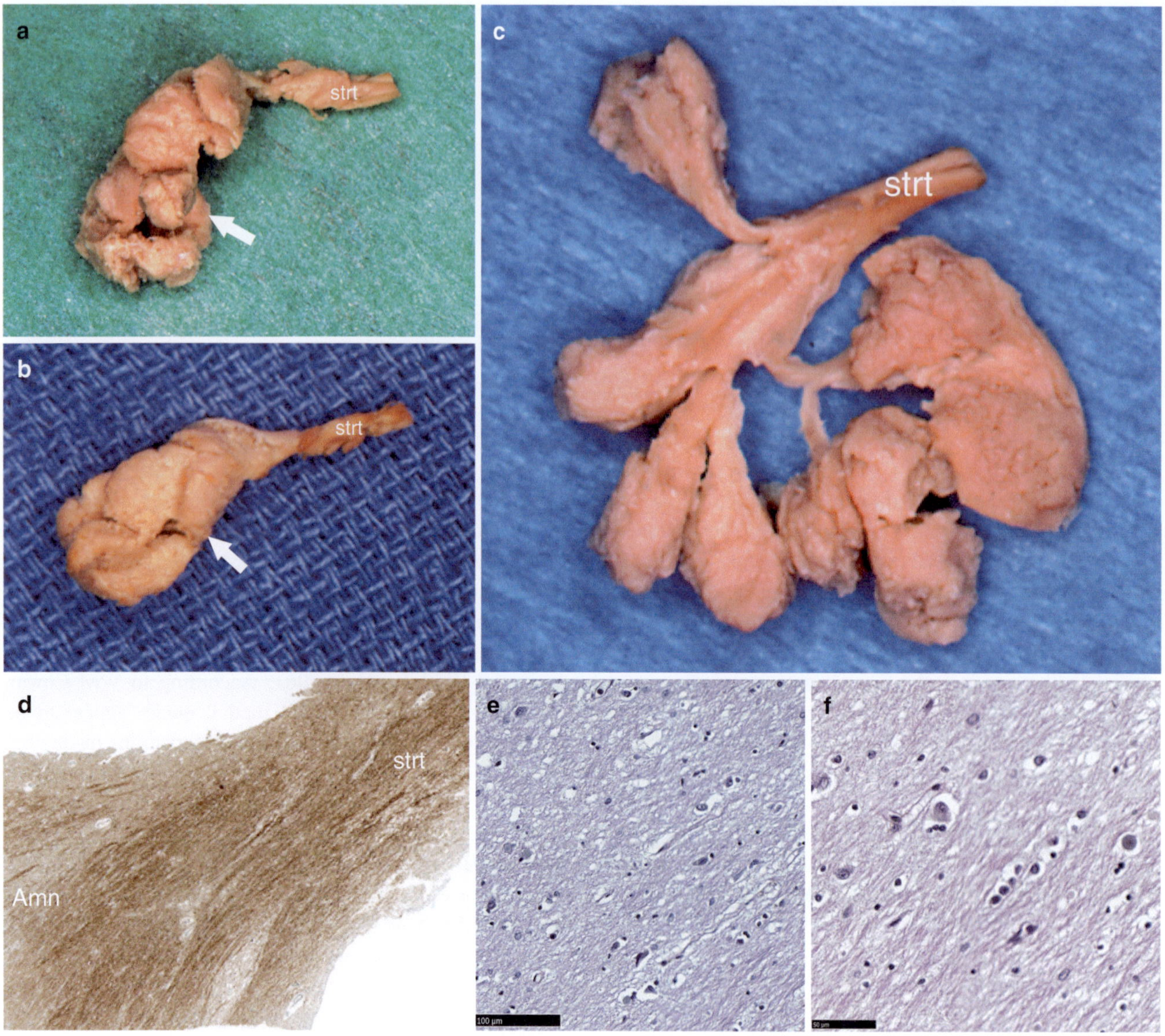

Fig. 6.1 The stria terminalis in the closest segment to amygdala (segment of origin or segment of ending, according to the considered fibres). (**a**, **b**) Two amygdalae (*Amn*) whose nuclei are dissected, with their own stria terminalis (*strt*). (**c**) Dissection and separation of the nuclei of an amygdala whose stria terminalis (*strt*) receives fibres' fasciculi from each nucleus (NB: 10 nuclei have been highlighted). (**d**) Junction between the amygdala and the stria terminalis (Loyez staining for highlighting the myelin sheaths of nerve fibres). (**e**, **f**) Luxol-fast-blue-PAS stain (×100, ×200) reveals some scattered neurons between the fibres and the glial cells. The *white arrows* on (**a**) and (**b**) show two intercalated nuclei

(Fig. 6.2). The choroidal web of the temporal horn of the lateral ventricle is attached to the above tenia and leaves the head of the hippocampus free and therefore not covered by the choroid plexi, unlike the other parts (body and tail) of this formation. The stria terminalis penetrates the opto-striated groove (or thalamo-striate sulcus), at the junction of the posterior thalamic convexity and concavity of the caudate nucleus. It therefore passes in the embryological joining area between the caudate nucleus (forebrain) and the thalamus (diencephalon). The stria terminalis is located in this groove successively in front of the thalamo-striate superior vein, at the level of the body of the caudate nucleus and below this vein, at the head of this nucleus. It is recalled that the superior thalamo-striate vein (terminal vein) drains the veins of the septum pellucidum and of the caudate nucleus and joins the homolateral superior choroidal vein to form one of the two internal cerebral veins, which are the roots of the great cerebral vein. The stria terminalis and its satellite vein will jointly form the bend with an anterior concavity of the thalamo-striate groove and thus reach the floor level of the frontal horn of the lateral ventricle, under the head of the caudate nucleus. They are covered, at this level, by the lamina affixa (lamina cornea), a thin epithelial layer, on the medial border of which are attached the choroid plexi of the lateral ventricle. Thus, the stria terminalis, over its arch-shaped

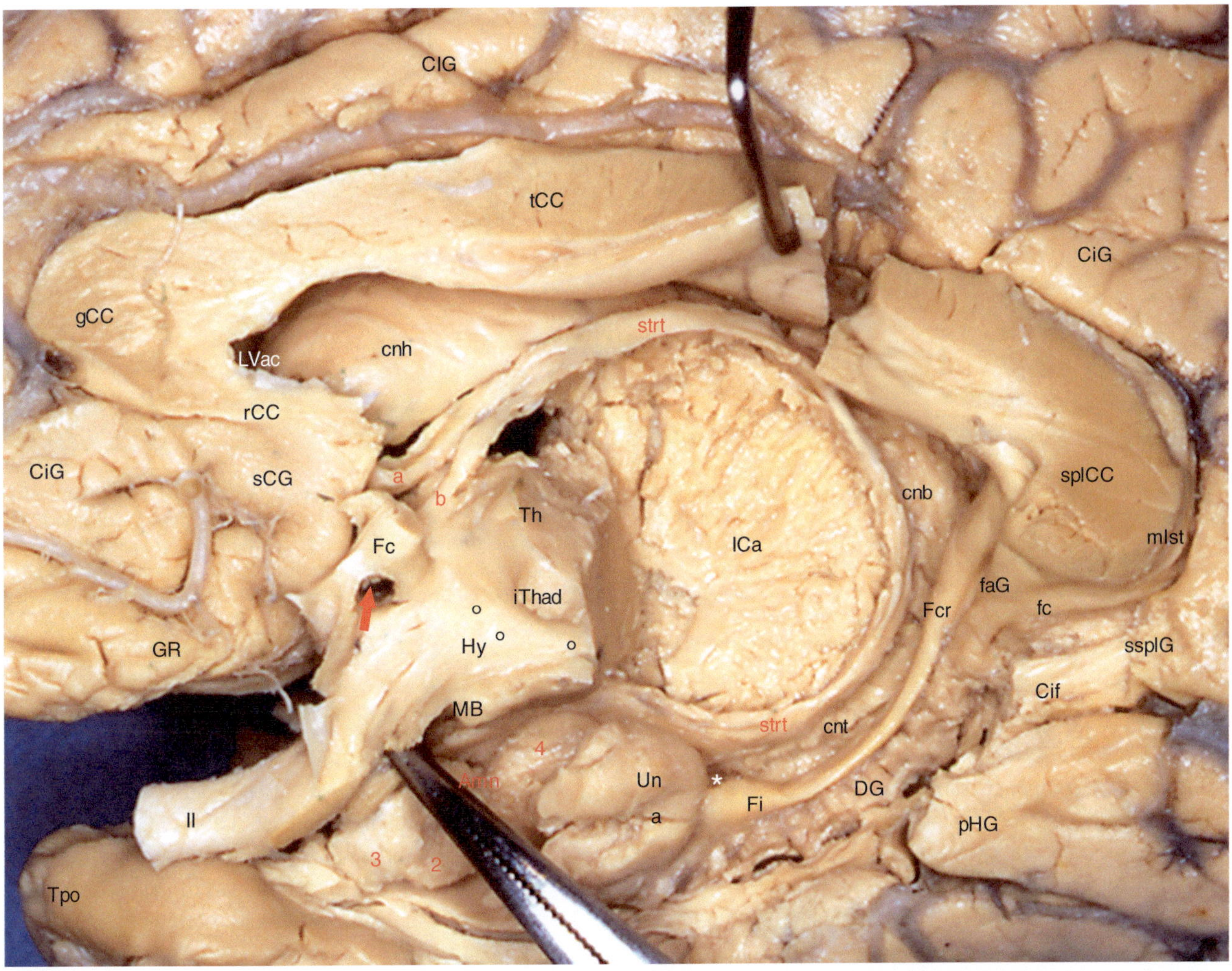

Fig. 6.2 Medial aspect of the right cerebral hemisphere, showing the rostral division of the stria terminalis. *Amn* amygdaloid nuclear complex, *2* baso-lateral, *3* baso-medial, *4* cortical, *CiG* cingulate gyrus, *cnb* caudate nucleus body, *cnh* caudate nucleus head, *cnt* caudate nucleus tail, *DG* dentate gyrus, *faG* fasciolaris gyrus, *Fc* column of fornix, *fc* fasciola cinerea, *Fcr* crus of the fornix, *Fi* fimbria, *gCC* genu of the corpus callosum, *GR* gyrus rectus, *sCG* subcallosum gyrus, *ssplG* subsplenial gyrus, *Hy* hypothalamus, *ICa* internal capsule, *iThad* interthalamic adhesion, *Lvac* lateral ventricule anterior cornu, *MB* mamillary body, *mlst* medial longitudinal stria, *pHG* parahippocampal gyrus, *rCC* rostrum of the corpus callosum, *splCC* splenium of the corpus callosum, *ssplG* subsplenial gyrus, *strt* stria terminalis, *a* ending fasciculus (supra-commissural component) destinated to the paraterminalis gyrus and to the accumbens and septal nuclei, *b* other ending fasciculus (hypothalamic component) destinated to the anterior commissure, the fornix and the hypothalamus), *tCC* truncus of the corpus callosum, *Th* thalamus, *Tpo* temporal pole, *Un a* uncal apex, *II* optic nerve, *Red arrow* it shows the commissural component of strt entering the Gratiolet's canal (the anterior commissure has been removed in order to see the fibres of this commissural component). *Little black circles* hypothalamic sulcus, *white asterisk* velum terminale of Aeby. The posterior portion of the diencephalon has been widely removed in order to show very well the stria terminalis. It is the same for the fornix's body. The rostral part of the septum pellucidum is tracted up by a little retractor. A forceps pulls lightly the hypothalamus's floor

trajectory, forms a 3/4 circle whose diameter is of 2.2 cm on average. Over this trajectory, the stria terminalis not only develops some thin branches, which target the caudate nucleus and the thalamus, but also a significantly sized branch which penetrates the latter at the level of the anterior third of the dorsal surface (Fig. 6.3). According to our chosen description approach, we can say that these branches contain input fibres targeting the thalamus. However, the opposite is also true and we can also say that these branches contain fibres from the thalamus, which are going to merge with the stria terminalis.

When the stria reaches the vicinity of the anterior commissure, it will be divided into three main pathways, which can be accessed for dissection (Fig. 6.2): the supra-commissural pathway (or precommissural component), the commissural pathway (commissural component) and the sub-commissural pathway (or retro-commissural component). These three pathways will develop as follows:

- The supra-commissural pathway passes above and in front of the anterior commissure and joins the septum to terminate in the septal nuclei (lateral nucleus and medial nucleus)

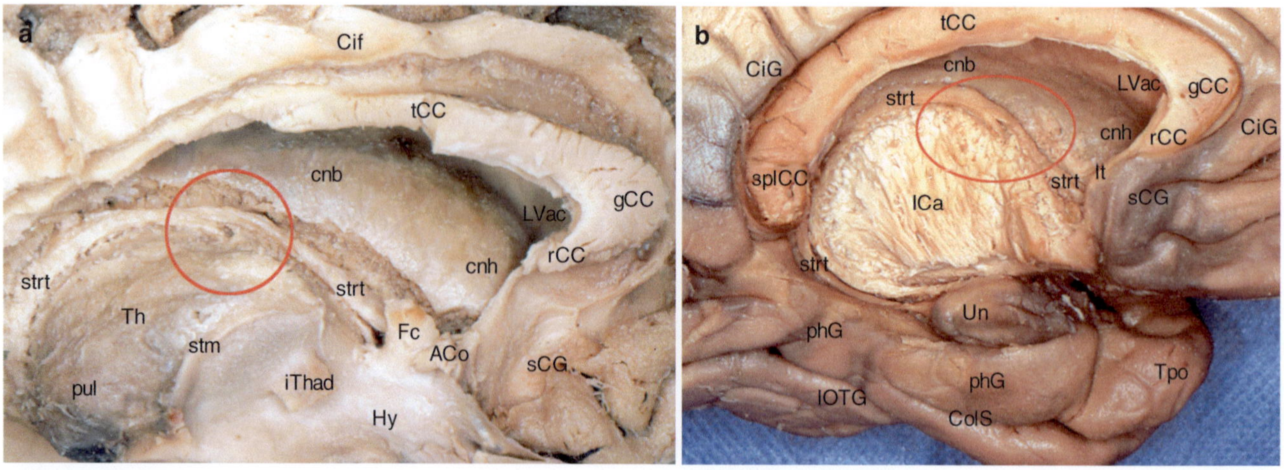

Fig. 6.3 The thalamic ramus of the stria terminalis, highlighted on two dissections of the medial face of the left hemisphere (the ramus is visible on the centre of each *red circle*). (**a**) Thalamus in position; (**b**) thalamus resected. *ACo* anterior commissure, *Cif* cingulate fasciculus, *CiG* cingulate gyrus, *cnb* caudate nucleus, body, *cnh* caudate nucleus, head, *ColS* collateral sulcus, *Fc* column of fornix, *gCC* genu of the corpus callosum, *Hy* hypothalamus, *ICa* internal capsule, *Ithad* interthalamic adhesion, *lOTG* lateral occipito-temporal gyrus, *lt* lamina terminalis, *LVac* lateral ventricle anterior cornu, *phG* parahippocampal gyrus, *pul* pulvinar, *rCC* rostrum of the corpus callosum, *sCG* subcallosum gyrus, *splCC* splenium of the corpus callosum, *stm* stria medullaris, *strt* stria terminalis, *tCC* truncus of the corpus callosum, *Th* thalamus, *Tpo* temporal pôle, *Un* uncus

(JS De Olmos 2004), and in the large nucleus accumbens septi, a key component of the pleasure and reward circuit. A part of the pathway reaches the cortex and terminates in the anterior perforated substance in the olfactory tubercle (G Paturet 1964) and in the paraterminal gyrus.

- Thc fibres of the commissural pathway (commissural component) (Fig. 6.4) join the anterior commissure, thus forming communication pathways between the two brain hemispheres including the two amygdalae.
- The sub-commissural pathway passes behind the anterior commissure and divides into several contingents:
  - The first component (fornix component) penetrates the homolateral fornix column and is routed up to the mammillary body.
  - The second component (stria medullaris component) follows a retrograde trajectory. "It is reflected on the external surface of the thalamus, outside the reticulated area" and follows the medullary stria of the thalamus to join, such as already shown by J Déjerine (1901), the nucleus of the habenula (Fig. 6.5).
  - The third component is formed by posterior fibres that will detach towards the rostral part of the thalamus and reach either the nuclei of this formation or those of the underlying hypothalamus (dorsal amygdalo-hypothalamic pathway connecting the centro-medial nuclear group of the amygdala to the preoptic and anterior nuclei of the rostral hypothalamus) (WR Adey and M Meyer 1952, WM Cowan et al. 1965).

The stria terminalis was recently studied in humans by fMRI (HG Kwon et al. 2011), involving a diffusion tensor tractography study, which provided some interesting images, which were, nevertheless, less accurate and less complete (some branches were not obvious) than anatomical dissections.

Fig. 6.4 The commissural component of the stria terminalis on two medial views (**a**) and (**b**) of the dissected right hemisphere. *Amn* amygdaloid nuclear complex, *2* baso-lateral nucleus, *3* baso-medial nucleus, *4* cortical nucleus, *5* central nucleus, *6* medial nucleus, *bmFa* baso-medial frontal artery, *cF* column of fornix, *chci* chiasmatic cistern, *Cif* cingulate fasciculus, *CiG* cingulate fasciculus, *cnb* caudate nucleus body, *cnh* caudate nucleus head, *cnt* caudate nucleus tail, *Fb* fornix body, *fc* fasciola cinerea, *Fcr* crus of the fornix, *Fi* fimbria, *gCC* genu of the corpus callosum, *Hb* hippocampal body, *Hh* hippocampal head, *Ht* hippocampal tail, *Hy* hypothalamus, *ICa* internal capsule, *iThad* interthalamic adhesion, *llst* lateral longitudinal stria, *lt* lamina terminalis, *LVac* lateral ventricle, anterior cornu, *LVb* lateral ventricle body, *LVic* lateral ventricle inferior cornu, *MB* mamillary body, *mD* margo denticulatus, *Mes* mesencephalon, *och* optic chiasma, *ot* optic tract, *perica* pericallosal artery, *r'* optic recess, *rCC* rostrum of the corpus callosum, *sCG* subcallosum gyrus, *Sepe* septum pellucidum, *strt* stria terminalis, *tCC* truncus of the corpus callosum, *Th* thalamus, *Tpo* temporal pole, *Un a* apex of the uncus, *UnF* uncinate fasciculus, **Fb* The *asterisk* shows the section's level of the fornix body, *II* optic nerve, *a1* anterior cerebral artery, *black circles* hypothalamic sulcus, *red curved arrow* anterior commissure, *red right arrow* shows the commissural component of the stria terminalis in the Gratiolet's canal; on **a**, the Cif is retracted to show the mD of the dentate gyrus; on **b**, the hippocampus is partially separated from the temporal lobe and retracted by an hook

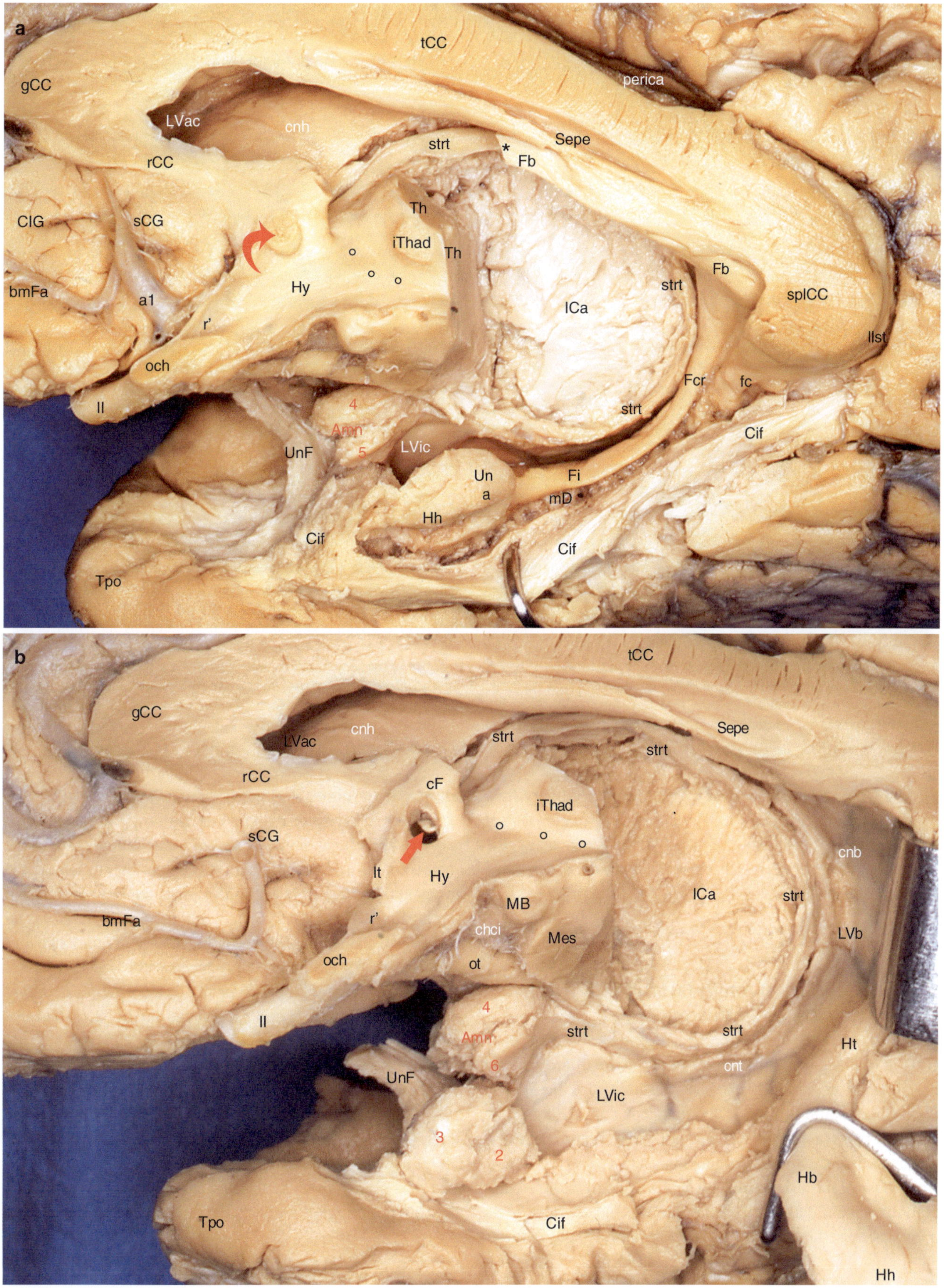
a
tCC
gCC
perica
LVac
cnh
rCC
strt
Sepe
*
Fb
ClG
sCG
Th
iThad
Th
bmFa
Hy
a1
r'
Fb
strt
ICa
splCC
llst
och
II
Fcr
fc
strt
4
Amn
5
LVic
Cif
UnF
Un
a
Fi
mD
Hh
Cif
Cif
Tpo
b
tCC
gCC
cnh
LVac
strt
strt
Sepe
rCC
cF
iThad
sCG
lt
Hy
cnb
bmFa
r'
MB
chci
Mes
ICa
strt
LVb
och
ot
II
4
Amn
6
strt
strt
cnt
Ht
UnF
LVic
3
2
Hb
Tpo
Cif
Hh

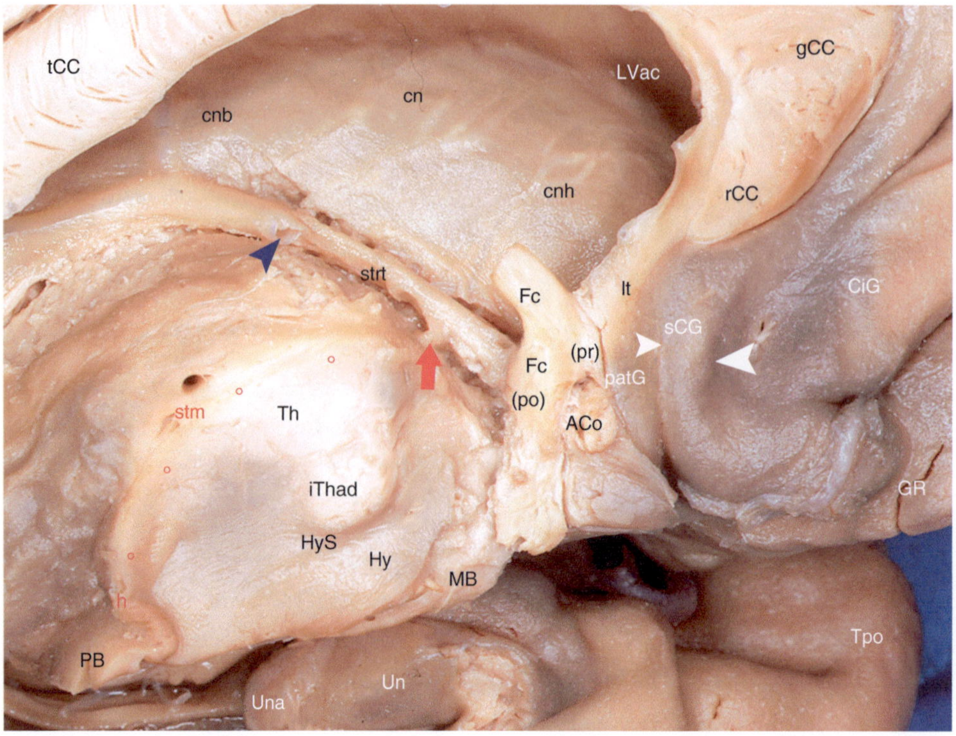

Fig. 6.5 Anastomosis between the stria terminalis and the stria medullaris thalami (dissection of the left hemisphere after a median sagittal section of the brain). *ACo* anterior commissure, *CiG* cingulate gyrus, *cn* caudate nucleus, *cnb* caudate nucleus body, *cnh* caudate nucleus head, *Fc (po)* fornix column (post-commissural), *Fc (pre)* fornix column (pre commissural), *gCC* genu of the corpus callosum, *GR* gyrus rectus, *h* habenula, *Hy* hypothalamus, *HyS* hypothalamic sulcus, *iThad* interthalamic adhesion, *lt* lamina terminalis, *LVac* lateral ventricle anterior cornu, *MB* mamillary body, *patG* paraterminal gyrus, *PB* pineal body, *rCC* rostrum corpus callosum, *sCG* subcallosum gyrus, *stm* stria medullaris thalami (the little *red circles* show this stria), *strt* stria terminalis, *tCC* truncus of the corpus callosum, *Th* thalamus, *Tpo* temporal pole, *Un* uncus, *Una* uncus apex, *blue arrowhead* shows the section of the thalamo-striate vena (the vein is resected rostrally to the section), *little white arrowhead* posterior paraolfactory sulcus, *large white arrowhead* anterior paraolfactory sulcus, *red arrow* shows the anastomosis between a branch of the stria terminalis and the stria medullaris thalami

At a global level, we can only note the extraordinary field of distribution of the stria terminalis, a major dorsal amygdalofugal pathway, which alone manages the communication between the cerebral amygdala and most of the major brain nuclei.

6.2 The Ventral Amygdalofugal Connection Pathway

It is a pathway consisting of fibres (which are mostly myelinated[5]) connecting, via a ventral trajectory, the amygdala to the rostral portion of the diencephalon. Although its characteristics (a wide, flat, spread out, short, ventral pathway, difficult to see, with few projections) are entirely different from those of the stria terminalis, its correspond to the second communication pathway (ventral pathway) of the amygdala and is as important as the first (dorsal pathway) but is faster because it is shorter and more direct. Similarly to the stria terminalis and despite its name, it has amygdalopetal fibres and amygdalofugal fibres. It connects the amygdala and the diencephalon. It emerges from the dorso-lateral part of the amygdala (several fibres originate from the baso-lateral and central nuclei) and firstly heads, almost horizontally, under the lenticular nucleus (Fig. 6.6), through the innominate substance in the direction of the diencephalon. Thus, it passes obliquely through the ceiling of the anterior perforated substance before widening into a fan shape and dividing into two pathways (R Nieuwenhuys et al. 2008): the amygdalo-hypothalamic pathway, direct and very short, which will terminate in the lateral preoptic area of the hypothalamus, and the amygdalo-thalamic pathway that will reach the thalamus. The latter, which is formed by the fibres coming from the baso-lateral nucleus of the amygdala, terminates in the magnocellular pars oralis of the dorsal medial nucleus of the thalamus[6] after having mingled with the fibres of the inferior thalamic

[5]The proportion of myelinated fibres is greater than in the stria terminalis.

[6]Parvocellular pars caudalis of the dorsal medial nucleus of the thalamus establishes reciprocal connections with the prefrontal cortex via the anterior radiations of the thalamus (J Bossy 1990).

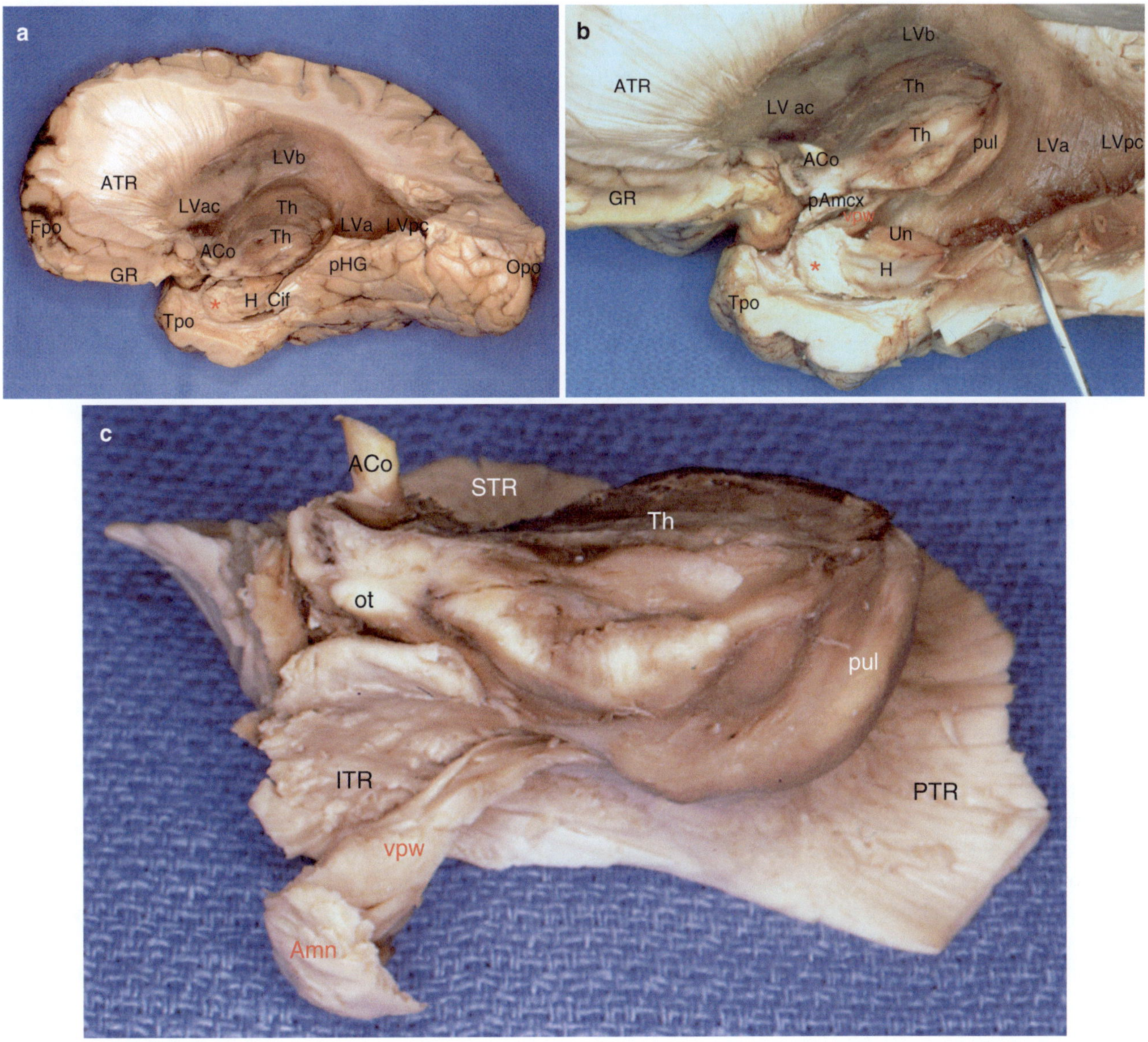

Fig. 6.6 Dissection on a right hemisphere, of the ventral amygdalofugal pathway (fibres arising from Amn and spreading to the thalamus). (**a**) Location of Amn (*red asterisk*) and thalamus. (**b**) Enlargement of **a**, showing the vpw; (**c**) dissection ex situ of the vpw (pushed back and down with the amygdala from which it arises). *ACo* anterior commissure, *Amn* amygdaloid nuclear complex (*red asterisk*), *ATR* anterior thalamic radiations, *Cif* cingulate fasciculus, *Fpo* frontal pole, *GR* gyrus rectus, *H* hippocampus, *ITR* inferior thalamic radiations, *LVa* lateral ventricle atrium, *LVac* lateral ventricle anterior cornu, *LVb* lateral ventricle body, *LVpc* lateral ventricle posterior cornu, *Opo* occipital pole, *ot* optic tract, *pAmcx* peri-amygdalar cortex, *pHG* parahippocampal gyrus, *PTR* posterior thalamic radiations, *pul* pulvinar, *STR* superior thalamic radiations, *Th* thalamus, *Tpo* temporal pole, *Un* uncus, *vpw* ventral amygdalofugal pathway

pathway. The entire system adopts an arch-shaped arrangement and forms the ansa peduncularis. It is joined by thalamocortical fibres to form the inferior thalamic peduncle. It should also be recalled that other fibres of the ventral amygdalofugal pathway, the anterior fibres, are directed rostrally to reach, on the one hand, the medio-frontal cortex (R Nieuwenhuys et al. 2008) and, on the other hand, the septum (amygdalo-septal pathway) and the diagonal band, the latter also being an additional connection pathway (see 7.1, Anterior Relations, in the Chap. 7). Highlighting the pathways of the ventral amygdalofugal pathway by dissection is difficult and uncertain.

- If the dissection is performed medially, on a half-brain obtained by performing a median sagittal section, these pathways can only be viewed after having mobilised the amygdala, by freeing it of the periamygdaloid cortex and of the anterior perforated substance. It is then possible, with large exposure and while taking care to keep the diencephalon, to observe a flattened strip of half a centimetre wide and two centimetres long, directed slightly obliquely and upwards and to the rear, which joins the prethalamic portion of the inferior peduncle of the thalamus and penetrates, with this structure, the thalamus under the inferior surface of its rostral portion (Fig. 6.6).

- If the dissection is performed laterally, still on a half-brain, the dissection proves to be longer but highlighting the ventral connection pathways is easier. This dissection consists, in fact, in successively discovering all of the pathways of the white substance according to the following stages: highlighting the insula after dissecting the superior longitudinal pathway, removing the insular cortex and highlighting the extreme capsule, ablating the extreme capsule and exposing the claustrum, ablating the claustrum and discovering the external capsule, removing the external capsule and discovering the lateral surface of the putamen, resecting the putamen and discovering the internal capsule. It is to be noted that the rostral dissection should, at the same time as the above-mentioned steps are being performed, allow the discovery of the uncinate and of the inferior occipito-frontal fasciculi. Therefore, the dissection becomes delicate and must take place in the area between the rostral fibres of the internal capsule and the summit of the convexity of the uncinate fasciculus. It allows to discover, near this fasciculus, the anterior commissure (sheathed inside the Gratiolet's canal), which will progress more or less vertically against the temporal half of the uncinate bend and develop towards the temporal lobe and then under the most rostral fibres of the internal capsule, the optic tract which is routed forward via a slightly oblique and upward trajectory. It is ultimately between the anterior commissure and optic tract that we discovered a pathway arranged more medially in a transverso-oblique plane: it is the ventral amygdalofugal connection pathway that has just passed through the sublenticular region and which will join the thalamus.

Such as noted, regardless of the approach firstly used to observe the ventral pathway, it is easy to understand that, by observing coronal sections, even if the amygdala and the diencephalon are close from a spatial viewpoint, their connection via the shortest and most direct route possible can only pass through the innominate substance of the sublenticular region, therefore within a sector which is particularly rich in nerve fibres (see Chap. 7).

6.3 The Anterior Commissure

This is one of the great commissures of the brain. Its specificity is to connect not only the cortical regions of the two hemispheres but also to establish a communication pathway between the two amygdalae and thus to permanently associate the emotional components processed at the level of the right and left brains.

The anterior commissure is a set of white fibres grouped as a transverse cord at the level of the middle part of this commissure and which develop into terminal amygdaloid, cortical and olfactory branches. The middle part, which corresponds approximately to a third of the commissure, is housed in a canal described by Gratiolet[7], in 1839, as the canal of the commissure and which has now become Gratiolet's canal. This canal is interrupted laterally, with a lateral opening for each hemisphere, at the level where the commissure crosses the temporal isthmus to form terminal branches.

In the axial view, the anterior commissure is shaped as a horseshoe, convex in the front region, or even as bicycle handlebars (see picture c, Fig. 6.7) whose ends terminate in a fan-type configuration. On the coronal sections, this formation has the shape of a perfect circular segment (see picture a, Fig. 6.7). The middle of the commissure is located immediately in front of the fornix columns and the rostral edge of the third ventricle (see pictures a and b, Fig. 6.7). This "white commissure" then curves gradually and passes under the lenticular nucleus before crossing the temporal isthmus, passing along the side of the amygdala and forming the temporal fan. The diameter of the commissure is practically that of an optic nerve: 4 mm on average (L Testut 1911). In the middle section, the cord fibres are twisted. It is also in this middle section that occur a few fibre exchanges between the anterior and posterior portions of the commissure, which have sometimes evoked a chiasma. However, most fibres follow a transverse route and are organised in three parts: a posterior part for the temporal region, a middle part and an anterior part for the olfactory region. The fibres of the anterior part are dedicated to the olfactory bulbs, olfactory tracts, prepiriform cortices and right and left anterior perforated substances and thus contribute equally to olfaction. These olfactory fibres have been described by Theodor Meynert (1888). According to this author, these anterior fibres can be direct, transverse or intertwined and are routed from an olfactory bulb to the contralateral temporal lobe. In addition to the two olfactory areas, the anterior fibres also connect the two stria terminalis (Figs. 6.2 and 6.4), the two BST and the two nuclei accumbens. The anterior part of the commissure is reduced compared to the rear part, which corresponds to

[7] LP. Gratiolet (1815–1865), anatomist, anthropologist and French zoologist, author (with Fr. Lorenzo) of "Anatomie du système nerveux, considérée dans ses rapports avec l'intelligence" (Anatomy of the nervous system, considered in its connections with intelligence) JB Ballière edit, Paris, 1839.

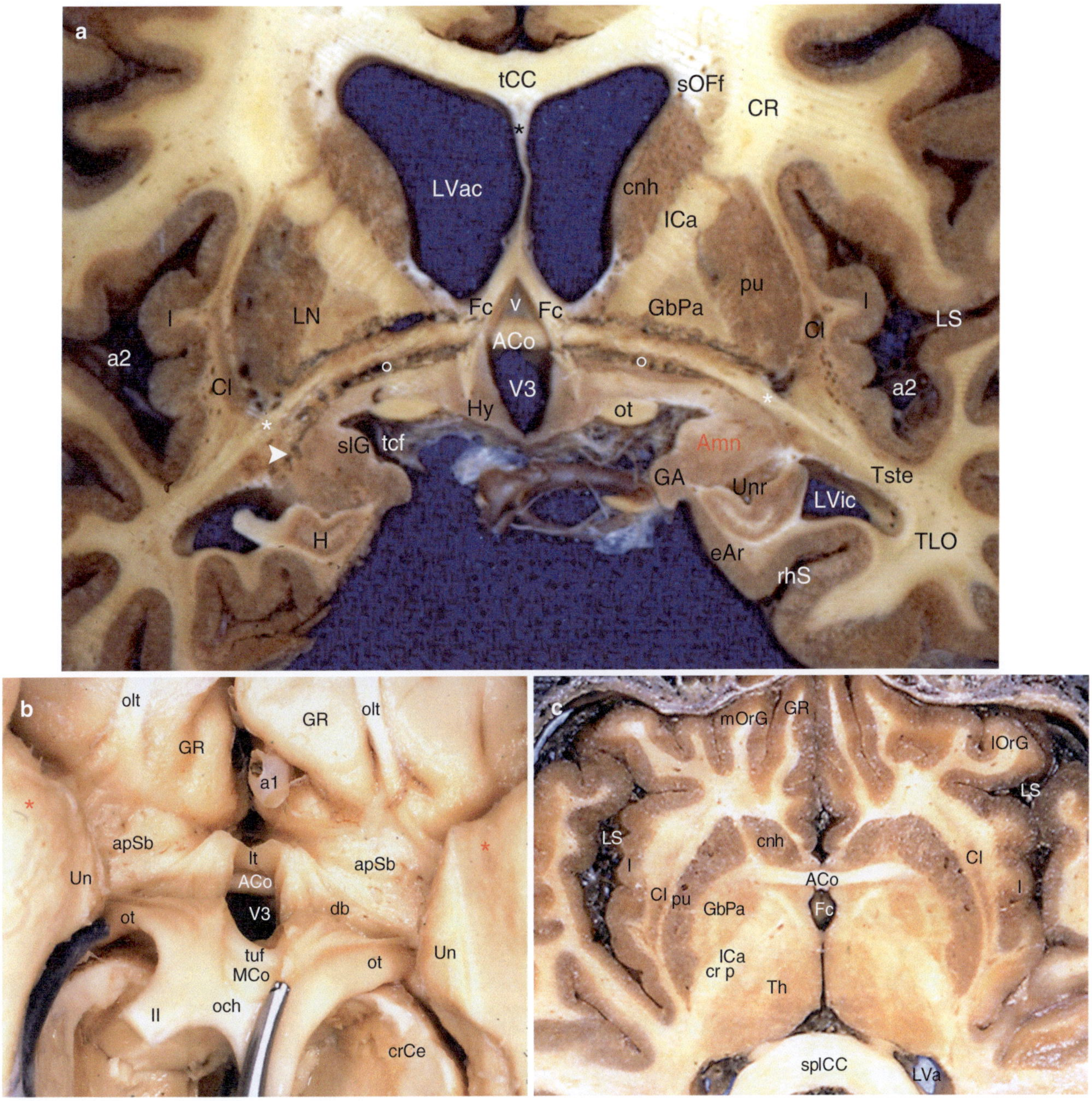

Fig. 6.7 Morphology and relations of the anterior commissure. (**a**) Dissection of the ACo on a coronal slice of the brain (posterior surface of the section). (**b**) Inferior view of the ACo on the basal brain's surface in the interpeduncular fossa (the optic chiasm has been rear retracted). (**c**) The aspect of the ACo (aspect in "bike handlebars") on a axial section of the brain through the thalamus. *ACo* anterior commissure, *Amn* amygdaloid nuclear complex, *apSb* anterior perforated substance, *Cl* claustrum, *cnh* caudate nucleus head, *CR* corona radiata, *crCe* crus cerebri, *db* diagonal band, *eAr* entorhinal area, *Fc* column of fornix, *GA* gyrus ambiens, *GbPa* globus pallidus, *GR* gyrus rectus, *H* hippocampus, *Hy* hypothalamus, *I* insula, *ICa (cr p)* internal capsule (crus posterior), *LN* lentiform nucleus, *lOrG* lateral orbital gyrus, *LS* lateral sulcus, *lt* lamina terminalis, *LVa* lateral ventricle atrium), *LVac* lateral ventricle anterior cornu, *LVic* lateral ventricle inferior cornu, *MCo* Meynert commissure, *mOrG* medial orbital gyrus, *och* optic chiasm, *olt* olfactory tract, *ot* optic tract, *pu* putamen, *rhS* rhinal sulcus, *slG* semi-lunaris gyrus, *sOFf* superior occipito-frontal fasciculus, *splCC* splenium of the corpus callosum, *tCC* truncus of the corpus callosum, *tcf* transverse cerebral fissure, *Th* thalamus, *TLO* temporal lobe, *Tste* temporal stem, *tuf* tuber fasciculus, *Un* uncus, *Unr* uncal recess, *v* vulva, *V3* third ventricle, *a1* anterior cerebral artery, *a2* middle cerebral artery, *II* optic nerve, *red asterisks* they locate the amygdalar areas, *black asterisk* septum pellucidum, *white circle* ACo in the Gratiolet's canal, *white asterisk* ACo out of the Gratiolet's canal, *white arrowhead* shows a ramus of the ACo penetrating in the amygdala

the involution of the rhinencephalon in humans and mammals, in general. The fibres of the middle part connect the two lentiform nuclei, the two insula, the frontal and parietal opercula, the fornix columns, the two stria terminalis, the two stria medullaris and the two hypothalami (J Bossy). The posterior fibres are the most numerous. They terminate medially in the amygdala, in the hippocampus and, laterally, in all the temporal gyri (Figs. 6.7 and 6.8) and more to the rear, in the occipital gyri. The temporal fibres "play, compared to the temporal lobes, the same role as the corpus callosum for the superior gyri of the hemispheres" (L Testut): they combine the two hemispheres for a single function. Fibres which reach the amygdala terminate in the nuclei of the baso-lateral group. Some of them only cross through the amygdala to reach the parahippocampal cortex and hippocampus (Crosby et al. 1962).

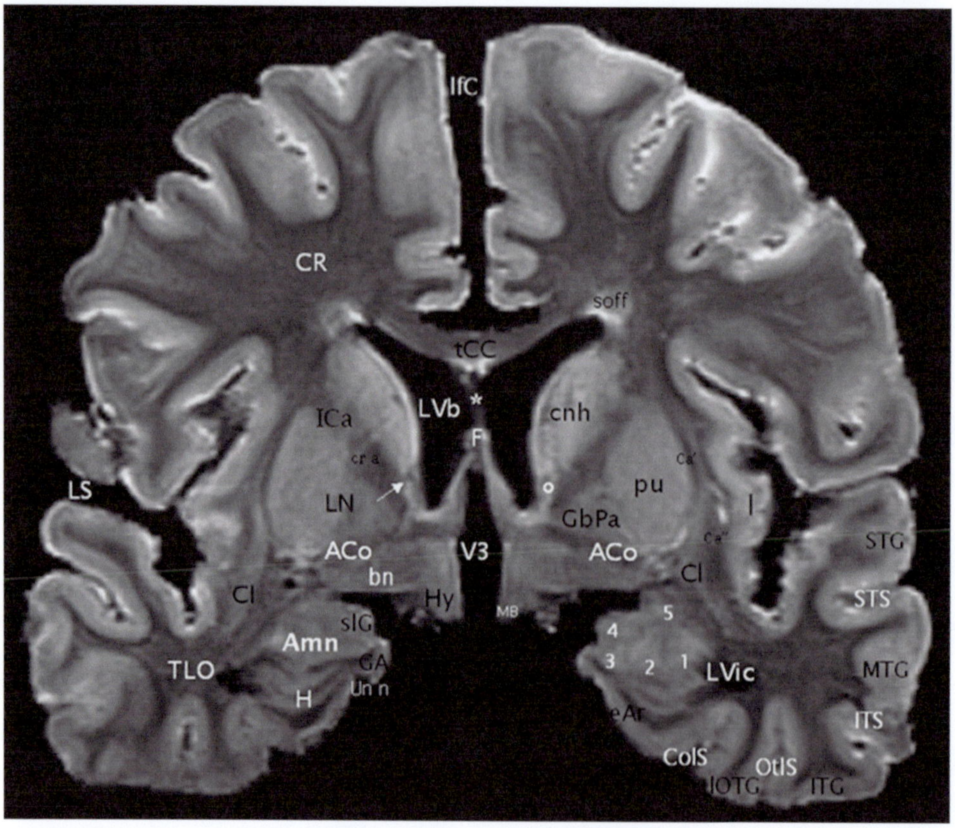

Fig. 6.8 The Aco, the amygdalae and the two BST on a MRI picture of a brain's coronal section. *ACo* anterior commissure, *Amn* amygdaloid nuclear complex, *1* lateral nucleus, *2* baso-lateral nucleus, *3* baso-medial nucleus, *4* cortical nucleus, *5* central nucleus, *bn* basal nucleus of Meynert, *Ca'* external capsule, *Ca"* extreme capsule, *Cl* claustrum, *cnh* caudatus nucleus, head, *ColS* collateral sulcus, *CR* corona radiata, *cr a* crus anterior (*ICa*), *eAr* entorhinal area, *F* fornix, *GA* gyrus ambiens, *GbPa* globus pallidus, *H* hippocampus, *Hy* hypothalamus, *I* insula, *ICa* internal capsule, *ITG* inferior temporal gyrus, *ITS* inferior temporal sulcus, *lfC* longitudinal fissure of the cerebrum, *LN* lenticular nucleus, *lOTG* lateral occipito-temporal gyrus, *LS* lateral sulcus, *LVb* lateral ventricle, body, *LVic* lateral ventricle, inferior cornu, *MB* mamillary body, *MTG* middle temporal gyrus, *OtlS* occipito-temporal lateral sulcus, *pu* putamen, *slG* semi-lunaris gyrus, *soff* superior occipito-frontal fasciculus, *STG* superior temporal gyrus, *STS* superior temporal sulcus, *tCC* truncus of the corpus callosum, *TLO* temporal lobe, *Un n* uncal notch, *V3* third ventricle, *septum pellucidum; *white circle BST; white arrow* stria terminalis

Relations of the Nuclear Amygdaloid Complex

7

The topographic location of this complex and its physiological significance explain that it must be permanently connected to a large number of other cortical and subcortical encephalic structures. The study of the relations of this formation is extremely complex and requires knowledge of both the brain and the brainstem[1].

We shall thus study successively the anterior relations, the lateral relations, the superior relations, the distant superomedial relations, the medial relations and the posterior relations. A paragraph will be dedicated to relations with the brainstem which we shall refer to as the postero-inferior relations.

7.1 Anterior Relations

Laterally and at the temporal floor, the amygdala is connected to the temporal pole. It is connected to the latter by a cylindrical looking bundle, constant during dissections, the fasciculus amygdalo-temporalis of Klinger (see medial relations).

Medially and at the junction between the temporal and frontal levels, the amygdala is surrounded by its periamygdaloid cortex (ambiens and semilunar gyri) and is directly connected to the entire olfactory system,[2] especially the prepiriform cortex and the anterior perforated substance, apSb (Fig. 7.1) It is to be recalled that the latter is situated behind the lateral and medial olfactory striae and the olfactory tubercle and that it is limited at the rear, by the optic tract, inside, by the diagonal band and the optic chiasma, and outside, by the endorhinal sulcus, which separates it from the uncus, and by the threshold to the insula or limen insulae, a deep sulcus forming "the bottom of the basal part of the lateral sulcus of Sylvius" (HM Duvernoy 1992). This perforated substance is comprised of paleocortex,[3] while the periamygdaloid cortex is made of periarchicortex. In humans, a third stria (referred to as the intermediate stria) is exceptionally linked to the two lateral and medial olfactory striae.[4] The triangular space between the lateral and medial striae is the olfactory trigone. Under the base of the trigone, the protrusion of the olfactory tubercle can be observed.

The tubercle[5] receives fibres of the lateral stria. The lateral olfactory stria becomes lost in the limen after having formed, just before the falciform fold (deep notch connecting the planum temporale to the orbital cortex and to the rostral part of the insula), a posterior thin bundle, which is difficult to dissect and which directly reaches the cortical nucleus of the amygdala (see Fig. 7.1). The medial olfactory stria is directed to the median region of the brain and reaches the paraterminal and subcallosal gyri (septal area 25) on the medial surface of the homolateral hemisphere, where is located, in particular, in the anterior olfactory nucleus. The anterior perforated substance (former grey substance of Soemmering[6]) is "perforated" by multiple small vascular orifices (the largest are lateral) via which penetrate the striated arteries dedicated to

[1]Some studies even refer to connections between the amygdala and the cerebellum, via the dorsal striatum (D Laricchiuta and L Petrosini 2014).

[2]Rhinencephalon is the name given by the former anatomists to the part of the brain which is dedicated to olfaction. Some authors extend this term to the entire limbic lobe.

[3]The paleocortex (previously considered as the former brain from a phylogenetic viewpoint, the most developed in inferior beings, which is used for the perception of smells and for the reactions of individuals following this perception) includes in humans, the olfactory bulb, tract, striae and tubercle, the prepiriform cortex, the apSb, the entorhinal cortex, the uncus, the diagonal band, the septal area, the surface of the limen and the inferior summit of the insular triangle.

[4]We also call up this stria the mean stria or stria intermedia or accessory stria of Mutel (Paturet 1964). It ends as a fan-shaped formation at the olfactory trigone.

[5]The olfactory tubercle whose cellular constitution is made by GABAergic projecting neurons (GF Alheid and L Heimer 1988) is considered from now on as being a part of the ventral striatum, as the accumbens nucleus and as the islands of Calleja.

[6]Samuel Thomas von Soemmering (1755–1830), German professor of anatomy and surgeon.

V. Di Marino et al., *The Amygdaloid Nuclear Complex: Anatomic Study of the Human Amygdala*,
DOI 10.1007/978-3-319-23243-0_7

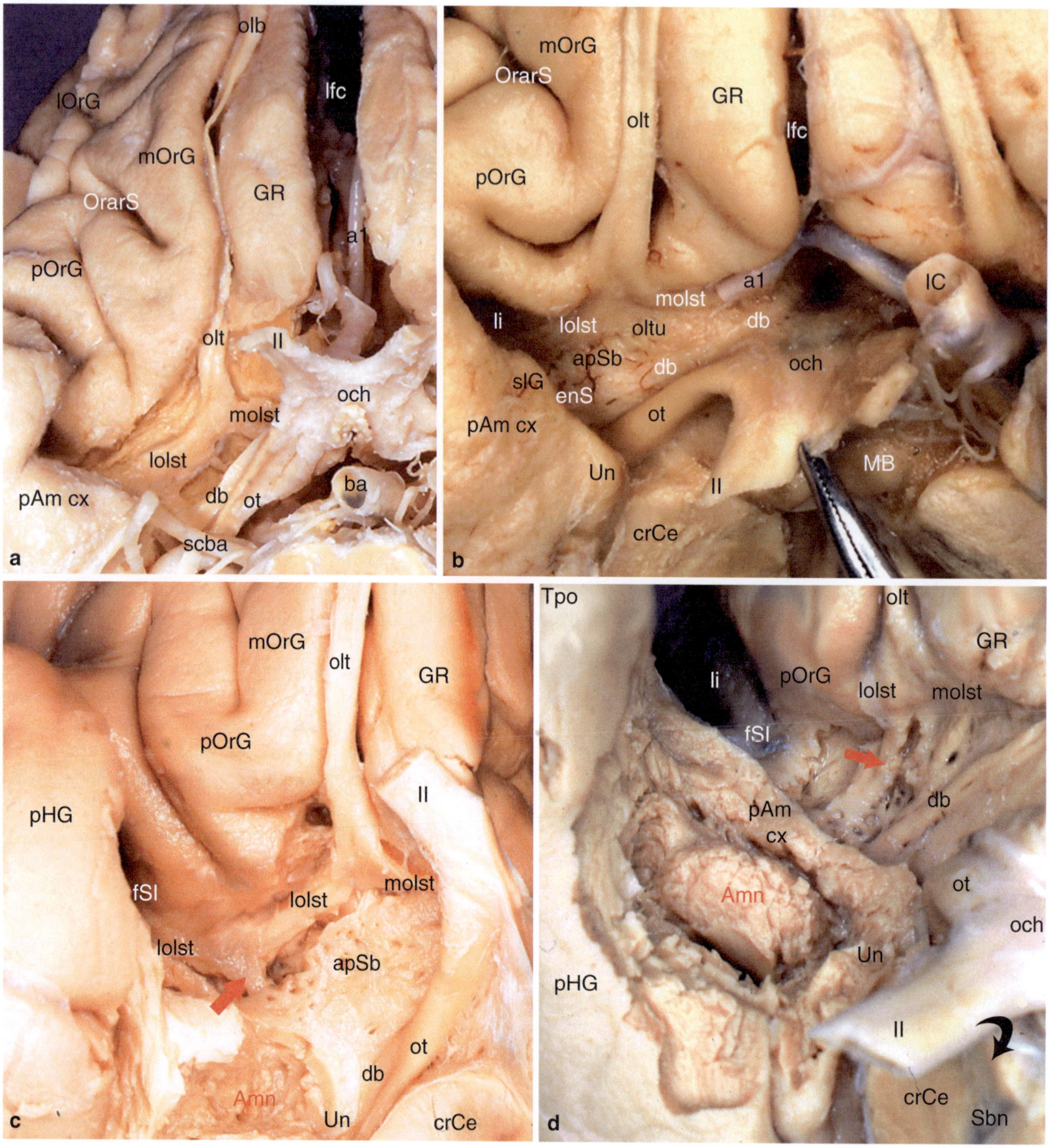

Fig. 7.1 Olfactory system and olfactory pathway to amygdala. (**a**) Dissection of the right olfactory system on an inferior view of the brain; (**b**) the right apSb on an inferior view of the brain. (**c**) Highlighting of the olfactory pathway (*red arrow*) to amygdala. (**d**) Highlighting of the olfactory pathway (*red arrow*) with dissection of the amygdala. *Amn* amygdaloid nuclear complex, *apSb* anterior perforated substance, *ba* basilary artery, *crCe* crus cerebri, *db* diagonal band, *enS* endorhinal sulcus, *fSI* falciform sulcus of the insula, *GR* gyrus rectus, *IC* internal carotid, *lfc* longitudinal fissure of the cerebrum, *li* limen insulae, *lolst* lateral olfactory stria, *lOrG* lateral orbital gyrus, *MB* mamillary body, *molst* medial olfactory stria, *mOrG* medial orbital gyrus, *och* optic chiasm, *olb* olfactory bulb, *olt* olfactory tract, *oltu* olfactory tubercle, *OrarS* orbital arcuatus sulcus, *ot* optic tract, *pAm cx* peri-amygalar cortex, *pHG* parahippocampal gyrus, *pOrG* posterior orbital gyrus, *Sbn* substantia nigra, *scba* superior cerebellar artery, *slG* semilunar gyrus, *Tpo* temporal pole, *Un* uncus, *a1* anterior cerebral artery, *II* optic nerve, *red arrows* olfactory pathway from lateral olfactory stria to amygdala, *curved black arrow* it shows the right optic nerve pushed back as on the picture (**b**)

the striatum. These rami, which developed from the anterior and middle cerebral arteries at their origin, are therefore vascular connections which are very close to the amygdala (see Fig. 9.1a).

From a topographic viewpoint, the apSb, which is only separated from the amygdala by the endorhinal sulcus and which is only separated from the orbitofrontal cortex by the narrow prepiriformis gyrus, therefore provides a connection between this amygdala and the orbitofrontal cortex (gyrus rectus inside the olfactory tract and orbital gyri outside this tract). Similarly, as it is connected to the limen, the apSb is directly connected to the planum temporale and the insular gyri, and therefore, these formations are connected, via this structure, to the amygdala!

Olfactory inputs (conscious projections), analysed in the olfactory bulb but also in the cortices bordering the posterior third of the olfactory sulcus (medial orbital sulcus), reach, via the lateral olfactory stria and the anterior perforated substance, the primary olfactory centres[7] (prepiriform cortex[8] and periamygdaloid cortex), the rostral part of the cortical nucleus of the amygdala and, in fine, the secondary projection area, the entorhinal area, area 28, underlying the gyrus ambiens (limbic area where the discrimination between pleasant and unpleasant smells takes place). The inputs also pass via the lateral olfactory stria and reach the limen, via which they reach the insular gyri. This most certainly explains the known links between taste and the smell of food, which interfere in the appreciation of dishes. It is necessary to recall the projections of the primary olfactory centres towards the medial, posteromedial and intermediate agranular insular areas (R Nieuwenhuys et al. 2008).

Furthermore, projection fibres depart from the medial olfactory stria and pass via the medial forebrain bundle (MFB) to reach the hypothalamus and the mesencephalon. Other fibres pass via the stria medullaris thalami to the habenula from which the habenulo-interpeduncular bundle allows the olfactory projection towards the interpeduncular nucleus and the reticular mesencephalic formation.

A major connection of the amygdala in this olfactory area is that of Broca's diagonal band, which plays several roles within the amygdaloid environment:

It represents the actual postero-internal limit of the apSb (more than the optic tract and the optic chiasm, which are more posterior and do not adhere to the apSb), it has connections to the amygdala itself and to the uncus, and it forms an incredible connection pathway to the septum and amygdala; thus it completes the amygdalo-septal connection implemented via the stria terminalis, by a septo-amygdaloid connection; and therefore forms an actual loop involving the development of amygdalofugal and amygdalopetal fibres.

Our dissections allowed to correctly locate the standard septo-amygdalo-uncal portion, mentioned by all the anatomists since Broca. However, we discovered and highlighted (Figs. 7.2, 7.3, 7.4, and 7.5) a posterior portion of this band, a narrower portion, which was perfectly dissectable and which followed the optic tract up to the pulvinar that it reaches a few millimetres above the top edge of the lateral geniculate body. Therefore, we can assert that the diagonal band is a complex, convex, outward and flattened structure, especially in its initial segment where it is stuck to the apSb, and comprises several parts, a short vertical septal part; an oblique transverse part at the rear and outward (often wrongly called the horizontal part), going from the septum to the unco-amygdaloid junction; and a narrower but more ribboned part, with a slight medial concavity, molded against the lateral surface of the mesencephalon, from the uncus to the thalamus, oblique at the rear and inward. Very rich cellular nuclei occupy three essential locations of the band: a rostral nucleus which is no other than the nucleus known as the "nucleus of the diagonal band", juxta-septal, a nucleus situated near the amygdala (Fig. 7.5) and the uncus, where the band changes direction (or, if we prefer, immediately before or at the junction of the two segments of the band, at the top of the curvature) and a terminal juxta-thalamic nucleus, at its posterior end.

Other observation: our dissections also revealed a double origin of the rostral portion of the band, which is formed not only by the standard septal bundle but also by a slightly arched and relatively thin bundle, coming from the area of the limen[9] (Figs. 7.3 and 7.4). Both bundles join before the band reaches the uncus and the amygdala.

Broca's band is known, at present, to transmit the olfactory inputs to the uncus and septal messages to the amygdala. With the new description defined thanks to our research work, we had to question the meaning of such a band. Sadly, we could not provide a satisfactory answer (oniric pathway? additional junction with the posterior part of the thalamus? Other?). Histochemical colourations and other anatomical dissections and research performed in neurosciences shall undoubtedly and shortly provide essential clarifications. It is to be recalled that Broca's band is very rich in cholinergic cells and that it is part of the MBF (magnocellular basal forebrain) such as Meynert's basal nucleus (see Chap. 11). It therefore belongs to a system which sends projections not only to the amygdala and hippocampus but above all to the cortex. Many authors have established correlations between

[7]The primary olfactory centres as well as the anterior olfactory nucleus of a hemisphere interconnect to the same structures of the contro-lateral hemisphere by fibres which pass via the anterior commissure and therefore form an actual right–left olfactory dialogue.

[8]The prepiriform cortex is a thin paleocortex layer which covers the lateral olfactory stria: it is the prepiriform gyrus. Topographically, it is the continuation of the posterior part of the posterior orbital gyrus.

[9]This fasciculus is perhaps that described by J Klinger and P Gloor (1960) as "amygdalo-insular fasciculus"?

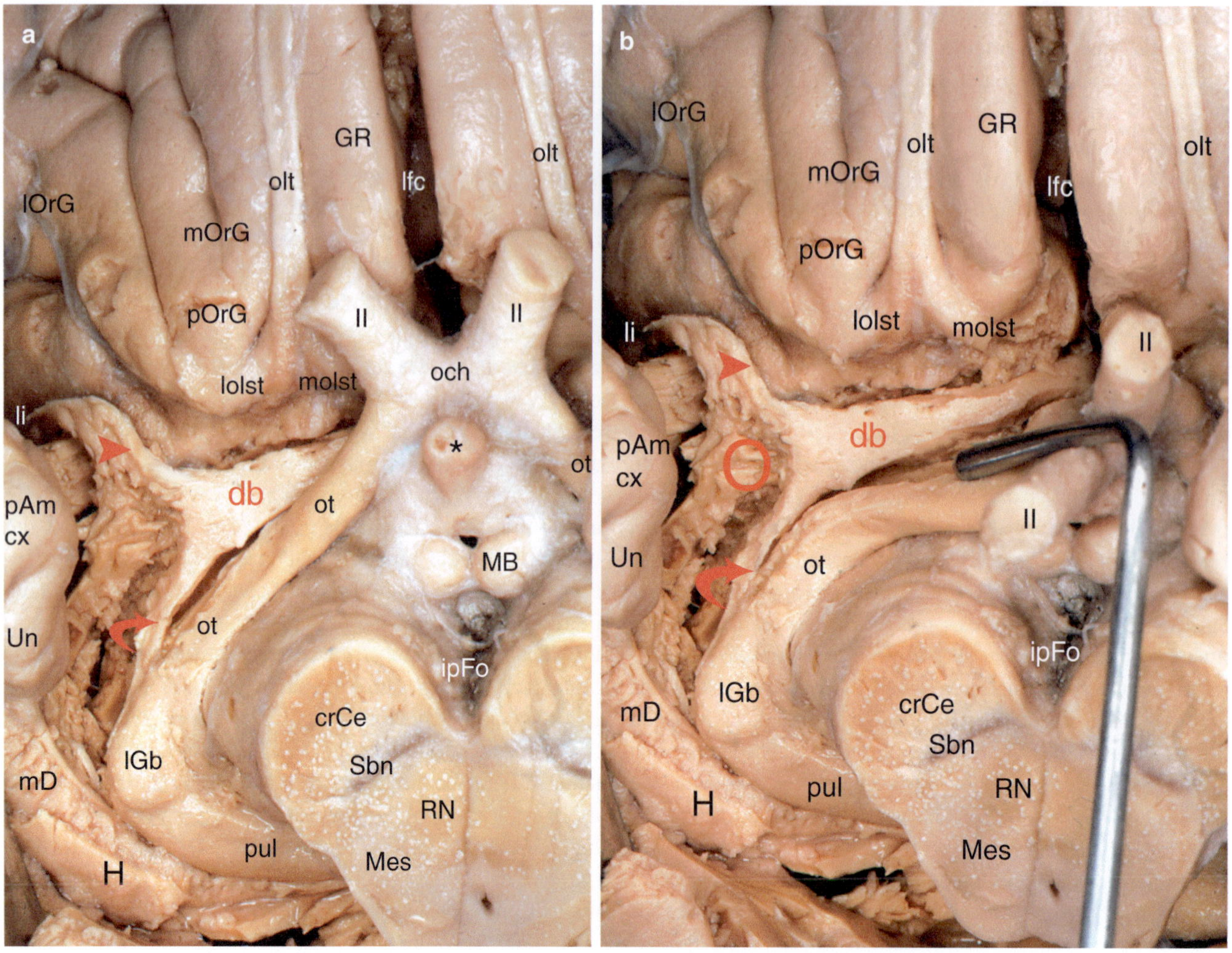

Fig. 7.2 Dissection of the right diagonal band on an inferior view of the brain (the temporal lobe has been tracted laterally. The *red circle* points the amygdalar area). (**a**) The optic chiasm is in anatomic position. (**b**) The optic chiasm has been retracted behind. *crCe* crus cerebri, *db* diagonal band, *GR* gyrus rectus, *H* hippocampus, *ipFo* interpedoncular fossa, *lfc* longitudinal fissure of the cerebrum, *lGb* lateral geniculate body, *li* limen insulae, *lOrG* lateral orbital gyrus, *lolst* lateral olfactory stria, *MB* mamillary body, *mD* margo denticulatus, *Mes* mesencephalon, *mOrG* medial orbital gyrus, *molst* medial olfactory stria, *och* optic chiasm, *olt* olfactory tract, *ot* optic tract, *pAm cx* periamygdalar cortex, *pOrG* posterior orbital gyrus, *pul* pulvinar, *RN* red nucleus, *Sbn* substantia nigra, *Un* uncus. *II* optic nerve, *red arrowhead* db pathway to the limen, *curved red arrow* db pathway to pulvinar, * the asterisk points the infundibular tip of V3

certain cases of senile dementia (including Alzheimer's disease) and the rarefaction of the cholinergic cells of the MBF. Other authors have referred, in relation to these pathologies, a degeneration of the axons of these cells in their terminal presynaptic part, which is responsible for the loss of the cortical cholinergic markers and the formation of senile plaques (JC Hedreen et al. 1984) and, in fine, the dementia.

7.2 Antero-Superior Relations (Relations with the Frontal Lobe)

The amygdala has, as we have observed, proximity relations with the temporal cortex, that of the apSB and the rostral poles of the cingulate cortex. However, other more distant relations form with a key part of the brain, the frontal lobe, and especially the basal part of the latter.

The cortex of this lobe comprises a dorsolateral surface, a medial surface and a basal or inferior surface based on an orbital cavity.

The dorsolateral surface (Fig. 7.6) on which, from the rear to the front, the precentral motor cortex (area 4 of Brodmann),[10] the premotor cortex (area 6)[11] and the prefrontal

[10] It is in the posterior part of the inferior frontal gyrus (astride Brodmann areas 44 and 45) of the dominating hemisphere that is located in Broca's area, the area of articulated language.

[11] In addition to the precentral area, a supplementary motor area (SMA) exists and is located above the cingulate gyrus, on the medial surface of the hemisphere (Fig. 7.7).

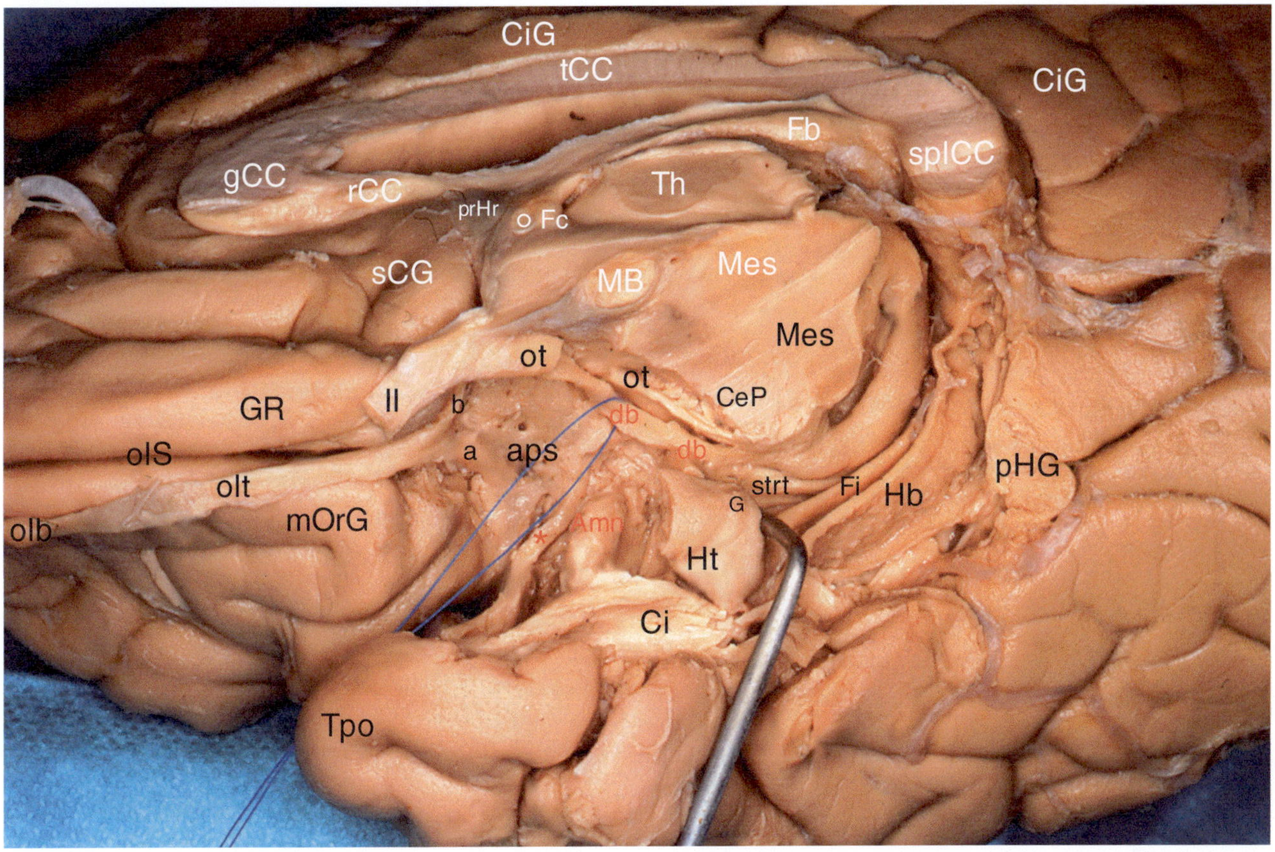

Fig. 7.3 Dissection of the diagonal band on a right cerebral hemisphere. (Infero-medial view of the brain). *The fimbria is slightly pushed downwardly by a retractor.* For an easy reading, the inscriptions of the sagittal section are *white* and those of the inferior area are *black* or *red*. *Amn* amygdala, *aps* anterior perforated substance, *CeP* cerebral peduncle, *Ci* cingulum, *CiG* cingulate gyrus, *db* (*red*) diagonal band (identified by a pull wire), *Fb* fornix (body), *Fi* fimbria, *Fc* fornix (column), *G* band of Giacomini, *gCC* genu of the corpus callosum, *GR* gyrus rectus, *Hb* hippocampus body, *Ht* hippocampus tail, *MB* mamillary body, *Mes* mesencephalon, *mOrG* medial orbital gyrus, *olb* olfactory bulb, *olS* olfactory sulcus, *olt* olfactory tract, *ot* optic tract, *pHG* parahippocampal gyrus, *prHr* pre-hippocampal rudiment, *rCC* rostrum of the corpus callosum, *sCG* subcallosum gyrus, *splCC* splenium of the corpus callosum, *strt* stria terminalis, *tCC* truncus of the corpus callosum, *Th* thalamus, *Tpo* temporal pole, *II* optic nerve, *a* lateral olfactory stria, *b* medial olfactory stria, ° anterior commissure, * peri-amygdaloïd cortex

cortex (areas 8, 9, 10, 45 and 46)[12] are observed. The prefrontal cortex alone represents a third of the cortical mass!

The medial or internal surface (Fig. 7.7) on which, on the one hand, the medial part of areas 6, 8, 9 and 10 and, on the other hand, rostral limbic areas 24, 25 and 32 are located.

The orbital surface, referred to as the orbital cortex (Fig. 7.8), which includes the inferior part of the frontal pole, the medial frontal gyrus, the orbitofrontal gyrus and the inferior part of the cingulate gyrus (areas 10, 11, 12, 25, 15 and 32).

Although the dorsolateral surface only has few connections to the amygdala as it is specifically involved in the motor and executive programmes, the orbital and medial surfaces of the prefrontal cortex are very closely connected to the amygdala and all structures governing the brain of emotions.[12] Thus, the prefrontal cortex receives direct projections not only from nearby cortices (motor cortex, cingulate cortex, sensorial cortex) but also from associative cortices (temporal, occipital, parietal) from the amygdala, hypothalamus and thalamus: it is therefore permanently informed of the individual's emotional, somatic, visceral and motivational conditions.

Such as we will see further in the document, the lateral part (lateral frontal gyrus) of the orbital cortex is dedicated to cognition, whereas the medial part (medial frontal gyrus), the orbitofrontal gyrus and the cingulate gyrus are specifically involved in the emotional circuitry. Due to the fact that all of these parts are interconnected and linked to diencephalic

[12]The dorsolateral prefrontal cortex controls the executive functions, the movement programme, favours attention, allows learning and is used for solving problems.

The fronto-median and anterior cingulate cortices are the basis of the initiative and entrepreneurial spirit.

The orbitofrontal cortex is used for calmly and without irritation dealing with situations, which may occur, and induces a behaviour compatible with a social life.

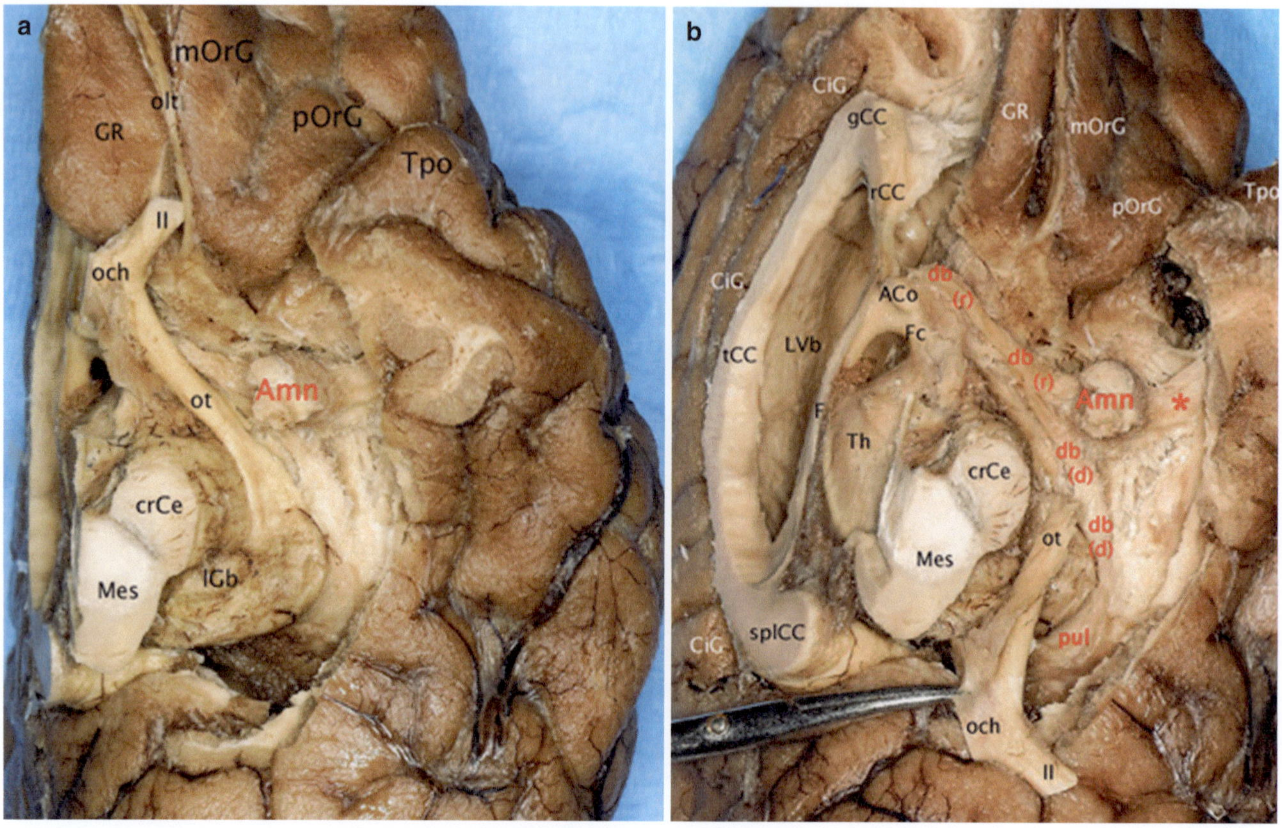

Fig. 7.4 Relationships of the amygdala with the diagonal band. (**a**) Inferior view of the left brain hemisphere. (**b**) Infero-medial view of the left brain hemisphere. (The left optical ways were pulled behind to clear the visibility of the diagonal band.) *ACo* anterior commissure, *Amn* amygdaloid nuclear complex, *CiG* cingulate gyrus, *crCe* crus cerebri, *db* diagonal band, *db r* rostral part, *db d* dorsal part, *F* fornix, *Fc* column of the fornix, *gCC* genu of the corpus callosum, *GR* gyrus rectus, *lGb* lateral geniculate body, *LVb* lateral ventricle body, *Mes* mesencephalon, *mOrG* medial orbital gyrus, *och* optic chiasm, *olt* olfactory tract, *ot* optic tract, *pOrG* posterior orbital gyrus, *pul* pulvinar, *rCC* rostrum of the corpus callosum, *splCC* splenium of the corpus callosum, *tCC* truncus of the corpus callosum, *Th* thalamus, *Tpo* temporal pole, *II* optic nerve, * *red asterisk* it shows the pathway connecting the amygdala and the superior temporal gyrus

structures (thalamus and hypothalamus), basal structures etc., the permanent interrelations existing between cognition, emotions and their corporal expressions, as well as the modulation possibilities of the emotional conditions of the prefrontal cortex, are considered.

7.3 Lateral Relations

The lateral relations of the amygdala are visible on the coronal and axial sections of the brain, but it is, above all, the coronal sections which show the amygdaloid topography the best. The following is very clear on these sections that the amygdala is located in contact with the temporal stem, a narrow junction area between the frontal and temporal lobes. Its lateral surface is adjacent to this temporal stem, which contains not only the grey substance belonging to the claustrum but also all of the fibres connecting these lobes, whether they are ascending or descending, and therefore has a major significance.

Although this stem has already been studied, even recently, by many teams, by using dissection and (or) radiography (J. Peltier et al. 2010, EL Kier et al. 2004, CY Choi et al. 2010), the topic subject to discussion remains the limit of this stem. Where is the limit between the frontal and the temporal lobe to be set? The question concerns both the lateral end of this limit (inferior circular sulcus of the insula, limen, sylvian valley?) as well as the medial end (roof of the inferior horn of the lateral ventricle, tail of the caudate nucleus, endorhinal sulcus?). According to our experience, it appears that concerning the sections providing the best view of this amygdala, this limit corresponds to the shortest line drawn on the fronto-temporal neck from the recess of the endorhinal sulcus.

The *relations of the amygdala with the isthmic white substance* are multiple:

- With *the uncinate fasciculus*, these connections are particularly remarkable (Figs. 7.9, 7.10 and 7.11). It is to be recalled that this arch-shaped fasciculus, a large fibrillar system which is junctional between the frontal

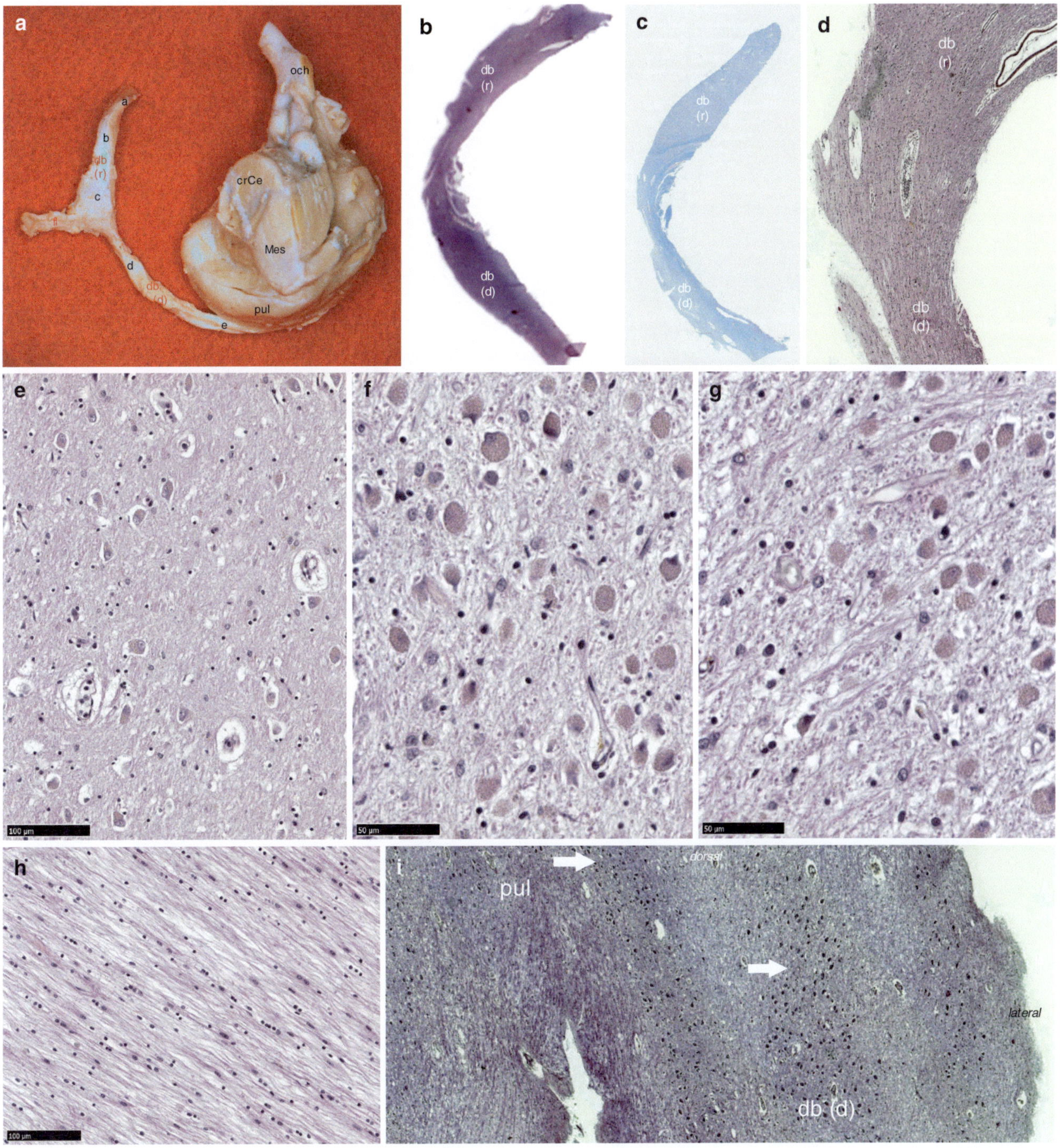

Fig. 7.5 Microscopic aspects of the diagonal band. (**a**) Dissection of the diagonal band (db) showing its constituent parts and its ending in the pulvinar; (**b**) macroscopic aspect of the db (classical staining); (**c**) same aspect as **b** with Luxol fast blue–PAS staining; (**d**) magnification of the curve between the rostral and dorsal parts of the diagonal band; (**e**–**h**) Luxol fast blue–PAS stainings (×200, ×400, ×400, ×200), at the levels a, b, c and d of the diagonal band, on picture **a**; (**i**) Luxol fast blue–PAS staining (×25) showing the ending part of the diagonal band in the pulvinar. *crCe* crus cerebri, *db* diagonal band, (*r*) rostral part, (*d*) dorsal part, *fl* fasciculus liminis, *Mes* mesencephalon, *och* optic chiasm, *pul* pulvinar, *white arrows* they show the clusters of neurons at the ending part of db and into the pulvinar. *Note the balloon-shaped aspect and the off-centred nuclei of the diagonal band's neurons*

and temporal lobes, comprises three parts: a frontal part, oblique at the top and front, which has developed, on the one hand, in the middle frontal lobe and the frontal pole and, on the other hand, in the orbital gyri and gyrus rectus; an intermediate part, the isthmus or infra-insular segment inside which the fibres combine, is curved and concave in the front region and covers the limen; and a temporal part, which is nearly horizontal

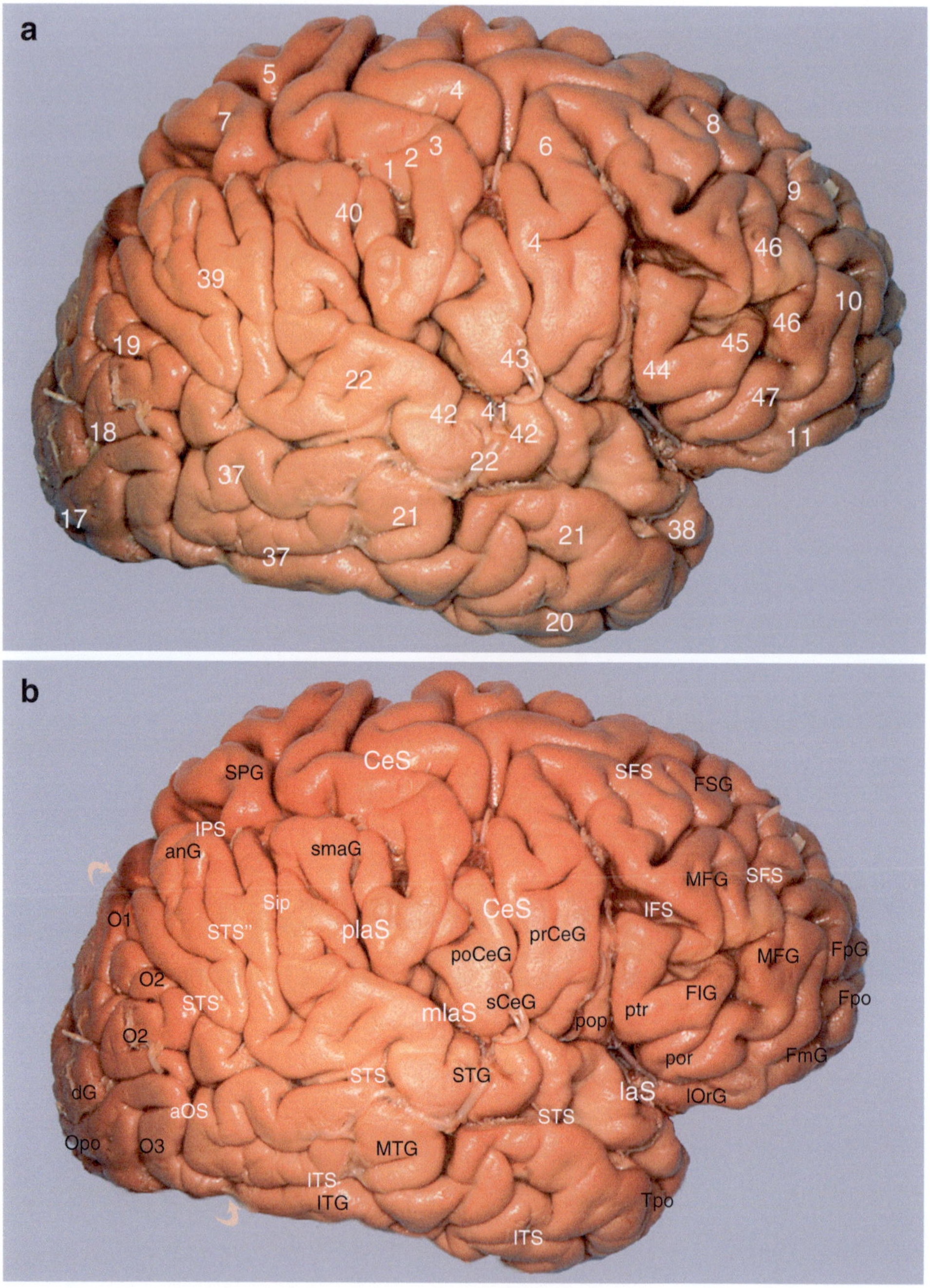

Fig. 7.6 Study of the amygdalo-cortical interrelations. Mapping of the cortical areas according to Brodmann (lateral views of the brain): (**a**) numbering of areas and (**b**) correspondences with gyri and sulci. *anG* angular gyrus, *aOS* anterior occipital sulcus, *CeS* central sulcus, *dG* descendens gyrus, *FIG* frontal inferior gyrus, *pop* pars opercularis, *ptr* pars transversalis, *por* pars orbitalis, *FmG* fronto-marginal gyrus, *FpG* fronto-polaris gyrus, *Fpo* frontal pole, *FSG* frontal superior gyrus, *IFS* inferior frontal sulcus, *IPS* intraparietal sulcus, *ITG* inferior temporal gyrus, *ITS* inferior temporal sulcus, *laS* lateral cerebral sulcus, *mlaS* medial lateral sulcus, *plaS* posterior lateral sulcus, *lOrG* lateral orbital gyrus, *MFG* medial frontal gyrus, *MTG* middle temporal gyrus, *O1* superior occipital gyrus, *O2* middle occipital gyrus, *O3* inferior occipital gyrus, *Opo* occipital pole, *poCeG* post-central gyrus, *prCeG* precentral gyrus, *sCeG* subcentral gyrus, *SFS* superior frontal sulcus, *Sip* sulcus intermedius primus of Jensen, *smaG* supra-marginalis gyrus, *SPG* superior parietal gyrus, *STG* superior temporal gyrus, *STS* superior temporal sulcus, *STS'* posterior part, horizontal segment, *STS"* posterior part, ascending segment, *Tpo* temporal pole, *curved superior arrow* parieto-occipital sulcus, *curved inferior arrow* temporo-occipital incisure

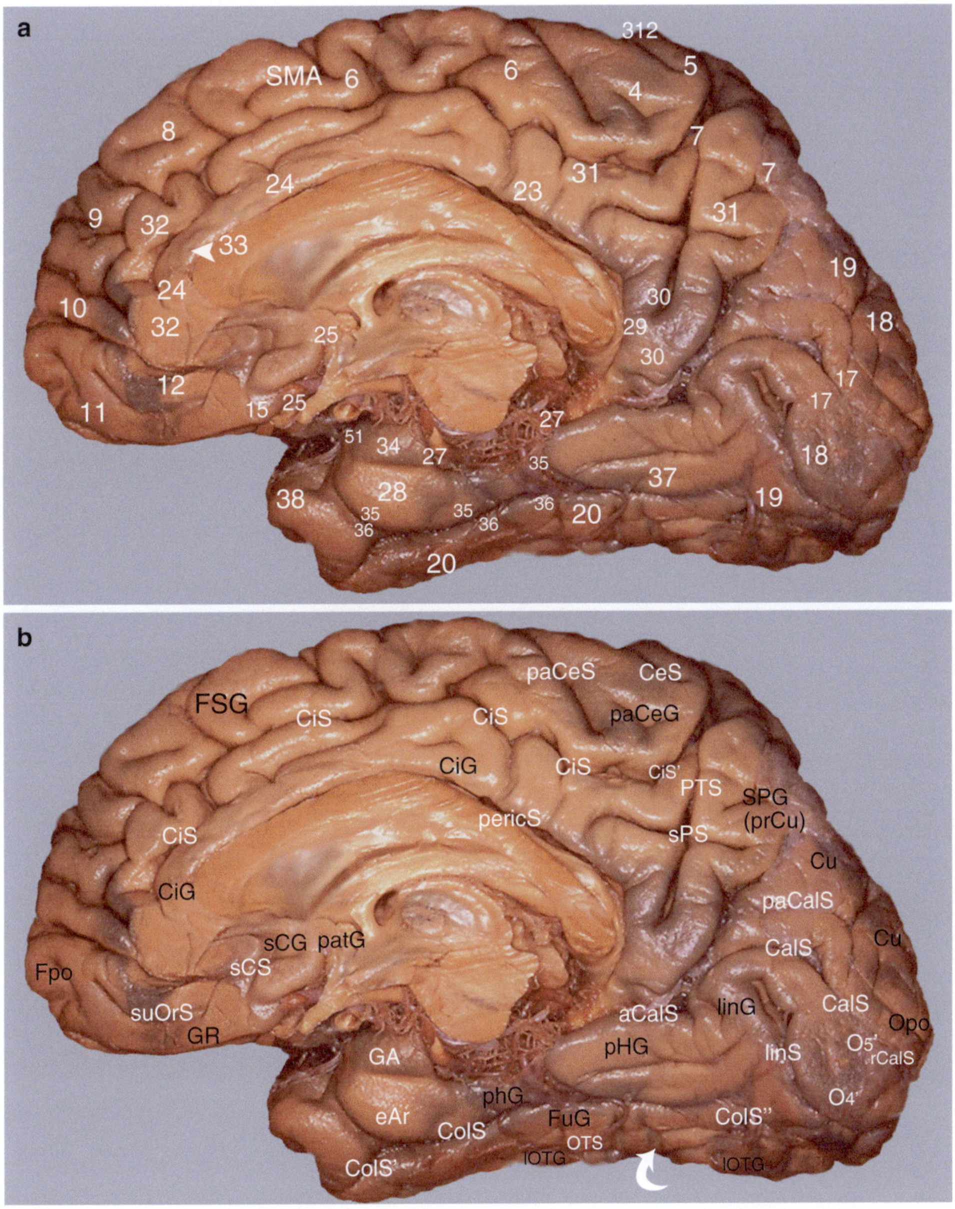

Fig. 7.7 Study of the amygdalo-cortical interrelations. Mapping of the cortical areas according to Brodmann (medial views of the right hemisphere): (**a**) numbering of the areas and (**b**) correspondences with gyri and sulci. *aCalS* antecalcarinus sulcus, *CalS* calcarine sulcus, *CeS* central sulcus, *CiG* cingulate gyrus, *CiS* cingulate sulcus, *CiS*' marginal part of CiS, *ColS* collateral sulcus, *ColS*' transverse anterior collateral sulcus, *ColS*" transverse posterior collateral sulcus, *eAr* entorhinal area, *Fpo* frontal pole, *FSG* frontal superior gyrus, *FuG* fusiform gyrus, *GA* gyrus ambiens, *GR* gyrus rectus, *linG* lingual gyrus, *linS* lingual sulcus, *lOTG* lateral occipitotemporal gyrus, *O4*' posterior part of FuG, *O5*' posterior part of linG, *Opo* occipital pole, *OTS* occipitotemporal sulcus, *paCalS* paracalcarine sulcus, *paCeS* paracentral sulcus, *paCeG* paracentral gyrus, *patG* paraterminal gyrus, *pericS* pericallosal sulcus, *pHG* parahippocampal gyrus, *PTS* parieto-temporal sulcus, *rCalS* retro-calcarine sulcus, *sCG* subcallosum gyrus, *sCS* subcallosum sulcus, *SMA* supplementary motor area, *SPG* superior parietal gyrus (*prCu* precuneus), *SPS* superior parietal sulcus, *suOrS* sus orbital sulcus, *white curved arrow* temporo-occipital incisures

and whose fibres have developed in the middle and Zinferior temporal gyri and the temporal pole. In 2004, EL Kier et al. showed, by means of MRI techniques studying previously dissected anatomical portions, that the frontal fibres of the uncinate fasciculus were routed in the external capsule and in the extreme capsule (a few fibres originate from the claustrum), prior to penetrating inside the middle part of the temporal isthmus. The uncinate fasciculus then nears the amygdala with which its intermediate part has developed close connections. The uncinate fibres border the lateral surface of the lateral nucleus of the amygdala and combine with them (Fig. 7.9). Others are distributed inside this lateral nucleus or penetrate it. This provides a comb

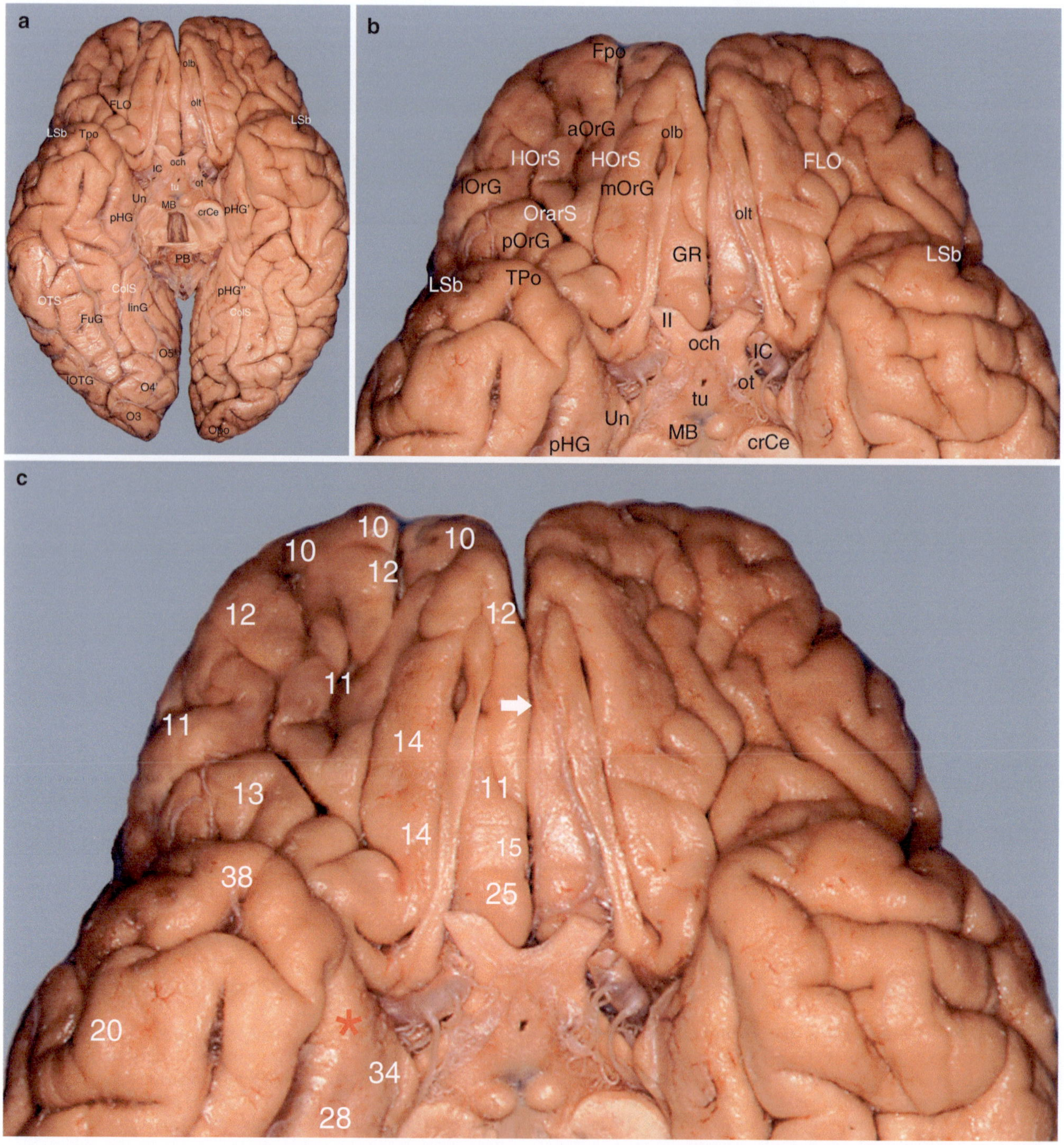

Fig. 7.8 Study of the amygdalo-cortical interrelations. Mapping of the cortical areas according to Brodmann (inferior views of the brain). (**a**) Basal aspect of the brain, (**b**) details of the frontal gyri and sulci, (**c**) numbering of the frontal areas. *aOrG* anterior orbital gyrus, *crCe* crus cerebri, *ColS* collateral sulcus, *FLO* frontal lobe, *Fpo* frontal pole, *FuG* fusiform gyrus, *GR* gyrus rectus, *HOrS* H orbital sulcus, *IC* internal carotid, *linG* lingual gyrus, *lOrG* lateral orbital gyrus, *lOTG* lateral occipitotemporal gyrus, *LSb* lateral sulcus (basal part), *MB* mamillary body, *mOrG* medial orbital gyrus, O3 inferior occipital gyrus, O4' posterior part of the fusiform gyrus, O5' posterior part of the lingual gyrus, *och* optic chiasm, *olb* olfactory bulb, *olt* olfactory tract, *Opo* occipital pole, *OrarS* orbital arcuatus sulcus, *ot* optic tract, *PB* pineal body, *pHG* parahippocampal gyrus, *pHG'* anterior part of pHG, *pHG''* posterior part of pHG, *pOrG* posterior orbital gyrus, *Tpo* temporal pole, *tu* tuber, *Un* uncus, *II* optic nerve, *white arrow* it shows the longitudinal fissure of the cerebrum (lfC), *red asterisk* it shows the amygdalar area and its proximity with the basal part of the frontal lobe

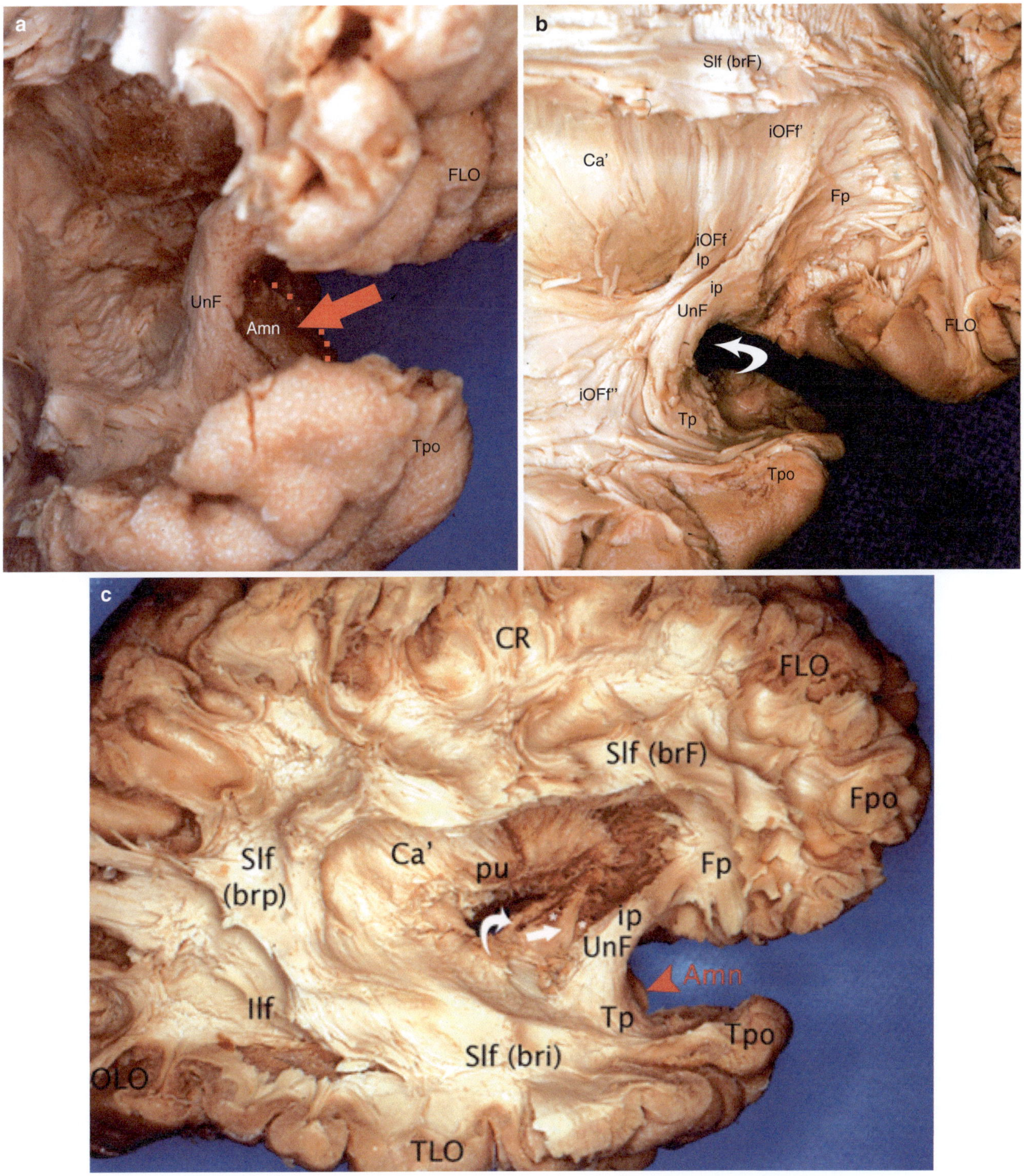

Fig. 7.9 Dissection of the lateral surface of the right hemisphere to illustrate the uncinate fasciculus and its relation to the amygdaloid nuclear complex. (**a**) 3/4 view in order to have a good visibility of the Amn joined with the medial aspect of the UnF; (**b**) dissection of the UnF. Note the IOFf adjoining to the isthmus of UnF. (**c**) Lateral dissection of the right hemisphere to localise Amn, UnF, ACo and ot. *Amn* amygdaloid nuclear complex (*red arrow* and *red arrowhead*), *Ca'* external capsule, *FLO* frontal lobe, *Fpo* frontal pole, *Ilf* inferior longitudinal fasciculus, *iOFf* inferior occipito-frontal fasciculus (*Ip* insular part, *iOFf'* frontal part, *iOFf'*' temporal part), *OLO* occipital lobe, *pu* putamen, *Slf* superior longitudinal fasciculus (*brF, brp, bri* brachium frontal, posterius and inferius), *Tpo* temporal pole, *UnF* uncinate fasciculus (*ip*, *Fp*, *Tp* isthmic frontal and temporal parts), *white curved arrow* on **b** it shows the limen, *right arrow* on **c** ACo (anterior commissure), *asterisks* amygdalofugal pathway, *white curved arrow* on **c** ot (optic tract)

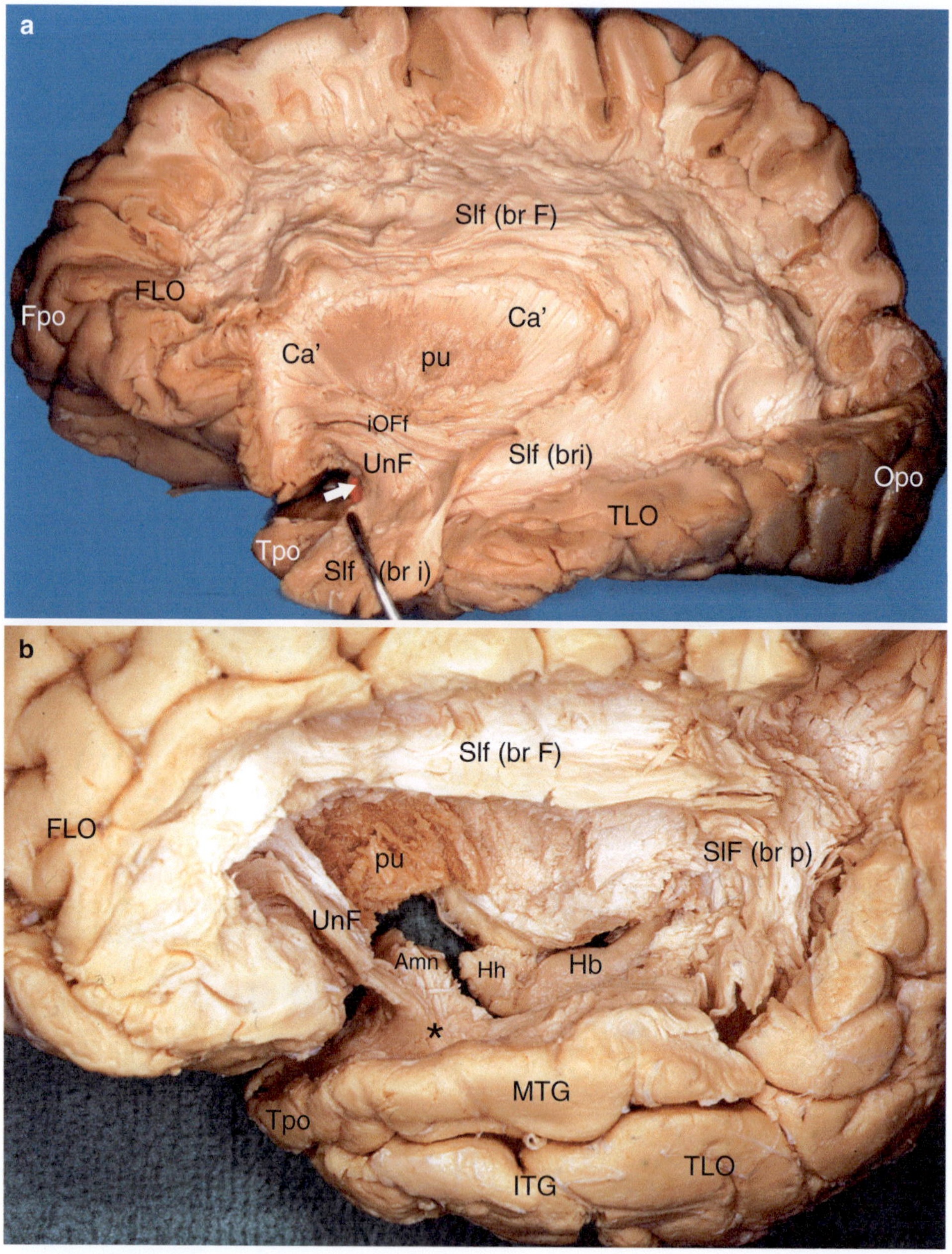

Fig. 7.10 Dissections of the lateral surfaces of two left hemispheres to precise the relations of the Amn with the uncinate fasciculus. (**a**) The *white arrow* shows the anterior edge of the amygdala (*red coloured*). (**b**) The superior temporal gyrus has been resected in order to show exactly the relations. *Amn* amygdaloid nuclear complex, *Ca'* external capsule, *FLO* frontal lobe, *Fpo* frontal pole, *Hb* hippocampus body, *Hh* hippocampus head, *iOFf* inferior occipito-frontal fasciculus, *ITG* inferior temporal gyrus, *MTG* middle temporal gyrus, *Opo* occipital pole, *pu* putamen, *Slf* superior longitudinal fasciculus, *br F* brachium frontalis, *br p* brachium posterius, *br i* brachium inferius, *TLO* temporal lobe, *Tpo* temporal pole, *UnF* uncinate fasciculus, *black asterisk* it shows the fibres of the uncinate fasciculus joining the white matter of the MTG

tooth-type aspect to its most lateral part. A recent microanatomical study performed on the fibres of the uncinate fasciculus (N Travers 2008) refers to similar connections to the baso-lateral nucleus of the amygdala. It must also be noted that certain fibres of this uncinate fasciculus seem to be originated from the peri-amygdaloid cortex, next to the cortical nucleus.

- With the *inferior occipito-frontal fasciculus*, the uncino-amygdaloid connections also concern the middle part of this fasciculus. The inferior occipito-frontal fasciculus is, in fact, a long fasciculus presenting a postero-superior concavity, adjacent to the uncinate fasciculus over the part of its route (D Peuskens et al. 2004). It connects the pre-frontal cortex of the frontal lobe to the posterior part of the temporal lobe (at the level of the lingual and fusiform gyri) and to the occipital lobe (up to the cuneus). This fasciculus, such as the uncinate bundle, comprises three parts: a frontal part, an occipito-temporal part and an insular intermediate part. The latter is adjacent to the uncinate isthmus, whose dorsal part it is positioned against, which is the reason why it was often believed that the inferior occipito-frontal fasciculus is the superior part of the uncinate fasciculus (J Sedat and H Duvernoy 1990). The persistent controversy concerning the capsule, inside which passes the inferior occipito-frontal fasciculus to each the temporal stem (extreme capsule for E Curran 1909, external and extreme capsules for JM Edeline and NM Weinberger 1992, for EL Kier et al. 2004, for J Peltier

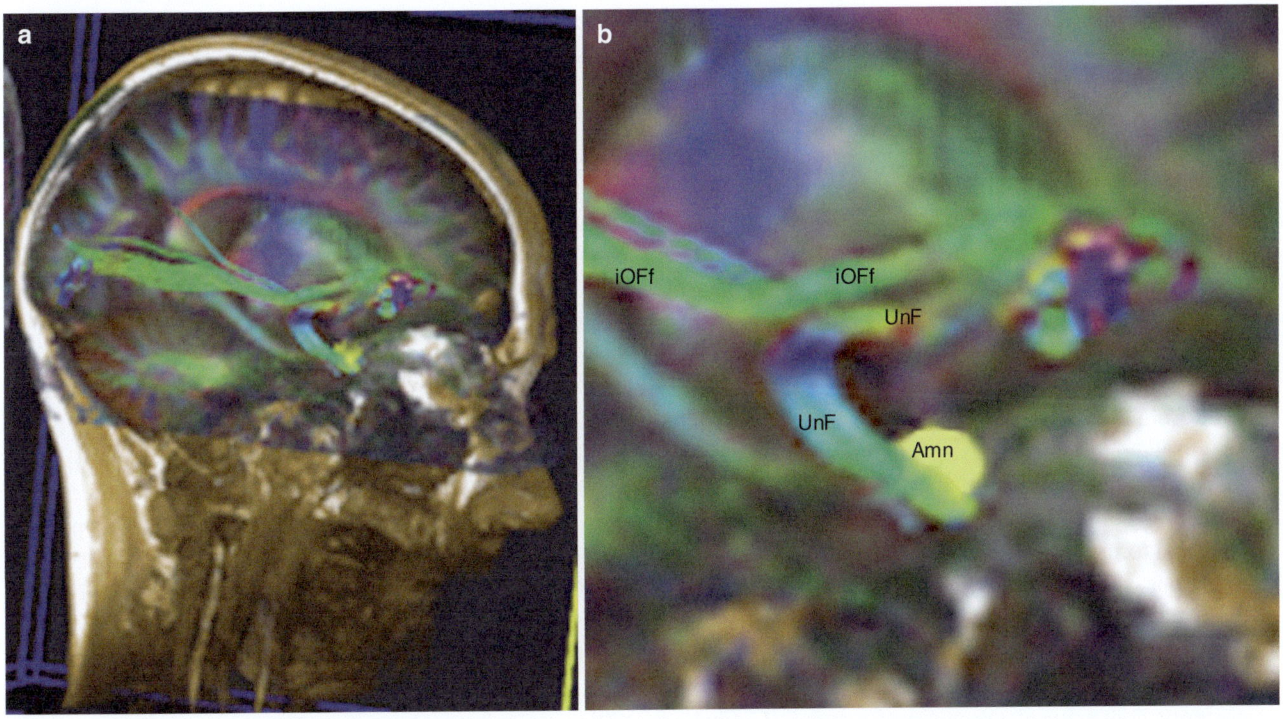

Fig. 7.11 Diffusion tensor imaging fibres tracking (DTI) of the uncinate fasciculus to show its relations with Amn. (**a**) Location on the head; (**b**) enlargement of the fasciculi seen on the previous picture **a**. *Amn* amygdaloid nuclear complex, *UnF* uncinate fasciculus, *iOFf* inferior occipito-frontal fasciculus

et al. 2010), is to be noted. It however remains certain that the inferior occipito-frontal fasciculus shares, with the uncinate fasciculus, contiguity connections to the amygdala and that some of its fibres are probably involved in the partition of the lateral part of the amygdala, which therefore provides the external part of its lateral nucleus with a streaked aspect.

- With the *Anterior commissure*, relations are close as part of the commissure fibres directly connect, as we have already observed (see Chap. 6), the two amygdalae, which they penetrate at the level of their respective basolateral nuclei, therefore forming the inter-amygdaloid component, which allows a direct connection between the left and right amygdalae. These amygdaloid fibres originate at the penetration of the lateral expansion inside the temporal stem. Other fibres, belonging to the inter-hippocampal component, originate at the same level and reach the hippocampus. The expansion of the commissure follows its route against the lateral surface of the amygdala, in a medial situation with respect to the uncinate and inferior occipito-frontal fasciculi. In this case, the expansion appears as "the diagonal of a rectangle whose two parallel edges would be the uncinate bundle and the optic tract" (J Peltier et al. 2010). When they reach the inferior edge of the amygdala, the expansions develop and disperse to form temporal components which essentially reach the middle and inferior temporal gyri.
- The amygdala does not seem to have any relation with the *corpus callosum*. However, as remarked by J. Peltier et al. (2010), "A short portion of the radiations of the CC seems to belong to the temporal stem", which therefore represents a potential relation with the amygdala as the amygdala is adjacent to the stem. Furthermore, the knowledge of these radiations is major: as shown by EC Crosby et al. (1962), the ventral fibres of the corpus callosum pass inside the extreme capsule, penetrate the temporal stem and bypass the cortex of the inferior sulcus of the insula to end in the superior cortex of the superior temporal gyrus. However, for the amygdala, this is only a proximity relation because the fibres of the anterior commissure will be interposed between the amygdala and callosal radiations, underneath which they pass to reach all of the parts of the temporal cortex.
- The amygdala has successively developed the following relations with the *inferior peduncle of the thalamus*:
  - Distant lateral relations, with its section which is developed on different parts of the temporal cortex (superior temporal, middle temporal and inferior temporal), parts corresponding to the inferior thalamic radiations which forward to the thalamus and the sensitive and sensory inputs from the temporal cortex.
  - Closer relations with the converging part of these radiations which form a pathway, the inferior thalamic peduncle, which will pass through the temporal stem

to reach the thalamus. Along this route, the peduncle receives from the amygdala the ventral amygdalofugal pathway and especially, in its anterior part, the amygdaloid fibres for the thalamus (see Chap 6).

During a lateral dissection of the hemisphere, it is possible to observe this ventral pathway between the optic tract and the anterior commissure, in a medial location with respect to the latter (see picture c, Fig. 7.9).

- The amygdala contracts contiguity relations with the *temporal loop* ("Adolph Meyer's loop, 1970", or "détour de Meyer" of French authors) (or geme temporalis of Putman), initial portion of the optic radiations which are freed from the lateral geniculate body. The anterior pathway of the optic tract fibres is briefly routed towards the roof of the inferior horn of the lateral ventricle. It then skirts round the end section of this horn to pass at a ventral level and form Meyer's loop or temporal knee of the optic radiations. It is this knee which is close to the posterior part of the lateral surface of the amygdala. It is separated therefrom by the most posterior fibres of the anterior commissure, which themselves are applied against the posterior convexity of the uncinate isthmus (U Ebeling and HJ Reulen 1988). Beyond the loop, the optic radiations move further away from the amygdala and pass under the inferior longitudinal fasciculus, in the sagittal stratum, with the most medial fibres of the inferior occipito-frontal fasciculus and a few fibres of the anterior commissure, in the direction of the occipital lobe.
  - Overall, it can be observed that the lateral fascicular relations of the amygdala explain the communication possibilities of this formation with the frontal and temporal areas. The medial fascicular relations will show similar possibilities with the entire limbic lobe.

The *lateral relations with the grey matter* concern only the peduncle of the lenticular nucleus and the ventral part of the claustrum. The latter, which belongs to the temporal stem from a topographic viewpoint, is contiguous with the lateral surface of the amygdala (see Fig. 4.9). Such as the lateral amygdaloid nucleus, the claustrum is dissociated by the fibres of the uncinate fasciculus and therefore appears fragmented. The claustrum portions which are the nearest to the amygdala are the most anterior and medial: pre-amygdalar claustrum and peri-amygdalar claustrum. The most lateral portions, which are relatively contiguous with the inferior limiting circular sulcus of insula, are the temporal claustrum, the limitans claustrum and the diffuse insular claustrum.

As for the peduncle of the lenticular nucleus, which sometimes is sus-isthmic but generally sub-isthmic, it can be developed and descend below the isthmic limit and appear as a triangular-shaped appendix of the putamen. It is located laterally and at the rear of the anterior commissure (see picture a, Fig. 4.11).

7.4 Medial Relations

7.4.1 Relations with the Cingulate Fasciculus

The most immediate connection of the amygdala concerns the inferior end of the cingulate fasciculus whose concave surface envelopes it such as "a chistera" in a game of "Basque pelota".

It is to be recalled that this white substance fasciculus known as the "faisceau du crochet" according to French anatomists, since Foville,[13] nearly surrounds the entire medial surface of the cingulated gyrus, from the anterior end, located inside the subcallosal gyrus, under the rostrum of the corpus callosum. The rostral part of the fasciculus bypasses the knee of the corpus callosum, while its posterior part bypasses the splenium of this formation, prior to passing inside the parahippocampal gyrus. It receives, along its entire circumference, white fibres which descend vertically from the cortical areas and then are curved and are incorporated therein. Some fibres ascend again towards one or several cortical areas that are more or less close (Fig. 7.12). Some fibres of the fasciculus are long and connect its antero-superior and inferior ends.

Furthermore, this cingulate fasciculus medially covers the entire hippocampus and more specifically the head of this formation and the uncus. With respect to the body of the hippocampus, this fasciculus occupies a position which is both medial and inferior. The fasciculus is relatively thick in the infero-medial part and appears to be divided into two parallel and contiguous layers, one is located against the hippocampus and the amygdala and the other is more medial (see Fig. 5.2).

Sometimes, some fibres of the end of the fasciculus insert themselves between the hippocampus and the amygdala. However, in most cases, the curve of the distal end of the fasciculus is only pre-amygdaloid. Such as observed on photos of dissections performed to highlight the medial part of the amygdala and the head of the hippocampus (see picture b, Fig. 5.1), following the removal of the cortices of the piriformis lobe and the parahippocampal gyrus, the resection of the end of this cingulate fasciculus is required in order to access the site. During this resection, we observed clear adherences between the amygdala and this fasciculus, which demonstrate the existence of communications between the amygdaloid nuclei and the periamygdaloid cortex, on the one hand, and the cingulate fasciculus and the cingulate cortex, on the other hand.

The connections of the cingulate fasciculus to the hippocampus, via the subiculum and the presubiculum, are also known. But the connections to the temporal pole are only

[13]LA Foville (1799–1878): Traité complet de la physiologie du système nerveux cérébro-spinal. Paris, Fortin Masson edit., 1844

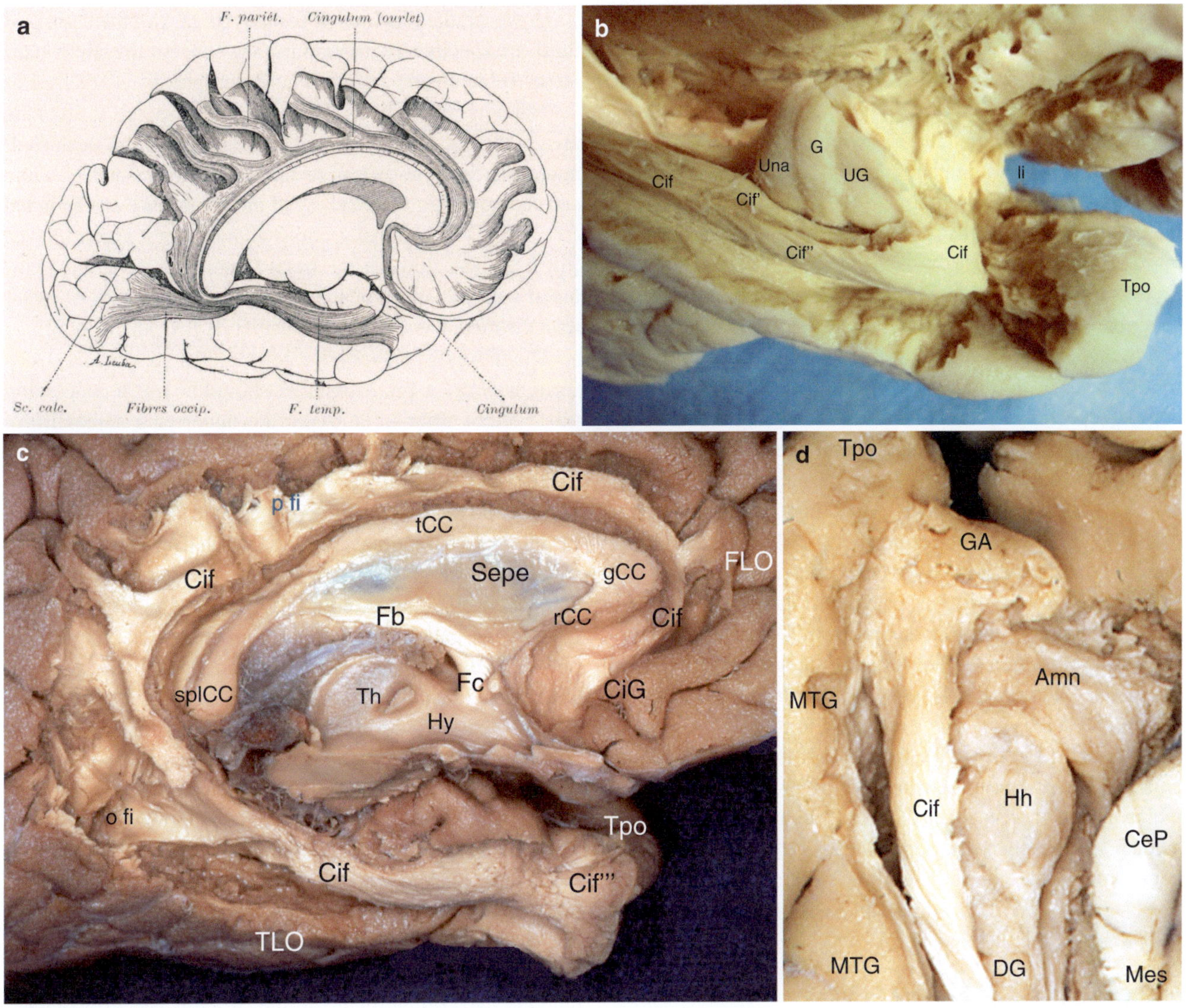

Fig. 7.12 Aspects of the cingulate fasciculus after dissection. (**a**) Original drawing of the ''faisceau de l'ourlet'' by LA. Foville. (**b**) and (**c**) dissections of the left hemisphere (**b**, to show the relations with the hippocampal head) (**c**, to show the entire fasciculus and its rostral expansion to the temporal pole) (**d**, inferior dissection of the right fasciculus after resection of the parahippocampal gyrus. Notice the expansion to the gyrus ambiens, in front of the amygdala). *Amn* amygdaloid nuclear complex, *CeP* cerebral peduncle, *Cif* cingulate fasciculus, *Cif'* juxta-hippocampic part of Cif, *Cif''* medial part of Cif, *Cif'''* rostral expansion of the Cif, *CiG* cingulate gyrus. *DG* dentate gyrus, *Fb* fornix body, *Fc* column of fornix, *FLO* frontal lobe, *G* band of Giacomini, *GA* gyrus ambiens, *gCC* genu of the corpus callosum, *Hh* hippocampus head, *Hy* hypothalamus, *li* limen insulae, *Mes* mesencephalon, *MTG* middle temporal gyrus, *ofi* occipital fibres, *pfi* parietal fibres, *rCC* rostrum of corpus callosum, *Sepe* septum pellucidum, *splCC* splenium of corpus callosum, *tCC* truncus of corpus callosum, *Th* thalamus, *TLO* temporal lobe, *Tpo* temporal pole, *UG* uncinate gyrus, *Una* uncus apex

rarely noted in the literature (J Klinger and P Gloor 1960).[14] However, we have constantly observed a more or less cylindrical bundle with a diameter of 3–4 mm, which leaves the amygdala at the rostral end of the cingulate fasciculus to reach the temporal pole. This bundle is obviously important and should be the subject of future neuroscientific studies as it directly links the rostral part of the amygdala to this temporal pole (see picture c, Fig. 7.12). Klinger had already called it the "fasciculus amygdalo-temporalis". The relations of the cingulate fasciculus, referenced by EC Crosby et al. (1962) in "Correlative anatomy of the nervous system", are also to be recalled. They not only concern the entorhinal cortex but also the frontal lobe and the paraolfactory area of the anterior perforated substance and therefore show the extraordinary connection possibilities between the amygdala and the hippocampus and other regions of the brain via this long and thick fronto-temporal cingulate fasciculus.

[14]J Klinger and P Gloor (1960) have also observed, when performing their dissections, fibres connecting the anterior surface of the amygdala to the gyri longi of the insula, which they had referred to as "amygdalo-insular fasciculus".

7.4.2 Relations with the Cortex

On the medial view of a half-brain, the amygdaloid body appears covered by a cortical layer referred to as the *peri-amygdalar cortex* (archicortex), within which are observed more or less protruding parts including, from top to bottom, the ambient gyrus, the semilunar gyrus, the prepiriform cortex and the cortex of the piriform lobe. This peri-amygdalar cortex is considered as part of the amygdala.

From a topographic viewpoint, if we observe the brain sections in 3D, it appears that the amygdala is incorporated inside the surface cortex. However, with the dissection, it is relatively easy to separate the amygdala and this surface cortex (especially as the white fibres of the cingulate fasciculus form a thin inter-amygdalo-cortical partition). Contiguity connections exist, on the one hand, to the cortex which covers the uncus, the uncinate gyrus; underneath the gyrus ambiens, to the entorhinal area; underneath the uncus, to the parahippocampal gyrus; to the other cortices of the temporal lobe (especially the anterior part of the lateral occipitotemporal gyrus); and, to the temporal pole. Contiguity connections also exist at a cranial level to the anterior perforated substance. However, this substance is located above the endorhinal sulcus and does not belong to the medial connections described in this paragraph. All of these cortical areas adjacent to the periamygdaloid cortex as well as the sulci which can be observed (rhinal sulcus, collateral sulcus, uncal sulcus) are directly connected to the amygdala and obviously contribute to the correct operation of the amygdalar complex, such as demonstrated by Neurosciences research work.

7.4.3 Peripheral Relations

Each amygdalar region has relations with the transverse cerebral fissure (large cerebral fissure of Bichat according to French authors) where develops the first centimetre of the middle cerebral artery, from which originate, within the subarachnoidal interpeduncular cisterna, the lateral striate arteries (which will vascularise the lentiform nucleus) (see Fig. 4.8). The neuroanatomic relations at this level concern, from front to back, the anterior perforated substance, underneath which is the distributed internal carotid artery, then the homolateral optic tract successively bordering the lateral surface of the hypothalamus and the floor of the third ventricle.

7.5 Superior Relations

These relations are those contracted by the amygdala to the basal forebrain, i.e. the basal part of the brain and its inferior edge. We have already studied the anterior relations of the amygdala with the olfactory area, especially with the apSb (anterior perforated substance). We will not refer to this topic again.

However, the relations of the amygdala with the basal part itself are specifically important: they concern the sublenticular or sublentiform region whose complexity is referred to several times in this book. This region corresponds to all of the area of the basal brain, located below the lentiform nucleus. It appears as a more or less rectangular narrow band of white substance, within which areas of grey substance are included (see Chap. 11). Coronal sections allow the best approach (see Fig. 6.9). The study of the basal forebrain is generally combined with the study of the septal region, which is no other than the medial end of the basal area (H Brockhaus 1942).

The more distant superior connections concern the lentiform nucleus. A brief reminder should be made concerning this nucleus prior to studying the underlying sublentiform region, which is less known.

The *lentiform nucleus* (or lenticular nucleus) is a grey nucleus located in the centre of the white substance, between the internal and external capsules, which have a pyramid-type shape and lateral base, which is triangular on sections. Its two components include the putamen, lateral, of telencephalic origin and the pallidum or globus pallidus, medial, of diencephalic origin. The globus pallidus comprises two portions, external and medial portions, and the latter is subdivided into a lateral segment and medial segment. The putamen and the pallidum are separated by a lateral medullary lamina. The two portions of the pallidum are separated by a medial medullary lamina. The two segments of the medial portion are separated by the accessory medullary lamina. The putamen forms, with the caudate nucleus which delimits it, the dorsal striatum. We have given the name of corpus striatum[15] to the lentiform nucleus + caudate nucleus system.

The immunohistochemical markers were used to differentiate the two compartments in the striatum, the striosomes,[16] dense cellular islands, and the matrix, area with a reduced density, in which they are included. The two compartments also vary at the level of their connections. The afferent fibres to the striosomes come from the orbitofrontal cortex and anterior cingulate cortex and are therefore connected to the limbic system. The efferences meet the pars compacta of the substantia nigra and follow the strio-nigral pathway. The afferences of the matrix come from the parietal, frontal and temporal associative cortices and are connected to the sensorimotor system. The efferences of the matrix reach the globus pallidus.

The globus pallidus is involved in the pathway of the fronto-striatal loops, whose tonic inhibiting role is starting to

[15] The neuronal masses, which extend from the base of the telencephalon to the tegmentum of the telencephalon, are referred to as the "ganglions of the base": they are the lentiform nucleus, the zona incerta, the subthalamic nucleus and the substantia nigra or locus niger.

[16] Anomalies have been observed on the striosomes in the case of certain neurodegenerative diseases, including Huntington's chorea.

be known.[17] During a recent study (2014) concerning the amygdalo-hypothalamo-striatal pathways, which are involved in approach/avoidance behaviours, D Laricchiuta and L Petrosini (2014) observed, on the fMRI, an activation at the level of the cerebellous cortex, which shows the participation of this organ and therefore with an additional striato-cerebellar connection for these pathways.

The sublenticular (or sublentiform) grey substance represented by the sub-commissural part of the lenticular nucleus, especially ventral pallidum (see the following paragraph), and by the small cellular islands forming Reichert's innominate substance, as we now know, is part of the continuum connecting the amygdala to the BST: the extended amygdala (see relevant chapter). The largest cellular island forms Meynert's basal nucleus whose neurons are rich in neuromelanin (Fig. 11.4). Most of the islands and especially Meynert's nucleus are cholinergic and act as key modulating relays between the amygdala and the cortex. The ventral striatum is also connected to the sublenticular grey substance. This striatum is essentially represented by the accumbens nucleus but also by the olfactory tubercle (L Heimer 1978).

The *accumbens nucleus* is a small oval 7–8-mm-long nucleus, with a large axis, which slightly ascends in the rostral direction and which is hidden underneath the cortex of the subcallosal gyrus and flattened (accumbere in Latin, which explains its name) against the lower and widened part of the septum. The large axis of its surface with a coronal section is oblique at the top and directed inwards (see Figs. 4.8 and 4.9). It is the main nucleus of the pleasure and reward pathway, discovered by chance by J Olds and P Milner, 1954, during their researches on the hypothalamus.[18] It comprises two parts, a central part, the "core", and a peripheral part, the "shell", which vary from an immunohistochemical viewpoint as well as from the viewpoint of their afferent and efferent connections (L Zaborszky et al. 1985). In our dissections (See Figs. 8.1 and 8.3), we were able to highlight this nucleus and observe the development, from its rostral pole, of white fibres which extend in the frontal lobe, where they pass against the medial surface of the anterior thalamic radiations. These fibres form the mesocortical portion of the medial forebrain bundle (mfb).

The olfactory tubercle is located behind the anterior olfactory nucleus. It covers the accumbens nucleus at a superficial level and represents, such as this nucleus, a wide projection field for the dopaminergic fibres coming from the ventral tegmental area. It also receives direct projections from the olfactory bulb (R Nieuwenhuys et al. 2008). It is in this olfactory tubercle that the greatest number of islands of Calleja can be found.[19]

The *sublenticular white substance* is extremely complex: it comprises elements as various as the white commissure, the ansa lenticularis, the inferior peduncle of the thalamus (seen previously), the postero-inferior peduncle of the thalamus (Arnold's temporo-thalamic bundle), the different components of the ventral amygdalofugal pathway, the fibres of the temporopontine fasciculus of Türck–Meynert and the Klinger's amygdalo-tegmental fasciculus.

- The *anterior commissure* obliquely passes through, in its Gratiolet's canal, the entire extent of the sublenticular region which it limits in its top section. In actual fact, it penetrates the basal part of the lenticular nucleus and thus isolates parts of this nucleus, which remain within the innominate substance and are known as the ventral pallidum and ventral striatum. Such as observed previously, the fibres of this commissure connect the two amygdalae inside which part of them ends, while the others develop inside the temporal gyri or form the commissural connection of the two olfactory systems (see Chap. 6).
- The *ansa lenticularis* is a bundle which develops from the three components of the lenticular nucleus and therefore from the putamen and lateral and medial parts of the pallidum. Its fibres emerge from the inferior surface of the lenticular nucleus, pass through the entire sublenticular area below the anterior commissure and combine to form a bundle which suddenly forms a bend (Gratiolet's ansa peduncularis) and reach the thalamus after having bypassed the ventral edge of the posterior crus of the internal capsule. When they reach area H, its fibres are mixed with those of the lenticular fasciculus[20] to end in the ventral anterior (ventralis anterior), ventral lateral

[17]Thus, when reading, the blinking phases alternate with non-blinking phases; the latter are only possible through inhibition of the former.

[18]The rats with an electrode implanted in this area of their brain and connected with a stimulating machine, which they can activate themselves by pressing on a small lever, feel such a pleasure as they autostimulate nonstop, forgetting even to eat.

[19]Such as recalled by J. Klinger et al., Meynert-type islands of the substantia innominata have been interpreted by some (especially Sanides 1957) as islands of Calleja, granule cell islands, referred to as "terminal insulae". In fact, the two entities are different due to the fact that Calleja-type islands are dopaminergic, whereas Meynert-type islands are cholinergic. Whatever the case, the two types of islands cohabit inside the substantia innominata. Calleja-type islands are not only found in the substantia innominata but also in the apSb and the olfactory tubercle (G Meyer et al. 1989), along the stria terminalis or even at the level of the accumbens nucleus (involved with the latter, in the reward system) or in the central nucleus of the amygdala. The islands of Calleja seem to intervene in the circulatory regulation of the ventral pallidum and the regulation of the systemic arterial pressure (vasodilating effect by the secretion of nitric oxyde) (J Rodrigo et al. 2002).

[20]The lenticular bundle comprises pallido-thalamic fibres which emerge from the superior surface of the pallidum. These fibres pass through field H2 of Forel, between the subthalamic nucleus and the zona incerta, before joining the fibres of the ansa lenticularis and reaching the thalamus. The thalamic bundle is formed by the fibres of the ansa lenticularis, by those of the lenticular fasciculus and by the dento- and rubro-thalamic fibres.

(ventralis intermedius) and centro-median (centromedianus) thalamic nuclei.

- The *postero-inferior peduncle of the thalamus* that Arnold had called the "temporo-thalamic tract" is only to be mentioned as it is involved in the sublenticular white substance. Due to its posterior location, it has no connection to the amygdala.
- The *inferior peduncle of the thalamus*,[21] in an anterior temporal location, is, however, an essential relation of the amygdala due to the fact that the ventral amygdalofugal pathway will follow it specifically to reach the diencephalic structures. Such as observed before, the fibres of this peduncle developed from the temporal gyri where they mix with the fibres of the anterior commissure. This inferior peduncle also receives, from inside this commissural component, fibres which have developed from the dorsolateral portion of the amygdala (J Klinger and P Gloor 1960).
- Two *amygdalar fasciculi*, the amygdalo-hypothalamic fasciculus and the amygdalo-septal fasciculus, will join the inferior peduncle of the thalamus to form the ventral pathway (or "ventral amygdaloid projection" of Klinger), the second major pathway connection of the amygdala (see relevant chapter). Both fibres are routed ventrally with respect to the lentiform nucleus and to the internal capsule.
  - The *amygdalo-hypothalamic* fasciculus emerges from the dorsal part of the amygdala and immediately follows a horizontal and medial direction, dorsal with respect to the Sbpa, thus allowing it to medially reach the anterior part of the lateral hypothalamus (Fig. 7.13) and then the preoptic region.
  - The *amygdalo-septal fasciculus* (Fig. 7.14a) initially follows the same sublentiform and sub-commissural route but then bends inwards to join Broca's diagonal band and reach the septal region.
- The fibres of the *temporopontine fasciculus*, temporal part of the parieto-temporopontine tract, will also pass through the sublenticular region and thus contract relations with the amygdala: These fibres develop from the parietal, occipital and temporal gyri and pass through the sublenticular segment of the internal capsule (Fig. 7.14b) (Klinger) to reach the cerebral peduncle. These fibres occupy the fifth lateral of the latter and end in the nuclei of the pons.
- The fibres of the *amygdalo-tegmental fasciculus* are particularly important due to the fact that they represent one of the major functional pathways of the amygdala, therefore allowing the latter to establish a direct physiological communication with the tegmentum and, via the latter, with the entire brainstem. The existence of this fasciculus, which could already be observed by dissection, has been certified, such as demonstrated by J Klinger and P Gloor, by means of degeneration techniques, experimental histological studies and electrophysiological observations. The original fibres develop from the temporal pole, the parahippocampal gyrus and the amygdala. They rapidly form a bundle which passes above the optic tract, then travels towards the pulvinar to position itself dorsally with respect to the temporopontine bundle (Fig. 7.14b) and then separates therefrom inside the midbrain tegmentum, at the level of the lemniscal layer.

[21] The ansa lenticularis and the inferior peduncle of the thalamus form the ansa peduncularis of Gratiolet.

The septal region (or ventral part of the septum) is the narrow interhemispheric cortical band, positioned underneath the septum pellucidum (dorsal part of the septum) when the two lamina, which form this wall, separate outwardly at the bottom.

It is located to the front of the lamina terminalis and therefore in front of the anterior commissure. It forms the medial limit of the basal forebrain. Many authors include, in its composition, the nucleus accumbens septi, especially due to the fact that this nucleus such as the septum provides the same hedonic effects when stimulated! This septal region is specifically important for our purpose for several reasons:

It contains several grey nuclei: not only the nucleus accumbens but also the nucleus of the diagonal band and the septal nuclei, the medial septal nucleus and the medial septal complex. All of these nuclei contain cholinergic neurons and belong to the cholinergic pathways, which are known to play a major role in the cortical activation process[22] (see Chap. 11).

It corresponds to the medial limit of the olfactory tubercle (which belongs, such as observed, to the ventral striatum).

It is closely connected to the amygdala by the precommissural component of the stria terminalis, by the diagonal band and by a fasciculus which only few books refer to (including J Klinger and P Gloor who provided their first description: the amygdalo-septal fasciculus (Fig. 7.14a), "bundle which is wider than the terminal stria and as wide as the anterior commissure"), which firstly passes horizontally under the anterior commissure and then straightens to reach the septal region.

It corresponds to the border between the anterior and posterior parts of the mfb.

"It contains a wide variety of neurotransmitters and neuroactive substances: 35 different neuropeptides in the septal neurons and septal fibres" (R Nieuwenhuys et al. 2008).

[22] There also exists, at the level of the roof of the mesencephalon, cholinergic neurones which activate the thalamus and then the brainstem. "These thalamic projections modulate the awakening and the awakening/sleeping cycle and are important for the genesis of the paradoxical sleep" (DL Felten and RF Jozefowicz 2003).

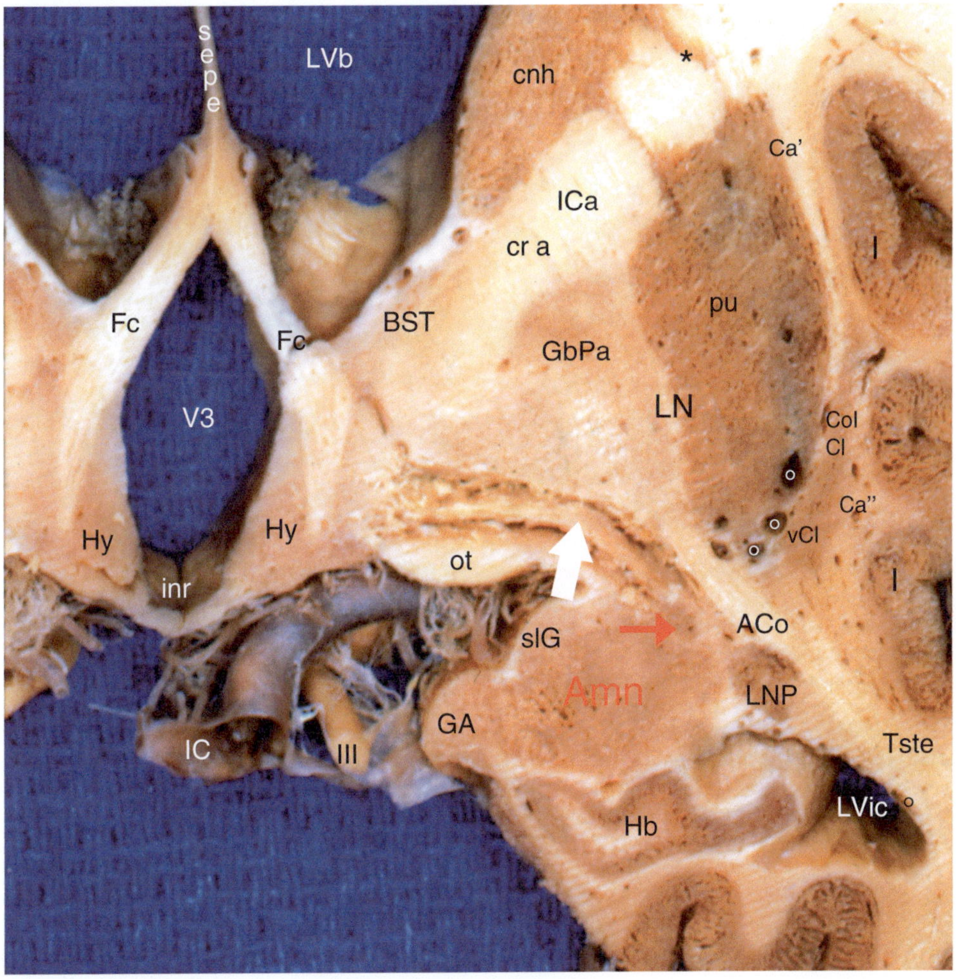

Fig. 7.13 The amygdalo-hypothalamic fasciculus (dissection of the fasciculus on a half coronal section of the brain). *ACo* anterior commissure, *Amn* amygdaloid nuclear complex, *BST* bed nucleus of stria terminalis, *Ca'* external capsule, *Ca"* extreme capsule, *cnh* caudate nucleus head, *ColCl* compact insular claustrum, *cr a* crus anterior of ICa, *Fc* column of fornix, *GA* gyrus ambiens, *GbPa* globus pallidus, *Hb* hippocampus body, *Hy* hypothalamus, *I* insula, *IC* internal carotid, *ICa* internal capsule, *inr* infundibular recess, *LN* lentiform nucleus, *LNP* lentiform nucleus peduncle, *LVb* lateral ventricle, body, *LVic* lateral ventricle, inferior cornu (temporal horn), *ot* optic tract, *pu* putamen, *sepe* septum pellucidum, *slG* semilunar gyrus, *Tste* temporal stem, *vCl* ventral claustrum, *V3* third ventricle, *III* oculomotor nerve, *black circle* caudate nucleus tail, *black asterisk* putamino-caudate bridge, *red arrow* central nucleus of the Amn, *white arrow* it shows the amygdalo-hypothalamic fasciculus; *white little circles* vessels of the striatum

7.6 Remote Supero-Medial Relations

Although they are less close than the sublenticular connections, these connections are as important as they can be used to see the proximity existing between the amygdala and the two diencephalic structures bordering the third ventricle, the hypothalamus and the thalamus (Fig. 7.15).

7.6.1 Relations with the Hypothalamus

In actual fact, the hypothalamus is not very far from the amygdala, which can be easily verified by observing the rostral coronal sections of the brain or even the standard median sagittal section. It is therefore easy to understand that the hypothalamus is one of the cogs of emotional expression for the amygdala, that relatively short axons connect it to the amygdala and that its long axons, which descend to the brainstem, are routed with similar axons from the amygdala. Our observation also shows that it is the nucleus of the lateral hypothalamus which is closest to the central nucleus of the amygdala. This explains why the ends of the amygdalo-hypothalamic bundle (Fig. 7.13) connect both nuclei.

The projections of the hypothalamus to the amygdala also concern other hypothalamic nuclei and especially the ventromedial, paraventricular and infundibular nuclei.

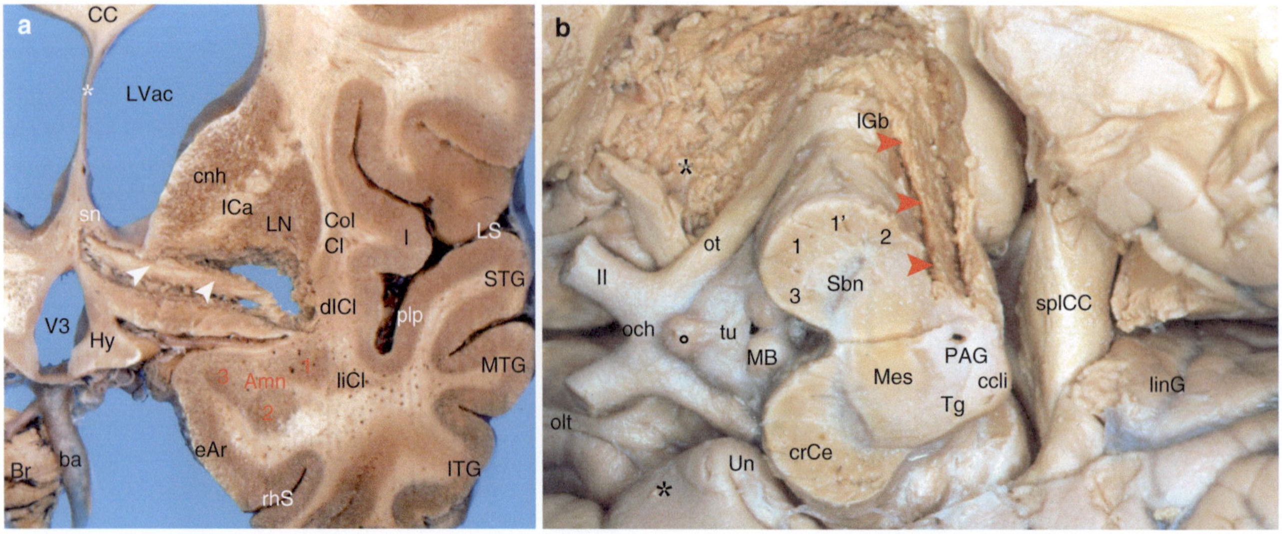

Fig. 7.14 Examples of two dissections of cerebral fasciculi, according to the Klinger's process. (**a**) Amygdalo-septal fasciculus. (**b**) Amygdalo-temporo-tegmental fasciculus. *Amn* amygdaloid nuclear complex, *1* (*red*) lateral nucleus, *2* (*red*) baso-lateral nucleus, *3* (*red*) baso-medial nucleus, *ba* basilary artery, *Br* brainstem, *CC* corpus callosum, *ccli* caudal colliculus, *cnh* caudate nucleus head, *ColCl* compact insular claustrum, *crCe* crus cerebri, *1* cortico-spinal tract, *1'* cortico-nuclear tract, *2* parieto-temporo-occipito-pontine tract, *3* fronto-pontine tract. *dlCl* diffuse insular claustrum, *eAr* entorhinal area, *Hy* hypothalamus, *I* insula, *ICa* internal capsule, *ITG* inferior temporal gyrus, *lGb* lateral geniculate body, *liCl* limitans claustrum, *linG* lingual gyrus, *LN* lentiform nucleus, *LS* lateral sulcus, *LVac* lateral ventricle, anterior cornu, *MB* mamillary body, *Mes* mesencephalon, *MTG* middle temporal gyrus, *och* optic chiasm, *olt* olfactory tract, *ot* optic tract, *PAG* periaqueductal grey matter, *plp* planum polare, *rhS* rhinal sulcus, *Sbn* substantia nigra, *sn* septal nuclei, *splCC* splenium of the corpus callosum, *Tg* tegmentum mesencephali, *tu* tuber, *Un* uncus, *V3* third ventricle, *white asterisk* septum pellucidum, *black asterisk* it marks the location of the amygdaloid nuclear complex, *black circle* it shows the hypophysis stalk, *white arrowheads* amygdalo-septal fasciculus, *red arrowheads* temporo-amygdalo-tegmental fasciculus

7.6.2 Relations with the Thalamus

There are less direct proximity relations than for the hypothalamus. However, this does not impact in any way the major axonal links between the amygdala and the thalamus that we will study in the chapter about amygdalopetal and amygdalofugal inputs. Furthermore, fascicular pathways directly connect the amygdala and the thalamus such as we have studied in the chapter concerning connections: dorsal pathway of the stria terminalis and ventral amygdalofugal pathway. Lastly, we cannot refer to the thalamus without mentioning the BST (see Chap. 12), another major component for physiological interrelations between the amygdala and the thalamus.

7.7 Posterior Relations

These relations are between the amygdala and the head of the hippocampus and the end of the inferior horn of the lateral ventricle. For contiguity topographical reasons with the wall of the lateral ventricle, we will also study the following in this chapter: relations with the tail of the caudate nucleus.

Relations of the amygdala with the head of the hippocampus are very close and show the complementarity of the functions of both anatomical structures. The head of the hippocampus is located behind and underneath the amygdala such that it forms a relatively postero-inferior relation thereof.

The relations between the two structures, observed on the medial surface of a hemisphere, are both, superficial and extra-ventricular, and deep and intra-ventricular.

- The *extra-ventricular relations* are relations between the cortex of the ambiens and semilunar gyri (peri-amygdalar cortex) and the anterior part of the uncal cortex (Fig. 7.16). We can observe, between the ambiens and semilunar gyri, the semi-annular sulcus which is sometimes referred to as the amygdaloid sulcus. This sulcus is determined by the print of the free edge of the tentorium cerebelli. The semilunar gyrus, below the endorhinal sulcus, which separates it from the anterior perforated substance, covers the cortical nucleus of the amygdala. The cortex of the uncus of the hippocampus, in front of the band of Giacomini,[23] is the cortex of the uncinate gyrus. As for the intralimbic

[23] The band of Giacomini, which is clearly visible on the medial surface of the uncus, is no other than the extension at the level of the hippocampal head, of the margo denticulatus, the visible part of the dentate gyrus (HM Duvernoy 1988). The visible part of the band on the medial surface of the uncus is referred to as "pars aperta". Its hidden part, on the inferior surface of the uncus, is the "pars occulta".

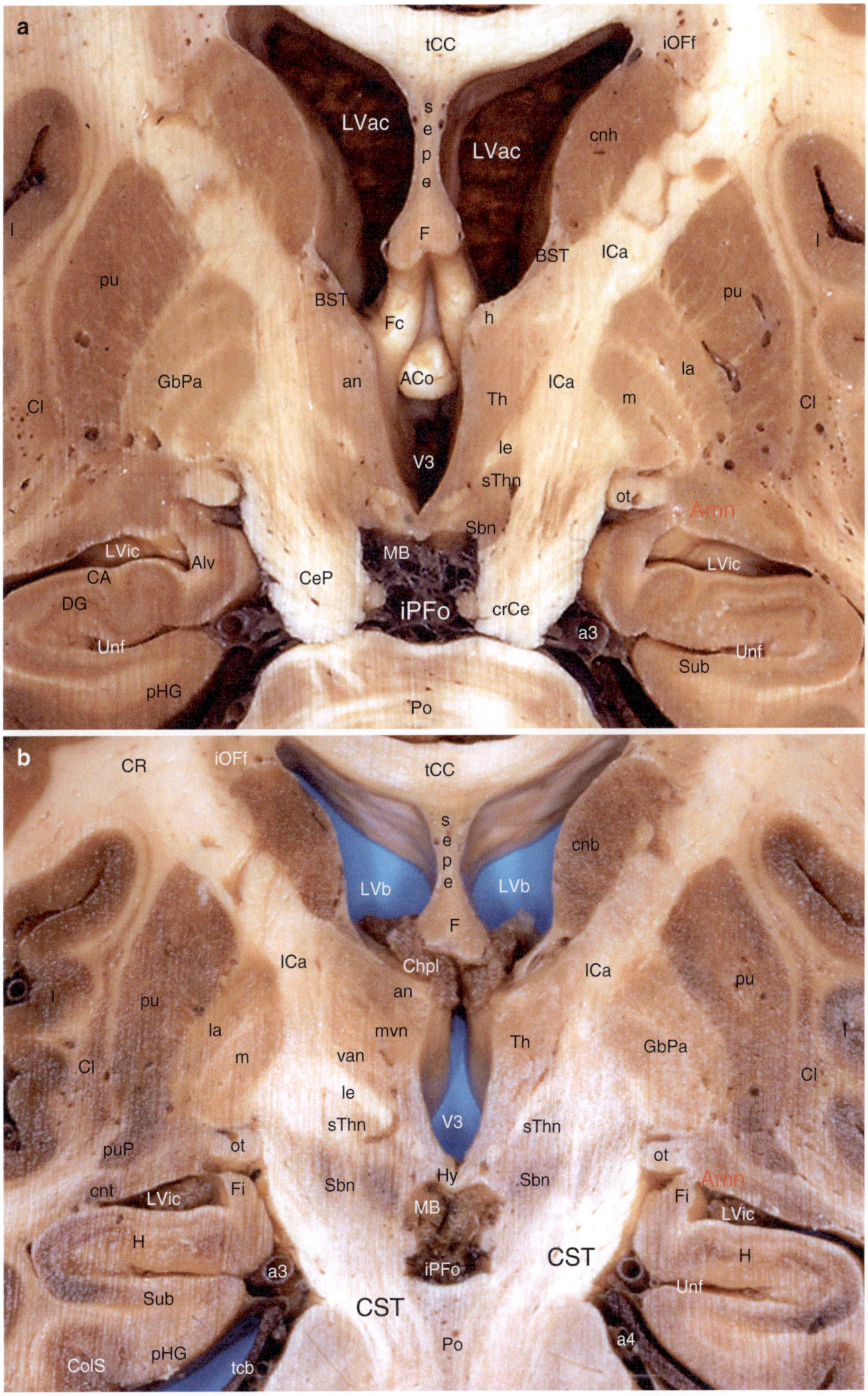

Fig. 7.15 Relationships of the Amn with the diencephalon on two coronal brain's sections. (**a**) Section behind the ACo; (**b**) section through the interventricular foramina. *ACo* anterior commissure, *Alv* alveus, *Amn* amygdaloid nuclear complex, *BST* bed of the stria terminalis, *CA* cornu ammonis, *CeP* cerebral peduncle, *Chpl* choroid plexus, *Cl* claustrum, *cnb* caudate nucleus body, *cnh* caudate nucleus head, *cnt* caudate nucleus tail, *ColS* collateral sulcus: *CR* corona radiata, *crCe* crus cerebri, *CST* cortico-spinal tract, *DG* dentate gyrus, *F* fornix, *Fc* column of fornix, *Fi* fimbria, *GbPa* globus pallidus, *la* pars lateralis, *m* pars medialis, *H* hippocampus, *h* habenula, *Hy* hypothalamus, *I* insula, *ICa* internal capsule, *iOFf* inferior occipito-frontal fasciculus, *iPFo* interpeduncular fossa, *le* lenticular fasciculus (field H2 of Forel), *LVac* lateral ventricle, anterior cornu, *LVb* lateral ventricle body, *LVic* lateral ventricle, inferior cornu, *MB* mamillary body, *ot* optic tract, *pHG* parahippocampal gyrus, *Po* pons, *pu* putamen, *puP* putamen peduncle, *Sbn* substantia nigra, *sepe* septum pellucidum, *sThn* subthalamic nucleus, *Sub* subiculum, *tcb* tentorium cerebelli, *tCC* truncus of corpus callosum, *Th* thalamus, *an, mvn, van* anterior, medio-ventral, ventral anterior nuclei, *Unf* uncal fissure, *V3* third ventricle, *a3* posterior cerebral artery, *a4* superior cerebellar artery

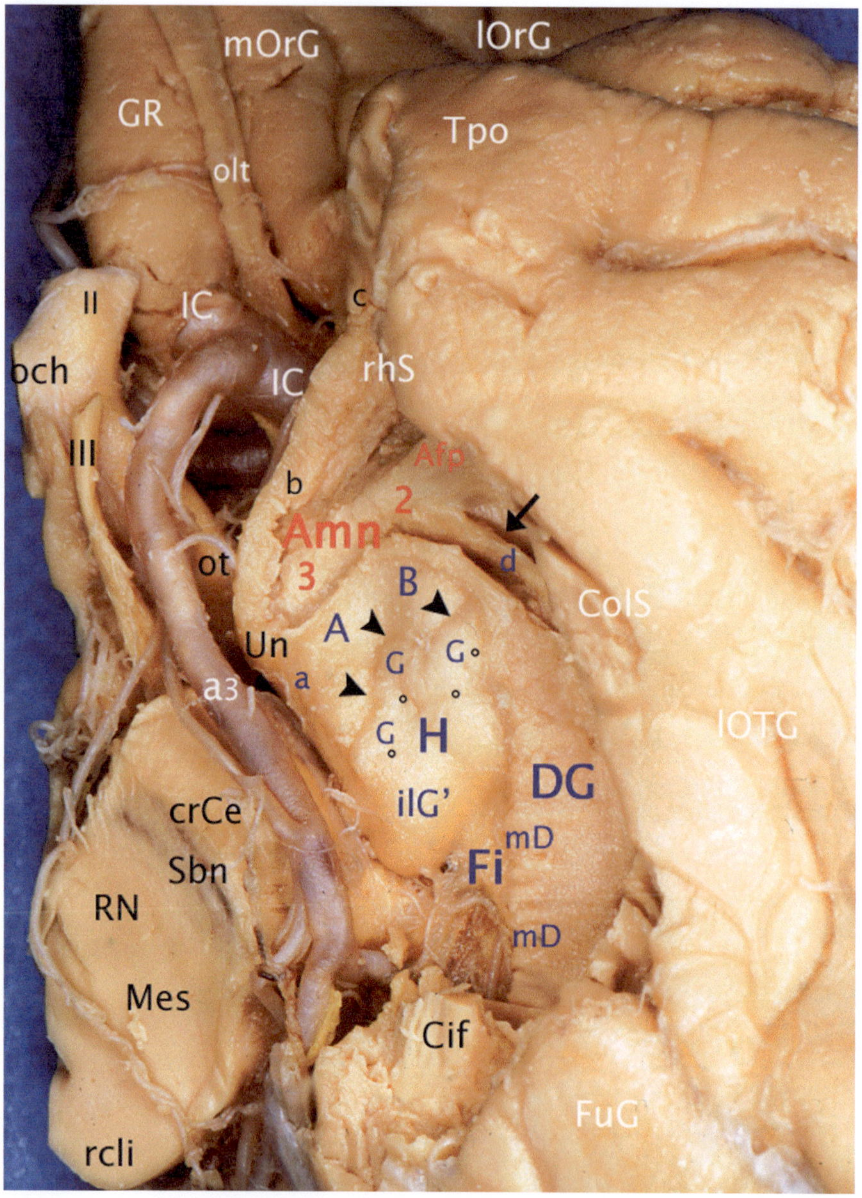

Fig. 7.16 Relations of the amygdala with the hippocampus. Dissection of the inferior surface of the left cerebral hemisphere (*the parahippocampal cortex and the cingulate fasciculus have been partially removed*). *Afp* amygdalofugal pathway, *Amn* amygdaloid nuclear complex: 2, baso-lateral nucleus; 3, baso-medial nucleus; *Cif* cingulate fasciculus, *ColS* collateral sulcus, *crCe* crus cerebri, *DG* dentate gyrus, *Fi* fimbria, *FuG* fusiform gyrus, *G* band of Giacomini (limbus Giacomini), *GR* gyrus rectus, *H* hippocampus, *a* hippocampal medial cortex, *b* periamygdaloid cortex, *c* temporal cortex, *d* hippocampal lateral cortex, *A* and *B* digitationes hippocampi, *IC* internal carotid artery, *ilG'* intralimbic gyrus (inferior aspect), *lOrG* lateral orbital gyrus, *lOTG* lateral occipitotemporal gyrus, *mD* margo denticulatus, *Mes* mesencephalon, *mOrG* medial orbital gyrus, *och* optic chiasma, *olt* olfactory tract, *ot* optic tract, *rcli* rostral colliculus, *rhS* rhinal sulcus, *RN* red nucleus, *Sbn* substantia nigra, *Tpo* temporal pole, *Un* uncus, *II* optic nerve, *III* oculomotor nerve, *a3* posterior cerebral artery with an aberrant origin (absence of the posterior communicating artery), *black arrow* inferior horn of the lateral ventricle, *black arrowheads* they follow the hippocampal sulcus, *little black circles* they indicate the prelimbic sulcus

gyrus, it covers, at the rear of Giacomini's band, the posterior part of the uncus up to its posterior apex.

It is interesting to also observe the inferior relations of the amygdala and hippocampal head, even if this requires the ablation of part of the para-hippocampic gyrus as well as the resection of the distal part of the cingulate fasciculus (see Fig. 5.3). We can therefore see the intralimbic gyrus, behind Giacomini's band, which is formed by fields CA 3 and CA 4 of Ammon's horn,[24] covered by the alveus (HM Duvernoy 2013).

- The *intra-ventricular relations* (Figs. 7.17 and 7.18) directly concern the amygdala and the head of the hippocampus. These relations are clearly visible, when the ventricle is open, as they are not masked by the choroid plexus, which do not reach the head of the hippocampus.

  The longitudinal axis of the amygdala is oblique at the top and at the rear such as that of the head of the hippocampus, which forms a relatively marked sharp angle with the large hippocampic corporeal axis (see Fig. 4.1). The amygdala therefore appears to be laying against the head of the hippocampus from which it is only partly separated by a narrow recess of the inferior horn of the lateral ventricle. In some cases, it determines a small cup-shaped notch on the hippocampal head (see Fig. 4.1).

  The relations occur between the baso-lateral part of the amygdala (lateral, baso-lateral and baso-medial nuclei) and the digitations of the hippocampal head, Retzius' digitations (G Retzius 1896), which amount to three or four and are orientated sagittally, except for the first which is vertical. It is to be recalled that one of the theories concerning the formation of these digitations involves the obstacle formed by the amygdala (see picture b, Fig. 3.3) during the anterior development of the hippocampus (Giacomini 1884). These digitations are formed by field CA 1.

  Junction bridges connecting the hippocampal head, at the level of the digitations, and the amygdala are frequently observed (Fig. 7.18). The number and diameter of these bridges vary. It is certain that these intra-ventricular bridges and the amygdalo-hippocampal appositions observed during the extra-ventricular dissection of the amygdala, on the superior and inferior edges of the ventricular recess (Fig. 7.18), are connections which turn the amygdalo-hippocampal system into an actual functional duo.

The relations of the amygdala with the end of the temporal inferior horn of the lateral ventricle also are posterior relations.

The distal end of the temporal horn inserts itself between the baso-lateral part of the amygdala and the hippocampal head, free of any choroid villosity. At this level, it forms the uncal recess, whose lumen is more or less narrow. This recess, which is easy to observe on sagittal sections (see Fig. 4.13), has two walls: a supero-lateral wall, which is slightly concave and lines the postero-inferior surface of the amygdala, and the other is infero-medial and convex, which lines the antero-superior surface of the hippocampal head, which protrudes in the ventricular cavity. This recess is more or less pronounced but generally leaves the medial parts of the amygdala and hippocampal head in a back-to-back configuration.

The relations of the amygdala with the tail of the caudate nucleus deserve to be studied here due to the fact that the tail itself is in close relation with the temporal horn of the lateral ventricle.

Such as observed on coronal sections, the tail of the caudate nucleus seems to be stuck against the supero-lateral wall of this temporal horn.

What connections does this tail have with the amygdala? Is it interrupted before reaching the amygdala? Is it interrupted when it comes in contact with the amygdala? Does it end inside the amygdala? Curiously, the authors' opinions vary and all types of cases are described. We will mention some of them as examples:

- Some refer to an amygdalar termination of the tail:

  "Caudally the amygdala fuses with the tip of the tail of the caudate nucleus" (TS Gray et al. 1989).

  "Where it terminates at the amygdaloid nuclei" (J Knierim et al. 2012).

  "The tail of the caudate nucleus terminates anteriorly in amygdaloid nucleus" (RS Snell 2010).

  "The tail of the caudate that … and extends rostrally to fuse with the amygdaloid body" (JR Augustine 2007).

  "La queue du noyau caudé se dirige en avant… et se termine dans le corps amygdaloïde" (The tail of the caudate N extends forward … and terminates in the amygdaloid body) (E Gardner et al., French Edition, 1979, adapted by J Bossy).

- Some, on the other hand, have specified that the tail of the caudate nucleus does not terminate in the amygdale:

  "Au niveau de son extrémité, la queue du noyau caudé est en contact avec le noyau amygdalien avec lequel elle n'affecte que des rapports de contiguïté" (At its end section, the tail of the caudate nucleus is in contact with the amygdaloid nucleus with which it only has developed contiguity connections) (G Paturet 1964).

  "A long tail which enters the temporal lobe and terminates in the vicinity of the amygdaloid complex" (R Nieuwenhuys et al. 2008).

[24]The structure of Ammon's horn ("cornu Ammonis") or "hippocampus proper" is complex. In 1934, Lorente de No named the four fields of this cortex, which is allocortex or archaeocortex: fields CA 1, CA 2, CA 3 and CA 4. CA 1 follows the subiculum. CA 2 is located between CA 1 and CA 3. CA 3 corresponds to the knee of Ammon's horn. CA 4 is located in the concavity of the inferior dentate gyrus.

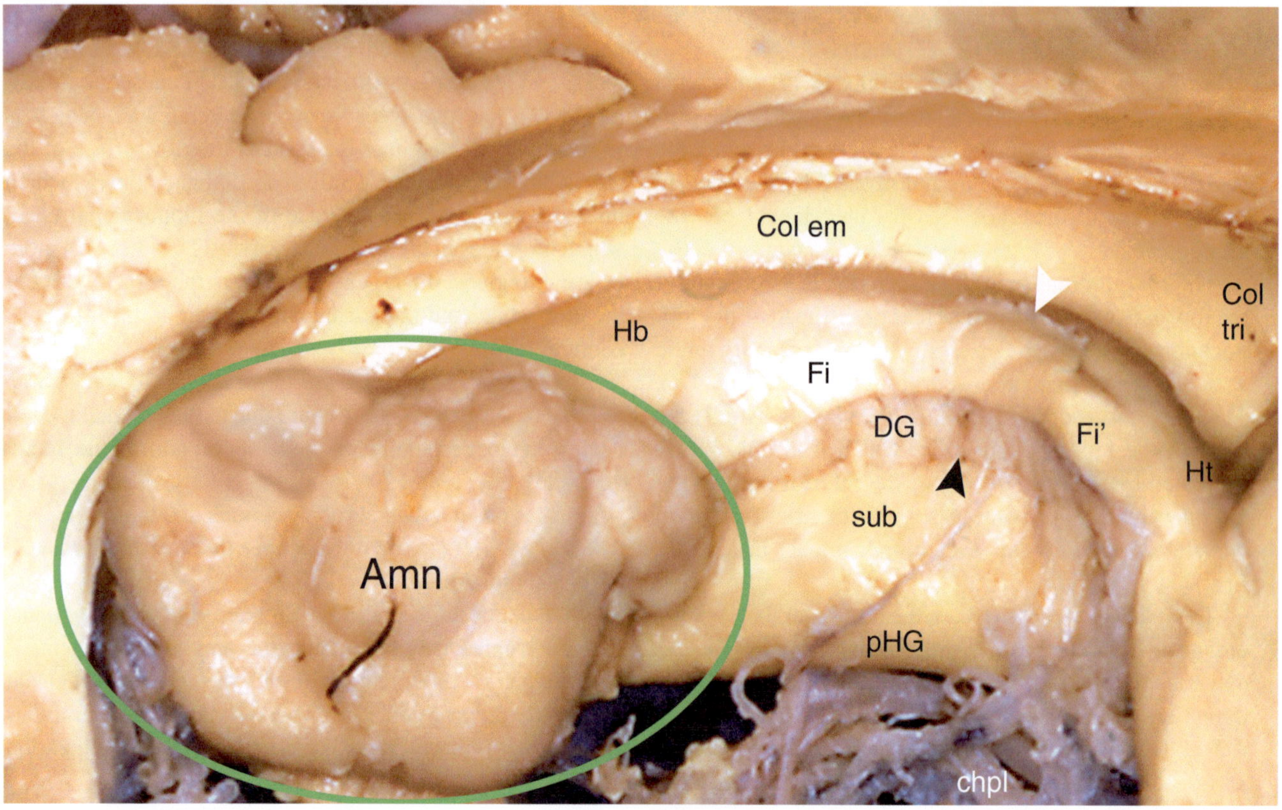

Fig. 7.17 Dorsal view of the amygdaloid nucleus in anatomical position, immediately above the head of the hippocampus (the temporal horn of the right lateral ventricle has been opened and the periamygdaloid cortex, removed). *Amn* amygdala (*green circle*), *chpl* choroid plexus, *Col em* collateral eminence of Meckel ("cuissart de Malacarne" of the old anatomists), *Col tri* collateral trigone, *DG* dentate gyrus (margo denticulatus), *Fi* fimbria, *Fi'* section of the fimbria, *Hb* hippocampus body, *Ht* hippocampus tail, *pHG* parahippocampal gyrus, *sub* subiculum, *black arrowhead* hippocampal sulcus, *white arrowhead* taenia of the fimbria

"Le corps amygdaloïde (amygdaloid nucleus) est une petite formation sriée située en avant de la queue du noyau caudé" (The amygdaloid body is a small striated formation located at the front of the tail of the caudate nucleus) (A Gouazé and J Laffont 2000).

"La queue du noyau caudé vient au contact du noyau amygdalien sans qu'il y ait fusion" (The tail of the caudate nucleus is in contact of the amygdaloid nucleus without any fusion) (Ch Foix and J Nicolesco 1925).

We confirm the affirmations of this last series of authors. Our most recent dissections have shown that:

The caudate nucleus does not terminate in the amygdala.

Both structures have neighbouring locations due to the fact that the end of the tail of the caudate nucleus terminates against the lateral surface of the distal end of the uncal recess, a recess which is capped, such as already observed, by the amygdala.

The only relations that can be mentioned are neighbouring, proximity or even contiguity connections but not fusion connections.

If difficulties arise concerning this termination of the caudate nucleus, they can be explained by the fact that this structure becomes thinner and is more or less discoloured towards its termination such that it can only be specified by performing a series of coronal sections of the termination area.

7.8 Relations with the Brainstem (Postero-Inferior Relations)

From a topographical viewpoint, the relations of the amygdala with the brainstem are remote connections due to the fact that the superior part of this stem, the mesencephalon, extends from the diencephalon and that the hypothalamus, inferior part of this diencephalon, is already relatively distant from the amygdala. Among the three parts of the brainstem (mesencephalon, pons and medulla oblongata), it is the mesencephalon which is the closest to the amygdala and which will therefore play a key role in the physiological amygdala-brainstem continuum (and vice versa).

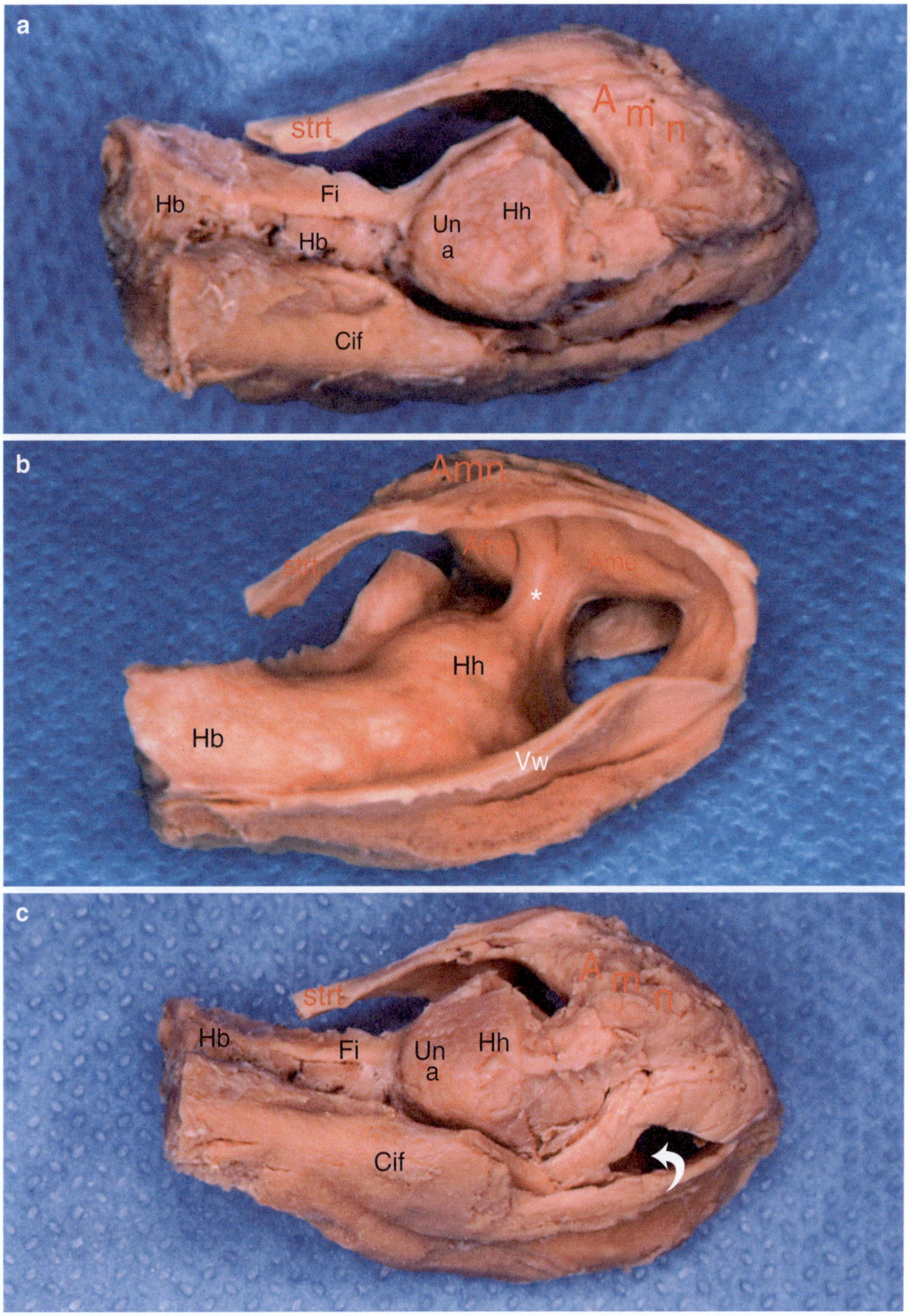

Fig. 7.18 Morphological aspects of the amygdala (dissection keeping the relations with the hippocampus). (**a**) Medial extra-ventricular view; (**b**) intra-ventricular view; (**c**) infero-medial extra-ventricular view. *Amn* amygdala, *Aem* amygdalar eminence, *Cif* cingulate fasciculus, *Fi* fimbria, *Hb* hippocampal body, *Hh* hippocampal head, *strt* stria terminalis, *Un a* uncal apex, *Vw* ventricular wall, *curved white arrow* it shows the cavity of the lateral ventricle, *white asterisk* it shows the intra-ventricular connections between the hippocampus and the amygdala

In fine, the amygdala will also be connected to the very numerous nuclei and loci of the brainstem, to the grey formations of the mesencephalon and to the major ascending and descending pathways which allow the formation of essential links and, especially of the longest among them, the amygdalo-hypothalamo-spinal connection.

Four major types of amygdala target nuclei are to be considered:

- *The nuclei of the autonomic nervous system*, attached to the nuclei of the cranial nerves of which the major parts are located in the floor of the fourth ventricle. They form the cranial parasympathetic.[25] At the level of the pons, we

[25]A second part of the parasympathetic system forms the pelvic or sacral parasympathetic. It is implemented by the hypothalamic or amygdaloid neurons which reach the raphe nuclei. The fibres of these nuclei descend in the hypothalamo-spinal bundle and target the neurons of the ventral horn of the sacral metamers of the spinal cord. The parasympathetic pathway develops from these neurons. It emerges from the motor roots of spinal nerves S2 to S4 and forms the pelvic splanchnic nerves (former Eckard's erector nerves). This parasympathetic pathway is also bineuronal. It is intended for the descending colon, the rectum, the bladder and the genitalia.

therefore obtain the superior salivary nucleus (attached to VII bis); at the level of the superior end of the medulla oblongata, the inferior salivary nucleus (attached to IX); and at the level of the inferior part of the medulla oblongata, the ambiguus nucleus (attached to VII bis, to IX and to X) and the dorsal complex of the vagus nerve (nucleus of the solitary tract and dorsal motor nucleus of the vagus nerve) studied in 1980 by JS Schwaber et al. (attached to X). It is to be recalled that each of these nuclei is the departure point of the first neuron of the parasympathetic pathway, which terminates at a ganglion of a cranial nerve, which is the departure point of a second neuron which terminates on a target organ (involving, in the order of the above-mentioned nuclei, the pterygo-palatine ganglion with the lacrimal gland; the nasal mucosa and the mucosa of the soft palate, as targets; the submandibular and sublingual ganglions with the submandibular gland and the sublingual gland as respective targets; the otic ganglion with the parotid gland as target). The nucleus of the vagus nerve is the departure point of the first neuron of the parasympathetic pathway of the stem, bineuronal and cholinergic.[26]

One nucleus is to be added to these nuclei, the only mesencephalic nuclei, the Edinger–Westphal nucleus (accessory oculomotor nucleus), centre of the pupillary light reflex, which adapts the diameter of the pupil to the intensity of the light that falls on the eye. The nucleus supplies a parasympathetic bineuronal pathway (with relay in the ciliary ganglion) which acts on the iris sphincter, causing pupil to contract (myosis), and also on the ciliary muscle, increasing the refractive power of the lens.

Reticular nuclei: It is to be recalled that the reticular substance is made of a series of neuronal centres which form a continuous nuclear column, layered along the brainstem, and whose entwined and interconnected axons, including many intercalated interneurons, form a grid between the interstices from which and pass other nerve fibres. In addition to its connections to the amygdala, the reticular substance forms the functional connection between the thalamus, the hypothalamus and the intermediate part of the grey substance of the spinal cord. The nuclei are organised according to three areas, the median (raphe nuclei), medial and lateral areas. For the raphe, we can observe at the level of the mesencephalon, the dorsal raphe nucleus, at the level of the pons, the nucleus raphes medianus and the nucleus raphes pontis and at the level of the medulla oblongata, the nuclei raphes pallidus, raphes obscurus and raphes magnus. Several of these nuclei contain group B neurons, which are serotoninergic (R Nieuwenhuys et al. 2008).

These serotoninergic neurons are divided into two groups: the first group in the mesencephalon and the rostral part of the pons and the second group in the caudal part of the pons and medulla oblongata. The quantity of serotonin contained in these neurons varies according to the nuclei under study:

For the medial area, at the level of the pons, we have the nuclei pontis oralis and caudalis and at the level of the medulla oblongata, the nucleus gigantocellularis and the central nucleus (medial subnucleus).

For the lateral area, we have, at the level of the mesencephalon, the pedunculopontine nucleus and the lateral parabrachial nucleus.

For the superior part of the medulla oblongata, we have the locus coeruleus and the medial parabrachial nucleus.

For the inferior part of the medulla oblongata, we have the lateral subnucleus of the central nucleus, the lateral reticular nucleus and the retroambiguus nucleus. The neurons of this lateral part belong for the major part to cell groups A and C, which are adrenergic (however, the pedunculopontine nucleus is formed of cholinergic Ch neurons!).

- *Loci of the greater limbic system.* These are layered cell areas, more or less assembled as nuclei, whose electrical stimulation triggers behavioural and autonomic responses. These loci have been specifically studied by WR Klemm and RP Vertes in 1990. We can specifically mention, at the level of the ventrolateral part of the medulla oblongata, Bötzinger's respiratory loci which generate the respiratory rhythm and the cardiovascular regulation locus (loci which recall the well-known "cardio-pneumo-enteric centre" of former physiologists!). At the level of the pons, reference has been made to Barrington's micturition nucleus, the pontine swallowing centre and the pontine attack site as well as to the vocalisation cell group Voc. At the level of mesencephalon, most loci are located within the periaqueductal grey (PAG). They are the defensive behaviour locus, the PAG centre for micturition and the PAG centre for vocalisation.
- *Grey formations of the brainstem* (Fig. 7.19). These formations are, with regards the mesencephalon, formations referred to with the abbreviations PAG, IP and VTA and for the brainstem, essentially the locus coeruleus.

PAG is the periaqueductal grey referred to above. It is the grey crown which surrounds the cerebral aqueduct (or aqueduct of the mesencephalon). With the tectum, the tegmentum mesencephali and the pes pedunculi (which includes the substantia nigra), it is the fourth part of the mesencephalon. It extends from the hypothalamic grey matter to that of the floor of the fourth ventricle.

[26] At the end of the parasympathetic pathway, the neuromediator is acetylcholine and the receptors of the target tissue are muscarinic. At the end of the orthosympathetic pathway, the neuromediator is noradrenalin, and the receptors of the target tissue are adrenergic.

As for visceral sensitivity, it transits via the solitary nucleus located in the inferior part of the medulla oblongata and the sensitive part of the dorsal nucleus of the vagus nerve.

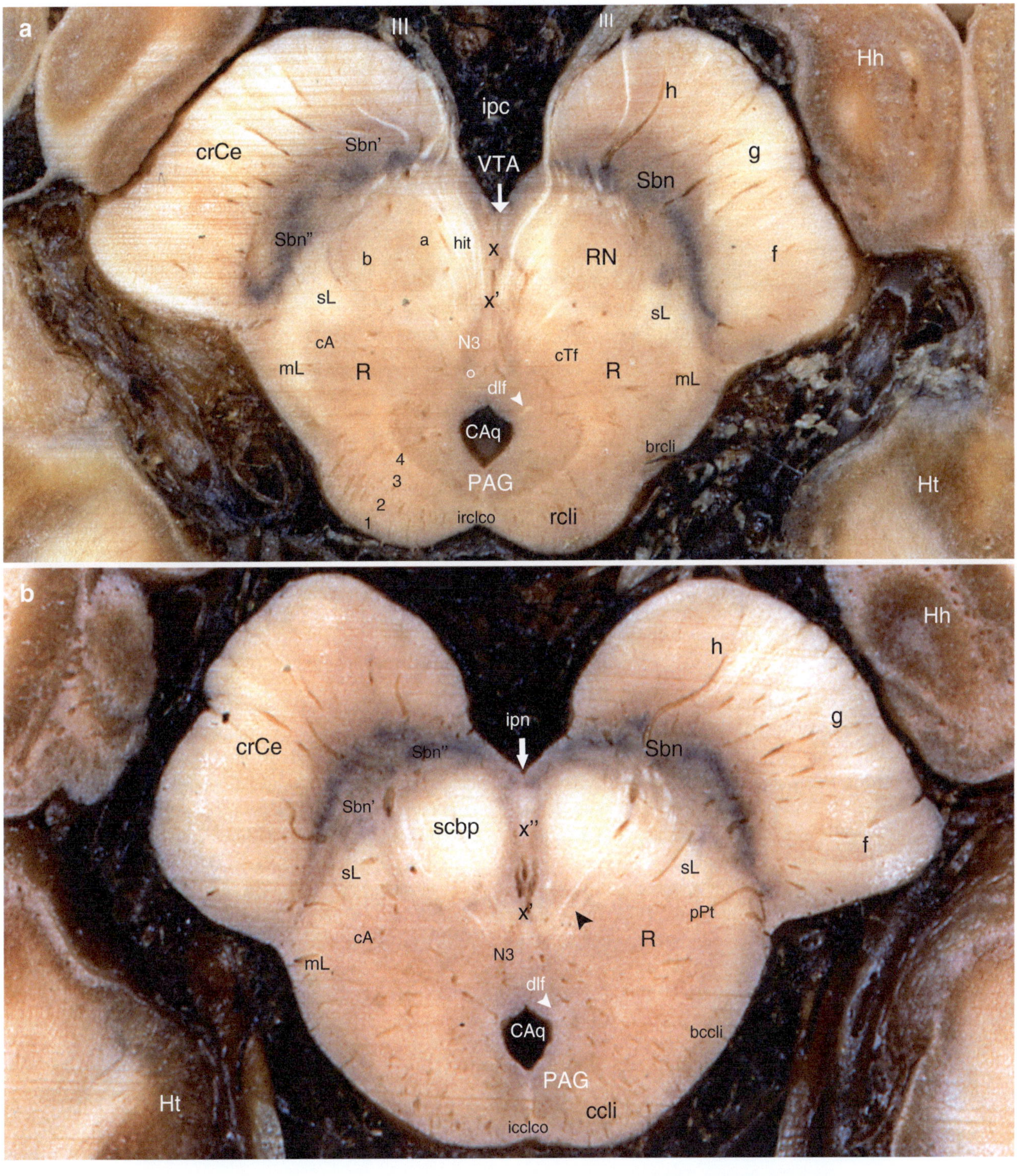

Fig. 7.19 Two axial sections of the midbrain ((**a**) section through the red nuclei; (**b**) section through the superior cerebellar peduncles) showing the main grey areas in relation with the pathways connecting the amygdala and the midbrain. *bccli* brachium of the caudal colliculus, *brcli* brachium of the rostral colliculus, *cA* cuneiform area, *CAq* cerebral aqueduct, *ccli* caudal colliculus, *crCe* crus cerebri, *f* parieto-temporo-occipito-pontine tract, *g* cortico-spinal and cortico-nuclear tract, *h* fronto-pontine tract, *cTf* central tegmental fasciculus, *dlf* (*white arrowhead*) dorsal longitudinal fasciculus, *Hh* hippocampus head, *hit* habenulo-interpeduncularis tract, *Ht* hippocampus tail, *icclco* inter caudal colliculi commissure, *ipc* interpeduncular cisterna, *ipn* (*white arrow*) interpeduncular nucleus, *irclco* inter-rostral-colliculi commissure, *mL* medial lemniscus, *N3* oculomotor nucleus, *PAG* periaqueductal grey, *pPt* pedunculopontine tract, *R* reticular formation, *rcli* rostral colliculus (*1* tractum zonale, *2* stratum opticum, *3* stratum lemnisci, *4* white layer), *RN* red nucleus (*a* magnocellular part, *b* parvocellular part), *Sbn* substantia nigra, *Sbn'* pars reticulata, *Sbn"* pars compacta, *scbp* superior cerebellar peduncle, *sL* spinal lemniscus, VTA (*thin white arrow*) ventral tegmental area, *x* rubro-spinal decussation, *x'* tecto-spinal decussation, *x"* superior cerebellar peduncle decussation, *III* oculomotor nerve, *black arrowhead* transpeduncular fibres of the oculomotor nerve, *white circle* oculomotor accessory nucleus (nucleus of Edinger–Westphal)

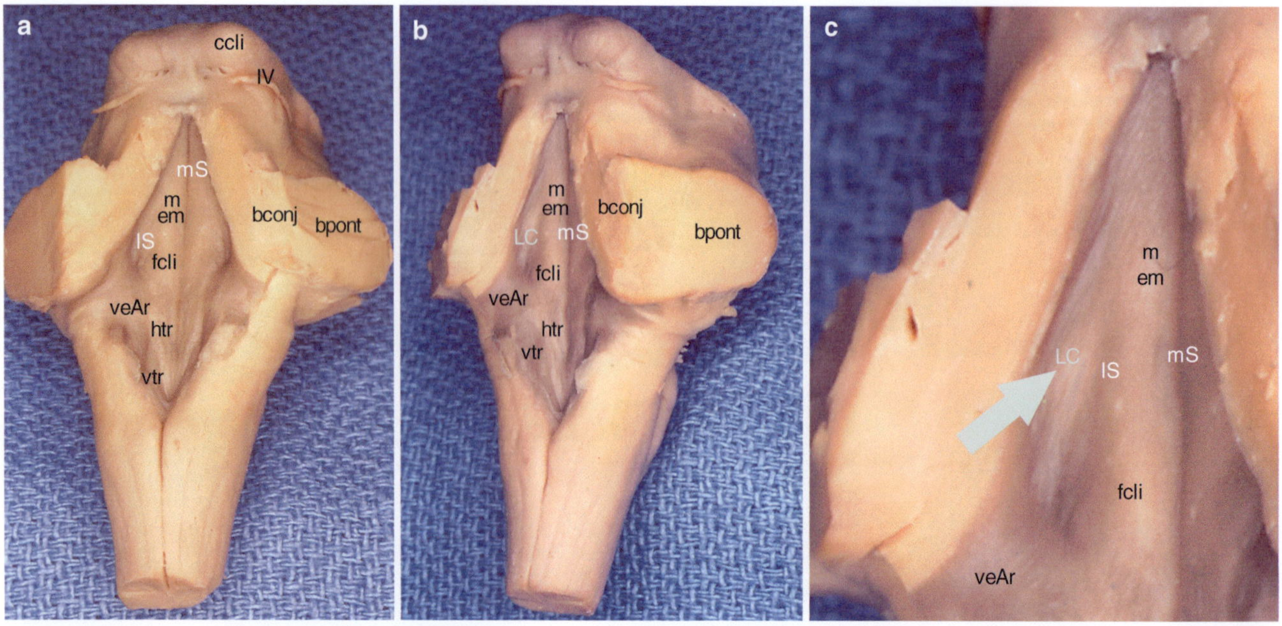

Fig. 7.20 Location of the locus coeruleus on the floor of V4. (The locus coeruleus is the area where are clustered the neurons of the noradrenergic A6 group projecting to the amygdala and to the entire central nervous system.) (**a**) Posterior view of the rhomboid fossa (floor of fourth ventricle). (**b**) 3/4 right view of **a**, showing the locus coeruleus area. (**c**) Enlargement of **b** to show the locus coeruleus' area. *bconj* brachium conjunctivum (superior cerebellar peduncle), *bpont* brachium pontis (middle cerebellar peduncle), *ccli* caudal colliculus, *fcli* facial colliculus, *htr* hypoglossal triangle, *LC* locus coeruleus, *lS* limitans sulcus, *m em* medial eminence, *mS* median sulcus, *vtr* vagal triangle, *ve Ar* vestibular area

IP is the interpeduncular nucleus (BJ Morley 1986). It is located in the floor of the interpeduncular fossa and also plays a key role within the mesencephalon. It is the final destination of the habenulo-peduncular tract (Meynert's retroflex bundle). It is connected via bidirectional pathways to the amygdala, to the hypothalamus and to the habenular nuclei. It contains cholinergic and GABAergic neurons. It plays a key role in sleeping mechanisms (H Shibata et al. 1986) and in the generalised emotional condition (BJ Morley 1986). Furthermore, recent work (R Zhao-Shea et al. 2013) has shown that this nucleus, via the activation of the GABAergic neurons that it contains, is responsible for the symptomatology of the withdrawal syndromes following smoking cessation.

VTA or ventral tegmental area is a group of cells located ventrally against the red nucleus and which belongs to the dopaminergic pathway and to the pleasure and reward pathway. Its cells produce dopamine which is released to the striatum (meso-striatal pathway). The VTA also targets the amygdala, the hippocampus, the cingulate cortex, the prefrontal cortex, the reticulate and the other grey mesencephalic formations. The VTA and the nucleus accumbens (mesolimbic pathway) are the targets of addictive drugs, which act by stimulating the release of dopamine (which induces the sensation of reward and psychomotor effects). All projection areas of the VTA target reciprocally towards it. The VTA also receives the particularly important projections from the BST (see Chap. 10).

The VTA is one of the components of the dopaminergic system whose other components are located at the level of the arched nucleus of the hypothalamus (dopaminergic tubero-infundibular pathway) and of the pars compacta of the substantia nigra.

The locus coeruleus: Its Latin terminology is explained by the fact that it is blue (Fig. 7.20). It is located in the rhomboid fossa, near to the rostral end of the limiting sulcus. Its neurons are adrenergic and noradrenergic and form group A6. Other adrenergic nuclei are observed in the brainstem, groups A1, A2, A4, A5 and A7. The adrenergic pathways and especially the locus coeruleus are "modulators of the excitability of other systems" (DL Felten and RF Jozefowicz 2003) and especially of the nociceptive pathways.

The major pathways of the brainstem will contain the axons allowing the amygdala to communicate with the nuclei and grey formations, which we have just examined, and therefore with the entire brainstem as well as with the spinal cord which extends therefrom. These pathways contain both amygdaloid descending axons and ascending axons, which are generally mixed with "hypothalamic axons". Two key pathways are routed through the brainstem: the dorsal longitudinal fasciculus and the medial forebrain bundle:

- The *dorsal longitudinal fasciculus of Schütz* (fasciculus longitudinalis dorsalis or posterior) (Fig. 7.21) is a relatively well-individualised pathway, which is mostly

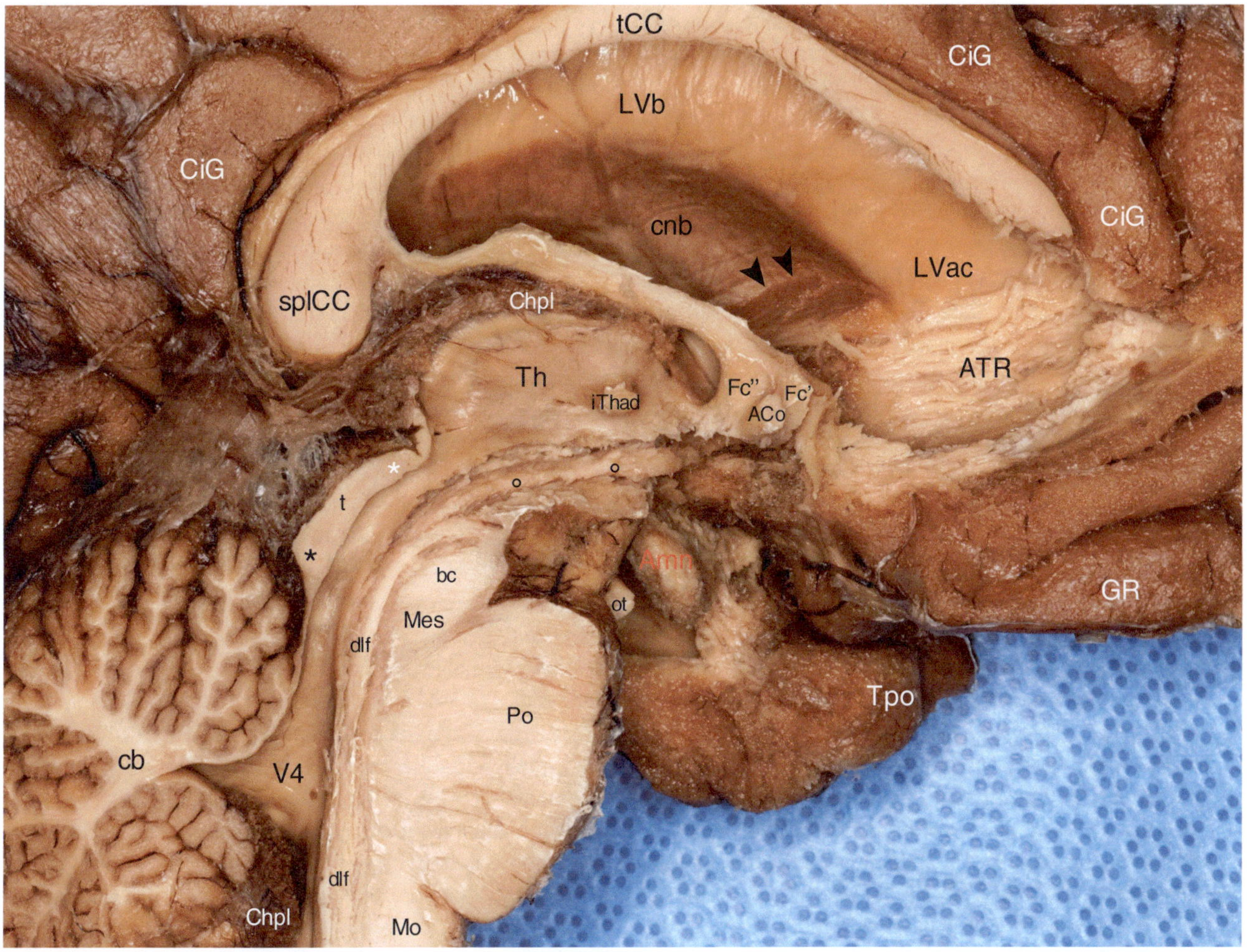

Fig. 7.21 Relationships of the Amn with the longitudinal dorsal fasciculus of Schütz. Dissection of the medial side of the left brain hemisphere (the head of the caudate nucleus and the anterior part of the corpus callosum have been resected). *ACo* anterior commissure, *Amn* amygdaloid nuclear complex, *ATR* anterior thalamic radiations, *bc* brachium conjunctivum (decussation of the superior cerebellar peduncles), *cb* cerebellum, *Chpl* choroid plexus, *CiG* cingulate gyrus, *cnb* caudate nucleus body, *dlf* dorsal longitudinal fasciculus (of Schütz), *Fc'* precommissural fornix, *Fc"* column of fornix, *GR* gyrus rectus, *iThad* interthalamic adhesion, *LVac* lateral ventricle, anterior cornu, *LVb* lateral ventricle body, *Mes* mesencephalon, *Mo* medulla oblongata, *ot* optic tract, *Po* pons, *splCC* splenium of the corpus callosum, *t* tectum of midbrain (*white asterisk* rostral colliculus, *black asterisk* caudal colliculus), *tCC* truncus of the corpus callosum, *Th* thalamus, *Tpo* temporal pole, *V4* fourth ventricle, *black arrowheads* they show the section of the caudate nucleus, *little black circles* they show the oblique course forward and inside of the fasciculus through the hypothalamus (dissected)

peptidergic and rich in somatostatin (W Kahle and M Frotscher 2005). It is formed by a network of ascending and descending fibres which connect, on the one hand, the amygdala, the septum and the hypothalamus and, on the other hand, the different nuclei of the brainstem (vegetative nuclei attached to the cranial nerves,[27] nuclei of the reticular formation) inside which they terminate and generate their nervous effects, especially autonomic, on target organs.

- Part of the axons continues down to the spinal cord, thereby forming a major amygdalo- and hypothalamo-spinal pathway which ends at the lateral cord and then passes in the intermedio-lateralis area of this cord. This corresponds to the point of departure of the first neuron of the bineuronal orthosympathetic pathway in the direction of the sympathetic ganglionic trunk where, for the head, neck, thorax and skin, the synapse with the second neuron is formed. It is at the end of the latter that the sympathetic effects[28] will occur.

The ascending fibres, which have developed from the various nuclei connected to the longitudinal dorsal path-

[27] This Schütz fasciculus receives, at the level of the mesencephalon, fibres from the habenulo-tegmental pathway. This is how the olfactory stimuli, which have reached the habenula via the medullary stria of the thalamus, generate salivary and gastric secretions through the action of fibres of the longitudinal dorsal pathway on the parasympathetic nuclei of the brainstem.

[28] For the abdominal and pelvic visceral the ssynapsis between the two neurons of the orthosympathetic pathway is made at the level of the abdominal preaortic ganglia and of the pelvic plexus.

way, are viscero-sensitive and inform the amygdala and hypothalamus of the vegetative condition of the subject.

In this longitudinal dorsal pathway, coexist fibres of various lengths, which terminate at the level of the numerous synaptic relays as well as at the interneurones.

This pathway, which is located underneath the ependyme (Fig. 7.20) and described in the descending direction, passes, at the level of the brainstem, via three neuronal areas which are part of these synaptic relays, the nuclei of the solitary tract, the parabrachial nuclei and periaqueductal grey substance. It then bends to take a more lateral route, inside the periventricular hypothalamus, and is slightly oblique at the top and in the front area. It terminates at the level of the septal region and of its nuclei, after having been joined by amygdaloid fibres.

- *The medial forebrain bundle* (see Fig. 8.1), medialis fasciculus telencephali (mfb or MFB according to the authors),[29] is the second pathway used by the fibres connecting the amygdala and the brainstem. It is also a peptidergic pathway (rich in VIP (vasoactive intestinal peptide), serotonin and enkephalin (W Kahle and M Frotscher 2005). It is less well individualised than the longitudinal dorsal pathway but similarly comprises fibres of various lengths, which form a sort of grid, orientated longitudinally and divided, at the level of the brainstem, into a lateral stream (the widest) and a medial stream (which only contains ascending fibres). The medial stream is close to the raphe nuclei which provide the numerous axons. Overall, the mfb receives, at the level of the brainstem, fibres mostly from the same nuclei and loci which we have mentioned for the longitudinal dorsal pathway, especially the parabrachial nuclei, the locus coeruleus, the vagal dorsal complex of the adrenergic loci and the pars compacta of the substantia nigra. From a topographical viewpoint, it passes in the brainstem at a less deep level than the longitudinal dorsal pathway. The two streams depart from the mesencephalon and join up, and the reunified mfb bends to laterally penetrate inside the hypothalamus, and by a route slightly oblique forwards at the top and inside, pass through the preoptico-hypothalamic area towards the septal region (see Fig. 8.1). It is connected to the ventral amygdalofugal pathway (via which it also sends amygdalopetal fibres to the amygdala) (JL Price and DG Amaral 1981), to the BST as well as to the nuclei in the septal region (nucleus accumbens and nucleus of the diagonal band).

Descriptions are also made of a fibrillary stream which reaches the olfactory bulb (R Warwick and PL Williams 1973). All of these terminations form the first terminal expansion of the mfb, termination of the "MacLean's mesolimbic pathway". This author has described another terminal expansion directed towards the anterior part of the cingulate cortex and towards the frontal cortex: it is "MacLean's mesocortical pathway", which forms what Nieuwenhuys has referred to as a "limbic forebrain-midbrain continuum" and whose integrity is indispensable for maintaining a harmonious neuropsychological condition.

In 2012, VA Coenen et al. were able to observe this bundle, using MRI and DTI FT (Diffusion Tensor Imaging Fibre Tracking), and so gave of it a little bit different description. According to these authors, the caudal origins of the mfb connect to the dentate nucleus of cerebellum. The pathway terminates inside the mesencephalon at the level of the ventral tegmental area (VTA) and then divides into two branches, an infero-medial branch (imMFB), which passes against the wall of the third ventricle to reach the lateral hypothalamus, the septum, the amygdala, the BST and the dorsomedial thalamus, and a supero-lateral branch (slMFB), which passes through the thalamus. This branch converges and is articulated with a second separate pathway, although it extends its direction forwards: these are anterior radiations of the thalamus, which will follow the median part of the anterior crus of the internal capsule and become part of the latter. Lastly, the anterior thalamic radiations (ATR) are routed towards the frontal gyrus and develop at the level of the cortex of the frontal pole and therefore towards the rostral anterior cingulate gyrus and the orbitofrontal cortex (Fig. 7.22). It should be noted that in this description, Mac Neal's two mesolimbic and corticolimbic pathways are represented by the imMFB and slMFB, respectively, with only a few differences.

However, beyond the descriptions which most definitely deserve a new approach, the comments made by the authors in relation to the MFB device are interesting: on the basis of the results obtained following the application of deep brain stimulations targeting formations connected to the MFB, cg25 (cingulate gyrus), the nucleus accumbens septi and the crus anterior of the internal capsule, which were performed during the treatment of psychiatric disorders (sever cases of depression and of obsessive compulsive disorder), the authors concluded that the MFB has a major importance in the management of the conditions. They made two observations: the stimulation of the slMFB has antidysphoric effects involving a renewed motivation, the restoration of energy and even euphoria, effects which are due to the activation of the limbic structures, especially the reward-seeking system (via the nucleus accumbens to which this pathway is connected). Inversely, the stimulation of the ATR triggers or increases a depressive disease with an intense tiredness syndrome, by triggering the panic emotional system, which probably results from the activation of cg25 to which the ATR is connected (we know that the activity of cg25 is increased in depressive patients). These

[29]The mfb and above all MFB abbreviations (which refer to the medial forebrain bundle) may lead to confusion with the MBF (which refers to the magnocellular basal forebrain), i.e. all of the cholinergic neurones of the basal region of the brain. We suggest that scientists should decide to use lower-case characters for one of these components and upper key characters for the other (or vice versa) in a definite manner.

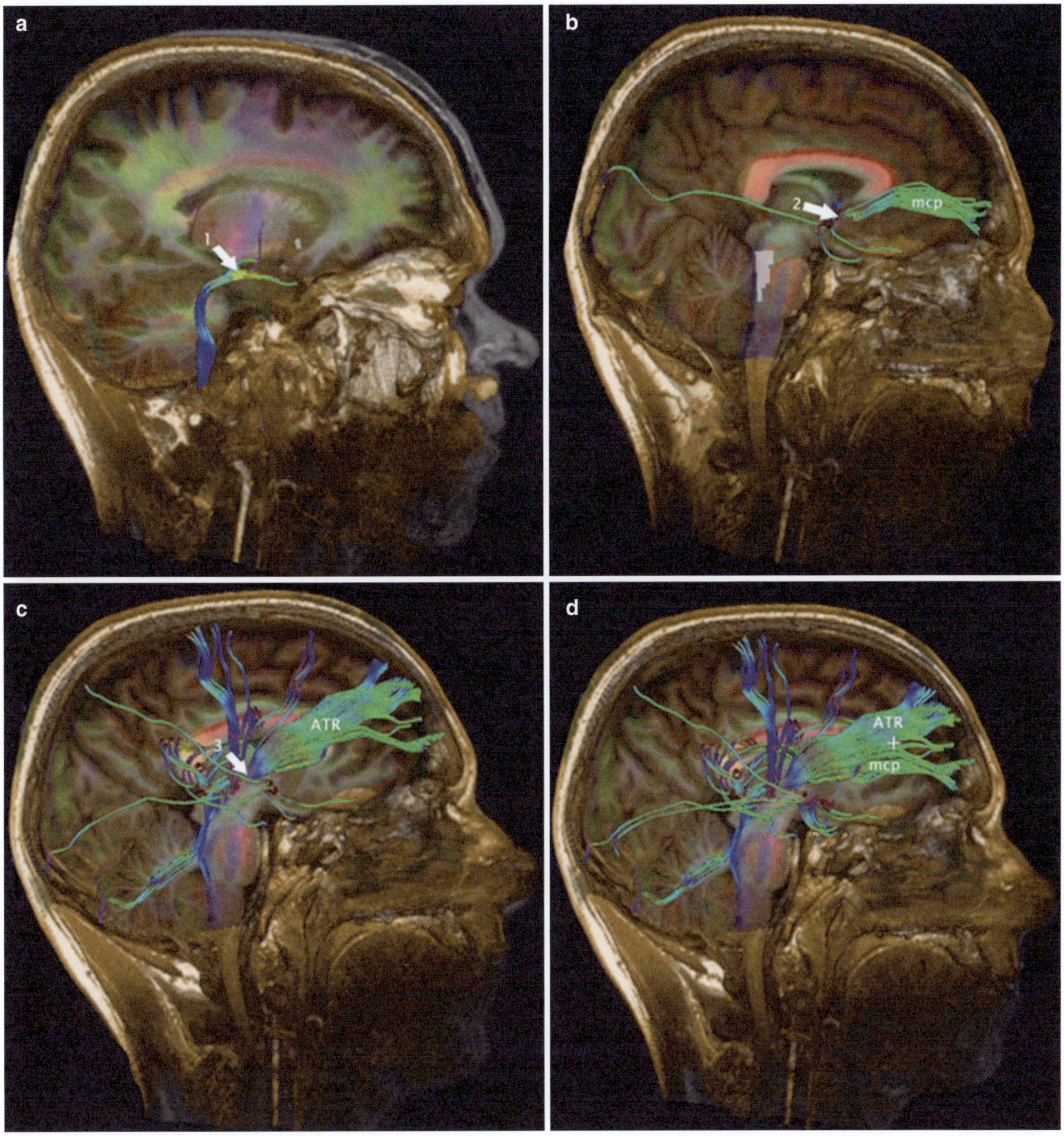

Fig. 7.22 Study by diffusion Tensor Imaging Fibers Tracking (DTI FT) of the left hemisphere showing the mesolimbic and mesocortical pathways, very important routes for the amygdalar neurons'axons. (**a**) The ROI (region of interest) is located by a *white arrow* on the mesencephalon (*white arrow 1*): we can see the longitudinal fibers of the brainstem behind the ROI and the "hypothalamic segment" of MFB (routed to the septal area) in front of the ROI. (**b**) The ROI is located on the accumbens nucleus (*white arrow 2*): we can see the fibers of the meso-cortical pathway (mcp) in front of the ROI. (**c**) The ROI is located on the anterior thalamic nucleus (*white arrow 3*): we can see the anterior thalamic peduncle (Anterior thalamic radiations: *ATR*). (**d**) Fusion of b+c (*ATR + mcp*) showing the rostral part of the MFB (medial forebrain bundle)

observations have led these authors to consider that the two parts of the MFB correspond to two systems whose opposite emotional effects must compensate each other so that the subject can have a perfect neuropsychological homeostasis. The authors also recall that the MFB is connected to the structures of the basal brain and to the BST (see Chap. 10). We have already seen that they play a major role in this balance.

8 Inputs and Outputs

The anatomical study of the relationship of the amygdaloid nuclear complex and its connection routes has given us already a very good idea of the structures with which this complex can exchange information and various stimuli, whether it receives their axons (carriers of the centripetal impulses "inputs") or it projects its own (carriers of centrifugal impulses or "outputs") to the said structures.

We must now make recourse to neuroscience to know from where the impulses entering the amygdala are coming and towards which structures are moving the impulses emerging from the amygdala that only immunohistochemical techniques and anterograde or retrograde tracing methods allow us to understand precisely. It is the results of studies using these methods we will review in this chapter, which therefore will be rather a review of the literature.

We shall examine successively the afferent nerve fibres in the amygdaloid complex and efferent fibres coming from this complex. The afferent fibres guide the centripetal amygdalopetal impulses, and the efferent fibres guide centrifugal amygdalofugal impulses.

8.1 Inputs

We distinguish between the afferent fibres, the fibres coming from the olfactory tract, the fibres from the hypothalamus, the fibres from the thalamus, the fibres from the basal part of the telencephalon, the fibres from the brainstem, the fibres coming from the hippocampus and the fibres from the neocortex.

8.1.1 Fibres from the Olfactory Tract

They arise from the olfactory bulb itself, the anterior olfactory nucleus (located in the groove occupied by the olfactory tract) and the lateral olfactory stria (Y Soudry et al. 2011). Let us recall that at the macroscopic level, this stria reaches the prepiriform cortex at the limen (several fibres reaching the primary olfactory cortex located in the insular angle in front of the agranular cortex of the insula) and gives a dissectible large contingent (see Fig. 7.1), detaching almost perpendicularly to reach the convexity which raises the cortical nucleus of the amygdala, at the level of the periamygdaloid anterior cortex. Research carried out by DL Rosene and CW Van Hoesen in 1977 and then by ST Carmichael et al. in 1994 that used, in macaques, a combination of anterograde and retrograde axonal tracers to study these inputs confirmed that the fibres of olfactory origin come to the cortical nucleus of the amygdala and the periamygdaloid cortex (semilunar gyrus and gyrus ambiens). Otherwise, the prepiriform cortex which also receives a macroscopic quota from lateral olfactory stria projects at the baso-lateral nucleus of the amygdala (TPS Powell et al. 1965). The olfactory bulb and prepiriform cortex also target the three parts (dorsal, ventral and ventromedial) of the entorhinal cortex (JE Krettek and JL Price 1977, online 2004) whose deep layers project to the amygdala (M Hoïstad and H Barbas 2008).

8.1.2 Fibres from the Hypothalamus

They have been studied by JP Aggleton et al. in 1980. They mainly come from the ventromedial hypothalamus and from other hypothalamic nuclei (including paraventricular and arcuate nuclei and the lateral hypothalamus) which are also involved in these inputs that go to the medial and central nuclei of the amygdala, leading visceral inputs. Note also that the anatomical dissection has already shown us bundles of nerve fibres connecting the lateral hypothalamus to the central nucleus of the amygdala (see Fig. 7.13).

V. Di Marino et al., *The Amygdaloid Nuclear Complex: Anatomic Study of the Human Amygdala*,
DOI 10.1007/978-3-319-23243-0_8

8.1.3 Fibres from the Thalamus

The thalamic inputs come from the dorsal and ventral regions of the thalamus. They were identified in 1979 by OP Ottersen and Y Ben-Ari in cats and in 1980 by WR Mehler in monkeys and also the same year by M Norita and K Kawamura (1980) and by JP Aggleton et al., and then confirmed in 1990 by JE LeDoux et al. (1990b).

Dorsal inputs For JP Aggleton et al. (1980) most of the afferent projections come from the rostral midline thalamic nuclei when no projection is found from the medio-dorsal nucleus. In 1982 Y Ben-Ari and E Tremblay specify that inputs coming from the central portion of the thalamus (intralaminar nuclei, in particular from the parafascicular nucleus) and the midline nuclear complex go to the central nucleus of the amygdala designated in 1979 by OP Ottersen and Y Ben-Ari, as " The major termination area of the thalamo-amygdaloid projections". DX Zhang, EH Bertram in 2002, obtained similar results using electrical stimulation methods of the median thalamus, but the excitation of the amygdala is associated with activation of the entorhinal cortex. They conclude that the midline thalamic nuclei play "a significant role in limbic physiology and may serve to synchronize the activity in this system".

The thalamic nuclei which receive inputs from the spinothalamic tract (JE LeDoux 1987) project to the baso-medial nucleus (JE LeDoux et al. 1990a).

Inputs coming from the parvocellular part of the VPM, the ventral postero-medial nucleus (nucleus that transmits the visceral taste information), go to the lateral nucleus of the amygdala.

Inputs coming from the medial division of the medial geniculate nucleus, the posterior intralaminar nucleus or the posterior medial complex also go to the lateral nucleus of the amygdala (JM Edeline and NM Weinberger 1992). Thus, during fear conditioned experiments, auditory stimuli (coming from the medial division of the medial geniculate body) go to the lateral nucleus of the amygdala, while contextual stimuli from the hippocampus go to the baso-lateral and baso-medial nuclei (NS Canteras and LN Swanson 1992; JE LeDoux 2000).

Ventral inputs The projections from the ventral nuclei of the thalamus go to the cortical, medial, central and lateral nuclei of the amygdala.

8.1.4 Fibres from the Basal Part of the Telencephalon

They have also been the subject of much research, particularly in 1980, in the rat (OP Ottersen) and in the monkey (M Norita et al.; WR Melher; JP Aggleton et al.) and in 2000 (K Semba et al.). Thus, Aggleton et al. who practised in macaques *macaca mulatta*, fluorescent markings with the enzyme horseradish peroxidase, could highlight inputs coming from the BST and the nucleus of the horizontal part of the diagonal band and ending at the medial and central nuclei of the amygdala. They also observed inputs from the innominate substance reaching the basal nuclei of the amygdala: These afferent fibres belong to large cholinergic neurons, which corresponds to the area in which is found the basal nucleus of Meynert. Other inputs from the ventral pallidum project to the lateral and baso-lateral nuclei of the amygdala (OP Ottersen 1980).

8.1.5 Afferent Fibres from the Brainstem

As demonstrated by the works of WR Mehler (1980), M Norita and K Kawamura (1980), JP Aggleton et al. (1980), OP Ottersen and Y Ben-Ari (1979), OP Ottersen (1981) and J Ciriello et al. (1994), all brainstem neurons send axons to the central nucleus of the amygdala which they approach through the ventral pathway. This is the case of neurons in the periaqueductal grey matter, the pars compacta of the substantia nigra, the ventral tegmental area, nuclei of the medulla oblongata (locus coeruleus, dorsal raphe nucleus, lateral parabrachial nucleus, nuclei of the solitary tract, interpeduncular nucleus) and the ventrolateral nucleus of the spinal cord. According to J Hanaway et al. (2001), all these fibres borrow in their ascending path, the medial forebrain bundle and dorsal longitudinal fasciculus (see Figs. 7.21 and 8.1).

The lateral parabrachial nucleus provides the largest contingent of these afferent fibres. It is involved in the severe, disabling tinnitus which results from interactions between auditory neural sites (cochlear and vestibular) and non auditory neural sites (parabrachial nuclei). These nuclei project the aberrant auditory stimuli towards the central nucleus of the amygdala. This one transfers its emotional responses to the insula which transforms them into conscious emotional feelings. We can see on fMRI the activations of the amygdala and the insula during the severe tinnitus and the regression of these activations when tinnitus was treated with high-frequency acoustic treatment (ML Lenhardt et al. 2007).

This is also from the parabrachial area and directly from the spinal cord that the central nucleus of the amygdala can receive nociceptive inputs (R Burstein and S Potrebic 1993). Since as in the amygdala, the nucleus is the effector of the lateral and basal nuclei of the amygdala, the messages transmitted to it by these nuclei can be modified by nociceptive inputs it receives, which can generate nociceptive outputs to the target organs.

The fibres come from the dorsal raphe nucleus, and fibres of the locus coeruleus (which are noradrenergic fibres) require not only the ventral pathway but also the terminal stria. As for inputs coming from substantia nigra and the ventral tegmental area, they explain the abundant dopaminergic innervation of the central nucleus (Y Ben-Ari 1981).

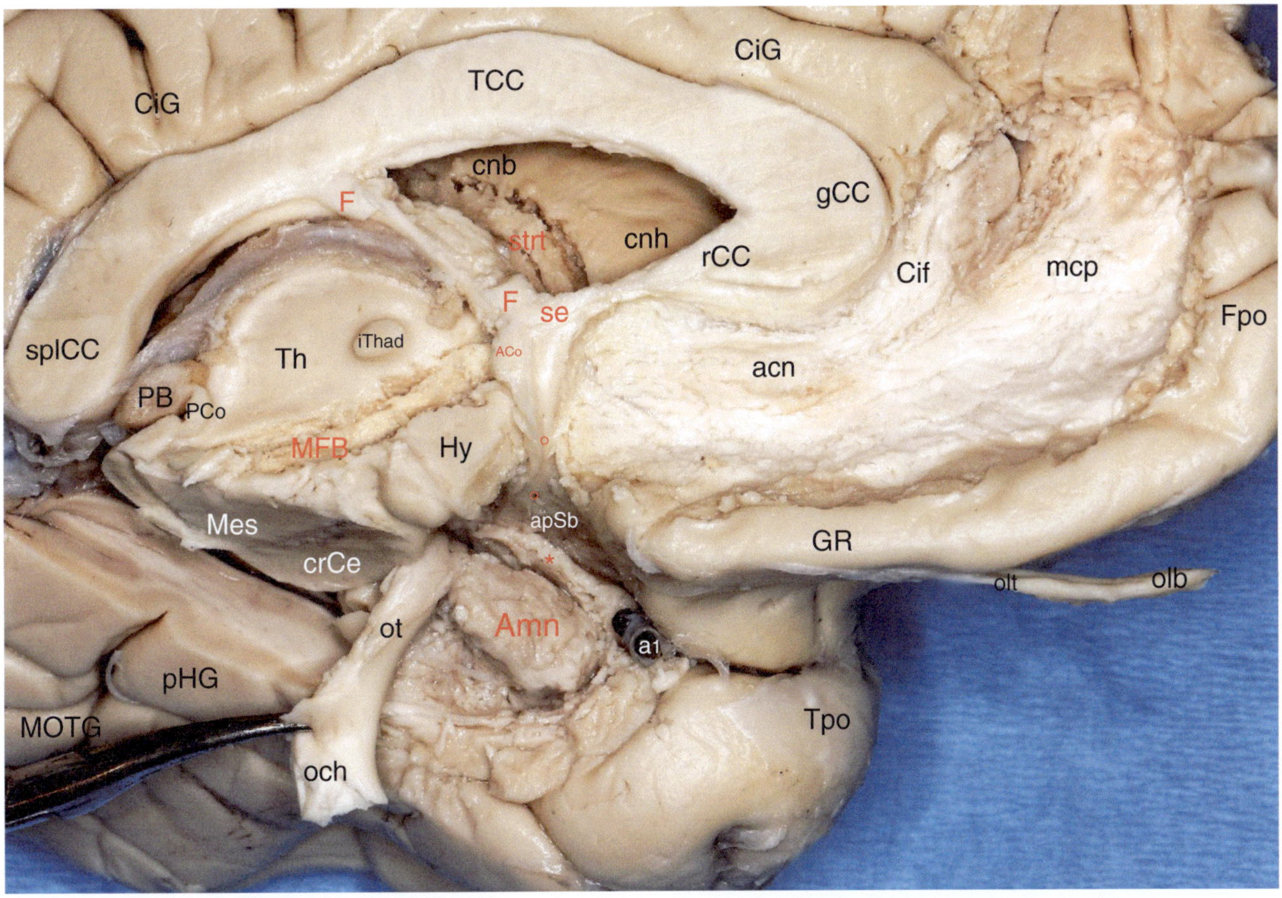

Fig. 8.1 The relations of Amn with the tracts and fasciculi converging to the septal area (components of the "mesolimbic system"). Medial dissection of the left hemisphere. *acn* accumbens nucleus, *ACo* anterior commissure, *Amn* amygdaloid nuclear complex, *apSb* anterior perforated substance, *Cif* cingulate fasciculus, *CiG* cingulate gyrus, *cnb* caudate nucleus body, *cnh* caudate nucleus head, *crCe* crus cerebri, *F* fornix, *Fpo* frontal pole, *gCC* genu of the corpus callosum, *GR* gyrus rectus, *Hy* hypothalamus, *iThad* interthalamic adhesion, *mcp* mesocortical pathway, *MFB* medial forebrain bundle, *MOTG* medial occipitotemporal gyrus, *och* optic chiasm, *olb* olfactory bulb, *olt* olfactory tract, *ot* optic tract (*moved down* and *backwards* by a forceps), *PB* pineal body, *PCo* posterior commissure, *pHG* parahippocampal gyrus, *se* septum, *splCC* splenium of the corpus callosum, *strt* stria terminalis, *Th* thalamus, *Tpo* temporal pole, *red asterisk* peri-amygdalar cortex (pAm cx), *little red circles* they locate the diagonal band, *a1* anterior cerebral artery. *Note*: *the oblique route of the MFB through the hypothalamus and the converging directions of the white fibre tracts to the septal area*

8.1.6 Hippocampal Inputs

The work of DL Rosene, GW Van Hoesen (1977) and RE Saunders et al. (1988) in rhesus monkeys has clarified that these inputs come from CA1 from the prosubiculum and essentially go to the baso-medial nucleus and the ventral cortical nucleus. The lateral and central nuclei also receive hippocampal inputs.

In 2000, the Finnish team of A Pitkänen et al., which has made anterograde and retrograde studies in rats, confirmed these data stating that the inputs to the amygdala come from the rostral half of the entorhinal cortex, from CA1, from the subiculum and from areas 35 and 36 of the perirhinal cortex. The nuclei that receive these inputs are the lateral, baso-lateral, baso-medial, central nuclei as well as the amygdalo-hippocampal area.

In 2008, M Höistad and H Barbas reviewed the temporal projections to the amygdala from the medial part of the temporal pole, the entorhinal and perirhinal areas, the granular and the dysgranular insula (the densest connections) and the agranular insula and the parahippocampal region (scattered connections (see for structure of the insula, Fig. 8.2)). They also confirmed that there are feedback connections between the entorhinal cortex, which projects from its deep layers to the amygdala and receives back projections in layers II–III, which ultimately target the hippocampus. This pathway could explain "how the amygdala can attach emotional value to environmental stimuli, participate in the sequence of information processing of emotions and modulate the formation of emotional memories".

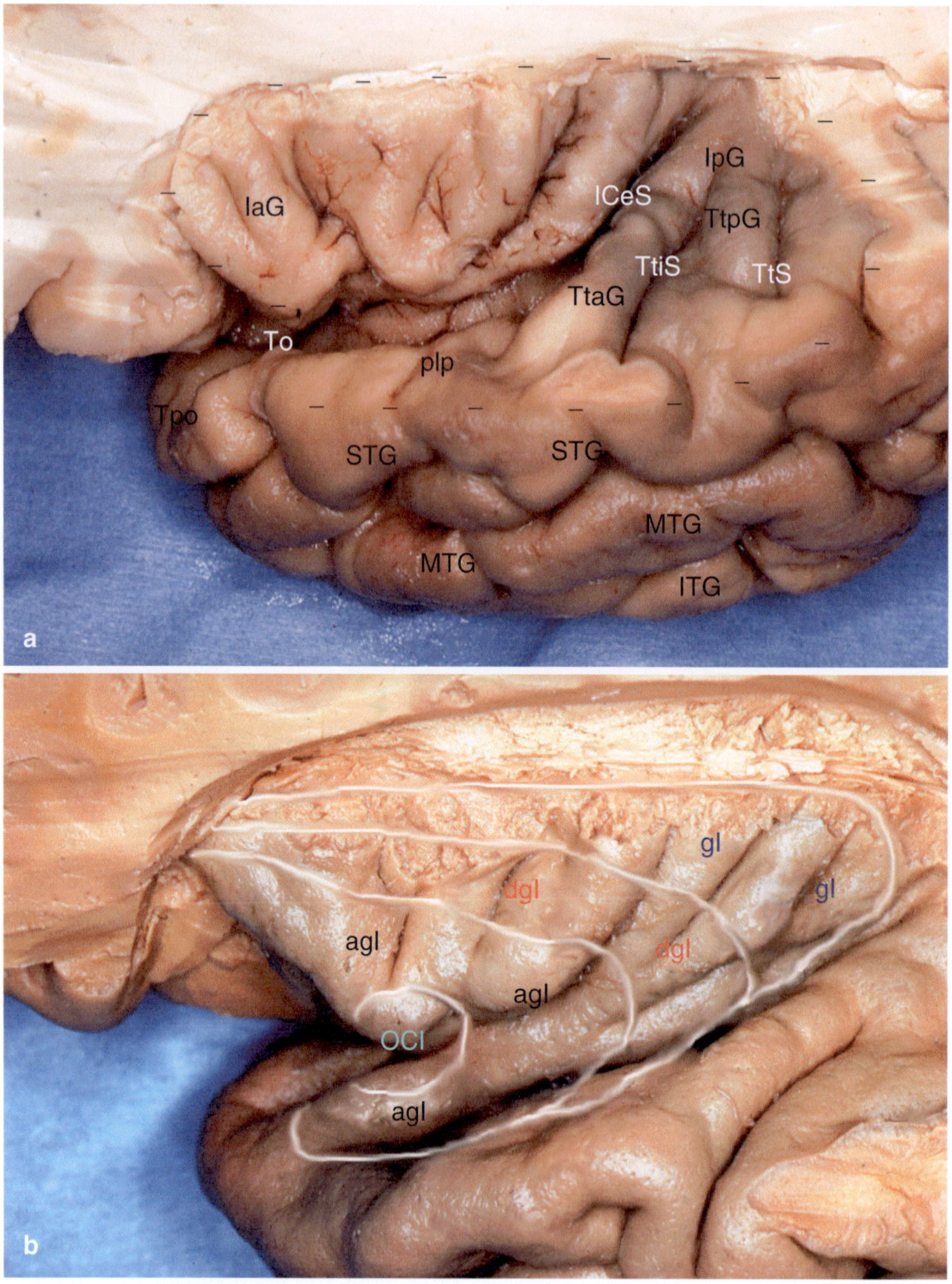

Fig. 8.2 An important relation of the amygdala: the insular cortex. (**a**) Supero-lateral view of the left human insula: descriptive anatomy. (**b**) Cortical structure of the insula according to MM Mesulam and EJ Mulson. *agI* agranular insula, *dgI* dysgranular insula, *gI* granular insula, *IaG* insular anterior gyri (gyri breves), *ICeS* insular central sulcus, *IpG* insular posterior gyri (gyri longi), *ITG* inferior temporal gyrus, *MTG* middle temporal gyrus, *OCI* primary olfactory cortex of insula, *plp* planum polare, *STG* superior temporal gyrus, *To* temporal opercule, *Tpo* temporal pole, *TtAG* temporal transverse anterior gyrus, *TtiS* temporal transverse intermedium sulcus, *TtPG* temporal transverse posterior gyrus, *TtS* temporal transverse sulcus. The *dashed line* follows the *ilcSI* (inferior limiting circular sulcus of insula of Reil)

8.1.7 Neocortical Inputs

They are particularly important and come from different associative cortical areas. They project for most to the baso-lateral nuclear group of the amygdala, which plays a major role for the receipt of inputs. GW Van Hoesen in 1981, L Stefanacci, DG Amaral in 2002 and JE LeDoux et al. (1990b) showed that the auditory inputs come from the anterior half of the superior temporal gyrus, that is to say, the rostral part of area 22 (extension of the lower section of the posterior long insular gyri or secondary acoustic area AII).

MJ Webster et al., in 1991 and JE LeDoux et al., in 1990a, have shown that inputs from the associative visual cortex come from the lower part of the temporal lobe, that is to say, areas 20 and 21.

We must also consider the very important role of the inferior longitudinal fasciculus (Fig. 7.21) which directly brings to amygdala the visual stimuli. RM Bauer, in 1982, reported the case of a patient whose cranial trauma had caused temporal haemorrhages responsible for the bilateral deterioration of inferior longitudinal fasciculus. This patient had lost any emotion aroused by a visual substratum and regretted having been brought to cancel a subscription of the magazine Playboy whose diligent reader he was before his accident and for which he had no more, from now on, any interest.

Note that we have seen in terms of macroscopic anatomy the bundles of fibres that connect the amygdala to the rostral part of each temporal gyrus, conducting resections of the successive gyri from top to bottom of the temporal lobe.

AG Herzog, GW Van Hoesen (1976) and then BH Turner et al. in 1980 and L Stefanacci and DG Amaral in 2000, showed showed that inputs from the multisensory association areas come from the rostral portion of the perirhinal cortex (areas 35 and 36), the cortex forming the dorsal bank of the superior temporal sulcus, the entorhinal cortex, the superior temporal gyrus, the parahippocampal cortex and the temporopolar area (area 38). Let us recall again that the macroscopic study showed us the obvious and important connections between the amygdala and the temporal pole (see Fig. 5.2).

EJ Mufson et al. (1981) studied inputs involved in gustatory and visceral functions: They come from anterior insular areas agranular areas (Fig. 8.2) that themselves receive impulses from the ventral postero-superior and ventral postero-inferior thalamic nuclei and from the parvicellular division of the ventral postero-medial nucleus of the thalamus. The agranular and dysgranular insular areas project to the lateral nucleus of the amygdala (L Stefanacci and DG Amaral 2000).

MM Mesulam, EJ Mufson in 1985; L Stefanacci, DG Amaral in 2002, studied the posterior part of the insula (Fig. 8.2): it contains a somatosensory association area (granular area), postero-superior, which is connected with the adjacent second sensory area (SmII) (located on the upper lip of the posterior segment of the lateral sulcus) which is involved in the process of nociceptive information.

In a more recent study (2008), M Höistad and H Barbas showed that the medial temporal pole, the entorhinal and perirhinal areas and the agranular and dysgranular parts of the insula have the densest connections with the amygdala, while the lateral temporal pole, the granular insula and the parahippocampal region have sparser connections.

All these inputs coming from the temporal regions (anterior superior temporal, anterior middle temporal and inferior temporal gyri, medial and lateral parts of the temporal pole) and from the insula go mainly to the lateral nucleus of the amygdala and little or not all to other nuclei (L Stefanacci and DG Amaral 2000). The temporal pole is the only one to also target the baso-medial nucleus (AG Herzog and GW Van Hoesen 1976).

We also know now thanks to experiments on conditioned fear (JE LeDoux 2000) that the cortex is not always involved in the immediate analysis of stimuli and the thalamus (receiver of the stimulus to the "medial division of the medial geniculate body") can directly transfer the stimulus (auditory case) to the lateral nucleus of the amygdala. This has been demonstrated by JS Morris et al. in 1999 by fMRI studies on humans: These authors have observed that during conditioning, amygdala activity changes are correlated with those of the thalamus as is not the case with the cortex which proves the importance of the direct thalamo-amygdaloid trajectory and the direct effector possibilities of the amygdala, in extreme emergency, without cortical modulation! But even in these cases, the stimulus is going to reach the brain analyser controller, some milliseconds later!

Inputs to the amygdala from the frontal lobe are of extreme importance. They are numerous and include the posterior orbitofrontal areas, areas 13 and 14; the medial prefrontal areas, areas 11 and 12 (HT Ghashghaei and H Barbas[1] 2002; H Barbas 2007); the anterior cingulate area, area 24 (DN Pandya et al. 1973); the prelimbic area, area 32; and the infra-limbic area, area 25. All these inputs from the frontal lobe target the baso-lateral and the baso-medial nuclei of the amygdala. Numerous authors have devoted their research to these inputs, identifying a number of details:

GR Leichnetz, J Astruc (1977) showed that the prefrontal cortex not only projects to the amygdala but also to the substantia innominata and the pallidum.

L Stefanacci and DG Amaral (2000) found found that the orbitofrontal cortex; the prefrontal medial cortex, areas 10, 11, 12, 13, 13a and 14; and the anterior cingulate cortex, areas 25, 24 and 32, mainly target the baso-lateral and baso-

[1]HT Ghashghaei et al. (2007) also observed projections towards the amygdala, from the lateral prefrontal cortex, but they are much more rare and scattered than those from the orbitofrontal and medial frontal areas. Moreover, these projections go preferentially to the baso-lateral nucleus of the amygdala.

medial nuclei and the orbitofrontal cortex and prefrontal medial cortex (but not the anterior cingulated cortex) and also project to the cortical and medial nuclei as well as to the periamygdaloid cortex.

NL Rempel-Clower, in 2007, observed in rats, as H Barbas did the same year in the monkey, that the orbitofrontal cortex also projects to the intercalated masses of the amygdala that are GABAergic and thus exerts inhibitory influences within the amygdala. H Barbas was able to clarify that, in fact, the orbitofrontal posterior cortex projects to the amygdala, through 2 "pathways" that have opposite effects on the central autonomous structures: The first path is the one mentioned above, which leads from the cortex to the intercalated mass and whose activation can ultimately disinhibit central autonomic structures during emotional arousal. The second system innervates the central nucleus of the amygdala that has inhibitory projections to the hypothalamus and autonomous structures of the brainstem and can cause inhibition of central autonomous structures, resulting in an autonomous homeostasis[1]. HT Ghashghaei and H Barbas in 2002 showed that the lateral prefrontal cortex "which has executive functions" specifically projects to the layer 5 of the orbitofrontal cortex. In 2007, the same authors found that the orbitofrontal neurons sending their axons to the amygdala are from this cortical layer 5. They so demonstrate that also the lateral prefrontal cortex has a controlling role over the inner workings of the amygdala (especially on the posterior half of the amygdala, the intermediate sector of the baso-lateral nucleus and the magnocellular portion of the baso-medial nucleus). It was also in 2007 that M Medalla et al. were able to demonstrate that at the level of the prefrontal cortex, there are two specialised classes of inhibitor neurons, belonging respectively to each of the two pathways mentioned above and expressing the duality and thus the complexity of inhibitory control: the area of inhibitory neurons 32, dedicated to emotional communication, and those in area 10 (see mapping Fig. 7.8), dedicated to working memory functions.

8.2 Outputs

The efferent fibres are not really modelled on inputs even if some of them use the same pathways as these.

We can classify these efferent fibres according to the parts of the brain to which the amygdaloid nuclei will send their outputs. There are also amygdaloid diencephalic projections to the hypothalamus and thalamus, projections to the brainstem, projections to the basal brain, projections to the hippocampus and projections to the neocortex.

Many of these projections borrow the bundles that were studied in terms of connections, and it is from such projections we will begin the study of efferent fibres.

8.2.1 Efferent Fibres Passing via the Stria Terminalis[2] (Dorsal Pathway)

1. Precommissural fibres of the stria have multiple destinations. They lead on the one hand to the septal and paraseptal nuclei (septal nuclei, diagonal band nucleus, nucleus accumbens) and to the rostral nuclei of the hypothalamus (preoptic and anterior). It is recognized that this contingent in the stria originates from the cortico-medial nuclear group of the amygdala.
2. Commissural fibres allow communication between two stria terminalis and thus between the nuclei of the two amygdalae. As we have seen in dissections, these fibres are positioned exactly in the centre of the anterior commissure (see Fig. 6.7). Before entering the commissure, a part of the contingent reaches the hypothalamic paraventricular, ventromedial and premammillary nuclei.
3. Post-commissural fibres terminate at the BST and the posterior hypothalamus: The fibres that go to the lateral BST come from the basal and central nuclei of the amygdala. The fibres that go to the medial BST come from the medial and cortical nuclei.

   A dorsal contingent of the stria may be attached to post-commissural fibres. It is the thalamic component of the stria which is detached from its dorsal concavity and penetrates the back of the thalamus to reach the mediodorsal nucleus of this formation. We were able to follow this through dissection, even within the thalamus (see Fig. 6.3). According to JP Aggleton, M Miskin 1984, only a small number of amygdalo-thalamic fibres progress into the terminal stria. We cannot agree to the term "small number" because the stria terminalis contingent destined for the thalamus has always seemed to be of a large diameter (see Fig. 6.3).
4. The fibres for the striatum use mostly the stria terminalis or its side branches to reach their targets. We have already seen the precommissural branch for the ventral striatum (nucleus accumbens, olfactory tubercle), considered to be the "limbic striatum compartment". But the stria terminalis also carries axons for the dorsal striatum (caudate and putamen nucleus), "non-limbic compartment" and yet essential. As Y Ben-Ari and E Tremblay (1982) recall since, it provides the amygdala, the possibility of inputs to the motor system and thus of having an effect on the latter. The terminal stria dissection allows to observe some branches for the caudate nucleus, fine for the most part: They touch the edge of this formation enabling axons from the amygdala to

[2]The accumbens nucleus receives not only inputs from the amygdala via the stria terminalis but also from hippocampus, via the fimbria (AE Kelley et al. 1982).

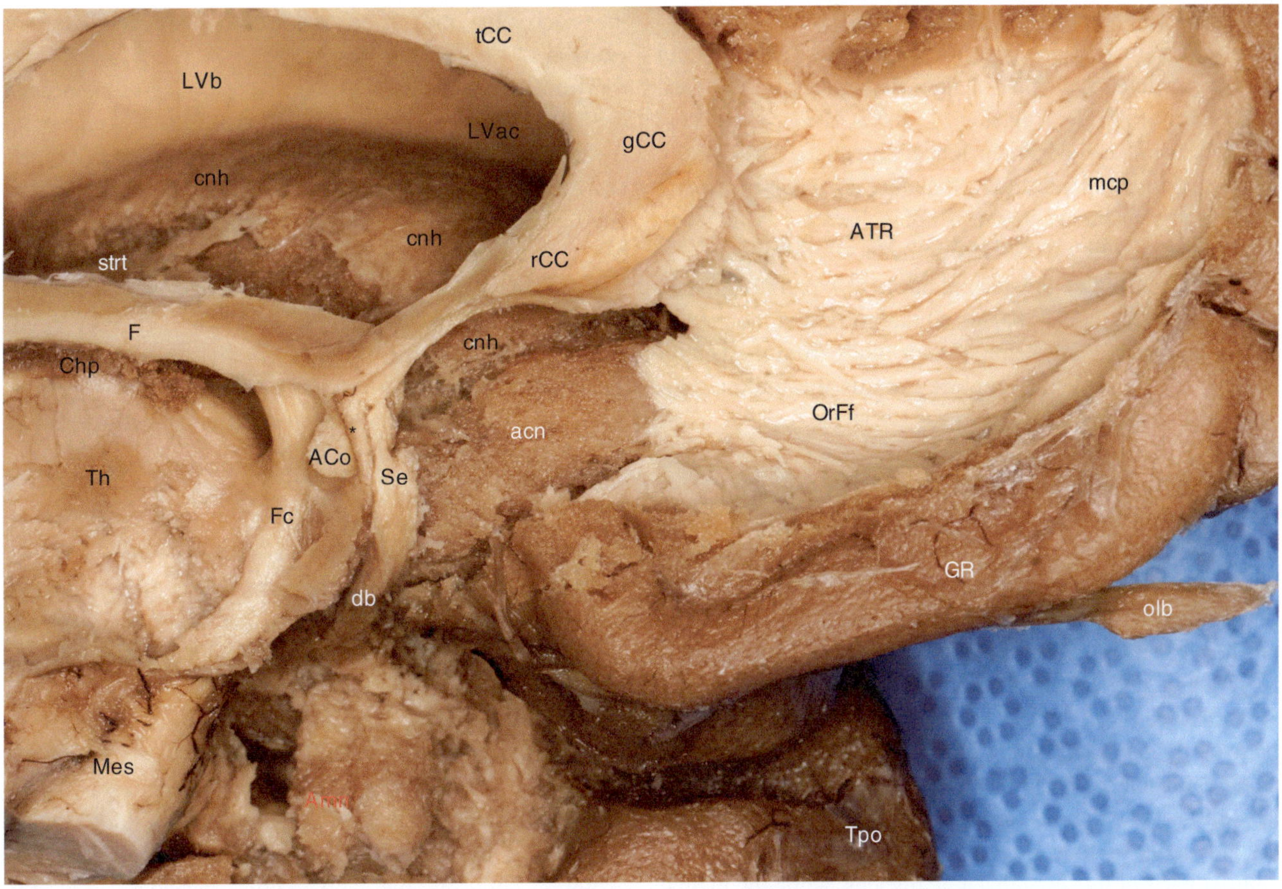

Fig. 8.3 Relationships of the amygdala with the nucleus accumbens septi (dissection of the medial side of the left brain hemisphere). *ACo* anterior commissure, *acn* accumbens nucleus, *Amn* amygdaloid nuclear complex, *ATR* anterior thalamic radiations, *Chpl* choroid plexus, *cnb* caudate nucleus body, *cnh* caudate nucleus head, *db* diagonal band, *F* fornix, *Fc* column of fornix, *gCC* genu of the corpus callosum, *GR* gyrus rectus, *LVac* lateral ventricle, anterior cornu, *LVb* lateral ventricle body, *mcp* mesocortical pathway of MacLean (limbic forebrain–midbrain continuum of Nieuwenhuys), *Mes* mesencephalon, *olb* olfactory bulb, *OrFf* orbitofrontal fibres, *rCC* rostrum of the corpus callosum, *Se* septum, *strt* stria terminalis (note the branch to the caudate nucleus), *tCC* truncus of the corpus callosum, *Th* thalamus, *Tpo* temporal pole, *black asterisk* prehippocampal rudiment

reach their striatal target (Fig. 8.3). Specifically the dorsal striatum also receives projections of the anterior baso-lateral nucleus that do not pass through the stria terminalis (GD Petrovich 2001). L Fudge et al. (2002) have shown that fibres which are all taken from the ventral striatum come from neurons of the baso-lateral and baso-medial nuclei of the amygdala. According to FT Russchen et al. (1985b), there is a specificity of origin of the fibres: fibres destined for the nucleus accumbens[2] are from the parvicellular part of the basal nuclear group and from the amygdalo-hippocampal area, while projections for the tail of the caudate nucleus come only from the magnocellular part of the basal nuclear group.

We were also able to demonstrate by dissection the connection between the terminal stria and the medullary stria of the thalamus (or stria medullaris). We see perfectly a direct posterior branch detaching from the stria terminalis prior to its development to continue directly with the medullary stria that will rejoin the habenula (see Fig. 6.5), the starting point of the habenulo-interpeduncular tract.

8.2.2 Efferent Fibres Passing Through the Ventral Amygdalofugal Pathway

They are created by the baso-lateral, baso-medial and central nuclei of the amygdala. They immediately use the ventral pathway and then target the structures of various regions: the septal region (particularly to the diagonal band nucleus), the sublenticular region (notably to the basal nucleus of Meynert), the BST, the thalamus and the lateral hypothalamic area.

8.2.2.1 The Amygdalo-thalamic Fibres

Amygdaloid projections go to the medio-dorsal nucleus and to the nucleus reuniens of the thalamus (Y Ben-Ari, E Tremblay 1982; JP Aggleton M Miskin 1984). Those destined for the medio-dorsal magnocellular nucleus essentially use the inferior peduncle of the thalamus and enter the rostral head of the thalamus before bending backwards to reach their target (see Fig. 6.6). They come mostly from the basal group (but almost all amygdaloid nuclei also participate) and terminate in the medial magnocellular portion of the medio-dorsal nucleus (JP Aggleton). Other fibres from the central nucleus of the amygdala project towards the midline thalamic nuclei (JL Price, DG Amaral 1981).

8.2.2.2 The Amygdalo-hypothalamic Fibres

In addition to the fibres connecting directly the amygdala and hypothalamus, which we mentioned earlier, there are many projections from the amygdala to the hypothalamus: the paraventricular nucleus, the dorso-medial nucleus, the perifornical region, the supramammillary and paramammillary nuclei and the lateral hypothalamus (JL Price, DG Amaral 1981). Among them there are the projections to the paraventricular nucleus that have been particularly studied using neuroscience techniques (TS Gray et al. 1989) in monkeys. It is clear from this research that the central nucleus of the amygdala supplies the caudal, lateral and medial parvocellular parts of this hypothalamic nucleus, while the medial nucleus of the amygdala supplies the rostral parvocellular sections of the same nucleus. The authors believe that these anatomical differences explain the amygdaloid modulation of neuroendocrine response to stress factors.

Other fibres coming from the basal nuclei and from the central nucleus of the amygdala will rejoin the medial hypothalamus and the lateral hypothalamic area (WJH Nauta, W Haymaker 1969). These projections show how the amygdala can act on hypothalamic pathways managing specific behaviours (nutritional, sexual, aggressive and defensive).

8.2.2.3 The Amygdalo-basal Fibres And Amygdala-BST Fibres

Amygdala-BST fibres are mainly distributed in three target organs: the diagonal band nucleus, the basal nucleus of Meynert and the BST. On the other hand, a study by FT Russchen et al. in 1985b, in monkeys, using autoradiographic tracing, found few amygdaloid axons reaching the innominate substance and the ventral pallidum. The origin of these amygdaloid fibres was the parvicellular basal nucleus, the caudal section of the magnocellular basal nucleus, the baso-medial magnocellular nucleus and the central nucleus. A new study by YT Cho et al., conducted in 2013, administering a tracer by two-way injection in the amygdala, has not only found a noticeably sharper projections to the striatum but also to the prefrontal and insular cortex, demonstrating at the same time the existence of cortico-amygdalo-striatal pathways.

We shall see successively the amygdalo-diagonal fibres and the fibres of the cortico-amygdalo-striatal pathways, reserving the study of the fibres that project to the nucleus of Meynert and the fibres that connect the amygdala and the BST in the chapter on the extended amygdala because these fibres either belong to the ventral amygdalofugal pathway and in the chapter cited above. Let us recall that in 1981, the links between the basal amygdaloid nuclei and the amygdala-BST were examined by JL Price and DG Amaral in their remarkable study using an autoradiography method for tracing axonal projections (In their publication, one can see images surimposed on whose which we present in the Chap. 11).

- The amygdalo-diagonal fibres connect the amygdala to the diagonal band nucleus. These are direct fibres that travel along the so-called horizontal part (actually more or less oblique) of the band to the lower septal region where the discrete thickening sits which corresponds to the nucleus of this band. It should be known that there is a real bond between the infero-medial pole of the amygdala and the lateral area of inflection of the diagonal band at or immediately after the latter has “received” its contingent coming from the limen insulae. This bond is more than just an apposition since there is an exchange of fibres between the amygdaloid periphery and the band’s fibres. Furthermore, the real constitution of the magnocellular diagonal band is complex (see Fig. 7.5) with not only a large part coming from the septum and forming the posterior limit of the anterior perforated material but also two other narrower portions, one coming from the limen and rejoining the band almost at right angles and the other continuing the initial trajectory of the band, curving slightly following a parallel path and overlying the optic tract and ending in the pulvinar, just above the lateral geniculate body (see Figs. 7.2, 7.3, 7.4, and 7.5). Such a constitution suggests amygdala projections which are not yet known towards the limen and probably the insula on the one hand and towards the posterior thalamus, on the other hand.
- The fibres of the cortico-amygdalo-striatal pathways

  They are better known thanks to the work of YT Cho et al. (2013). They include projections from various prefrontal cortical insular regions towards the baso-lateral and baso-medial nuclei of the amygdala. From these nuclei other neurons target the entire striatum “area extending well beyond the classic ventral striatum”. Depending on the degree of differentiation of the cortical layer from which the fibres reach the amygdala, the authors distinguish three main pathways to the striatum:
  - A primitive pathway from the least differentiated cortical regions (agranular insula, areas 25 and 32 of the

medial prefrontal cortex) and reaching through all nuclei of the amygdala, the ventral striatum.
- An intermediate pathway coming from more differentiated insular regions (dysgranular and granular) and areas 24 and 14 of the medial prefrontal cortex, projecting to the rostro-medial and centro-medial areas of the parvocellular portion of the baso-lateral nucleus, the magnocellular portion of the baso-medial nucleus and the magnocellular portion of the baso-lateral nucleus. From these nuclei, efferents pass beyond the classic ventral striatum to the caudo-ventral putamen and the ventromedial caudate nucleus.
- A developed pathway, originating in the most differentiated cortex (the entire orbitofrontal cortex and the most developed segment of the medial medio-frontal cortex, medial area 10) only projecting to the magnocellular regions of the baso-lateral and baso-medial nuclei. From these nuclei the efferents project to the caudal part, the dorsolateral body, the knee and the tail of the caudate nucleus. The authors state that they always found this pathway in conjunction with the other two pathways.

8.2.3 Fibres Projecting to the Brainstem

Many of them belong in their initial part of their journey to the ventral amygdalofugal pathway which they use at the exit from the amygdala before changing direction to reach the various levels of the brainstem and spinal cord (N Mizuno et al. 1985). They are an important contingent of which the neuronal cell bodies are in the central nucleus of the amygdala (JL Price and DG Amaral 1981). Thus, the central nucleus is, as has been documented by Y Ben-Ari and E Tremblay (1982), "the source of a massive long downward projection system which, through the ventral amygdalofugal system, bypasses the hypothalamus to directly innervate autonomic and other centres". The axons pass through the lateral hypothalamus and then reach the brainstem. Many of them join the medial forebrain bundle and the dorsal longitudinal fasciculus before reaching the brainstem (see Figs. 7.19 and 8.1).

The projection nuclei of these fibres (JG Veening et al. 1984; JG Price and DG Amaral 1981; AL Jongen-Relo and DG Amaral 1998) are:

At the level of the midbrain: the nucleus peripeduncular, the substantia nigra (pars compacta), the ventral tegmental area and the periaqueductal grey (or midbrain central grey)

At the pons: the cuneiform nucleus: the upper central nucleus, the raphe magnus nucleus, the locus coeruleus, the subcoeruleus nucleus, the parabrachial nuclei and the rhombencephalic reticular formation

At the level of the medulla oblongata: the pallidus and obscurus raphe nuclei, the medial and lateral parabrachial nuclei (MM Moga and TS Gray 1985), the nucleus ambiguus and the vagal complex (nucleus of the solitary tract and dorsal motor nucleus of the vagus nerve) studied in 1980 by JS Schwaber et al., the cuneiform nucleus and the rhombencephalic reticular formation

Most neurons in the central nucleus of the amygdala are projected not only to the parabrachial nucleus, the dorsal vagal complex (JG Veening et al. 1984; OA Lyubashina 2004) which give rise to parasympathetic cranial protoneurons and the nucleus ambiguus (AL Jongen-Relo, DG Amaral 1998) but also to the locus coeruleus and Barrington's nucleus that will be relays on the amygdalo-spinal pathway. This later joins the spinal ventral horn of the spinal cord's sacral metameres which will produce the protoneurons of the parasympathetic sacral chain. These then leave the spinal cord to progress to the ventral spinal nerve roots from S2 to S4 and then to the erector nerves that emerge from these roots.

In 1985, MM Moga and TS Gray had shown by the combination of fluorescent retrograde tracing and immunofluorescence that the lateral part of the central nucleus of the amygdala is a major source of the corticotropin-releasing factor, neurotensin and somatostatin, while neurons of the medial segment only contribute to the bundle which connects the central nucleus of the amygdala to the parabrachial nucleus.

In 1984, JG Veening et al. also researched, using a fluorescent retrograde tracer and by means of immunohistochemistry, the presence of peptides in cell bodies and the axons of these projections. They observed that projecting neurons do not contain neuropeptides and predominate in the medial nucleus of the amygdala, while peptide-containing perikarya (neurotensin, somatostatin and corticotropin-releasing factor) predominate in the lateral segment of the central nucleus of the amygdala.

In 1985, MM Moga and TS Gray obtained somewhat different results for the combination of fluorescent retrograde tracing and immunofluorescence: the lateral segment of the central nucleus of the amygdala is a major source of corticotropin-releasing factor, neurotensin and somatostatin ("neurotransmitters that very likely guarantee regulation of the parabrachial nucleus"). But it is more, according to these authors, the origin of the amygdalo-parabrachial bundle and the neurons of the medial segment of the nucleus only contributing to this bundle. The authors conclude that the bundle of fibres that go from the central nucleus of the amygdala to the parabrachial nucleus of the bridge is peptidergic and can contribute to respiratory or cardiovascular responses related to stress.

AL Jongen-Relo and DG Amaral, for their part, used retrograde tracing to show that a large number of neurons in the amygdala's central nucleus are GABAergic: half of the neurons in the medial part and 30 % of the neurons in the lateral part.

Through the reticular structure nuclei on which they project and which we have cited (see Chap. 7), the neurons from the central nucleus of the amygdala are connected to a large variety of visceral nuclei independent of the brainstem (AL Jongen-Relo and DG Amaral) themselves connected to the spinal cord through the brainstem's major pathways. The various components of the vegetative system, orthosympathetic (sympathetic) or parasympathetic, may also be activated one by one, by the amygdala, depending upon the case and the context.

The parasympathetic cranial system is then implicated by the axons that reach the lateral and medial parabrachial nuclei. These then project to the parasympathetic nuclei connected to the cranial nerves and in particular to the dorsal vagal complex, which is the main source of the parasympathetic pathways.

For the sympathetic system, in the majority of cases, the axons coming from amygdala and the hypothalamus follow, like the previous ones, the dorsal longitudinal fasciculus and the medial forebrain bundle, but they extend beyond the brainstem and regroup in a hypothalamo-spinal tractus. They also reach the spinal cord. It is in its intermedio-lateral zone (between the C8 and L2 metameres) where the bineuronal orthosympathetic pathway with adrenergic nerve endings is born.

Recall that the hypothalamo-spinal tract also contains noradrenergic axons from the parabrachial structure, the raphe magnus nucleus (J Hanaway et al. 2001), the locus coeruleus and groups A1, A2, C1 and C2 of the medulla oblongata, all noradrenergic. These axons go directly towards their target organs, including the adrenal medulla that will in turn secrete catecholamines: adrenaline and noradrenaline.

Of the amygdala fibres that send their axons to the brainstem, we must also discuss those that comprise the amygdalotegmental pathway (see Chap. 7).

8.2.4 Output to the Cortex

8.2.4.1 Projections to the Hippocampus

Unlike input neurons that reach the amygdala directly from the hippocampus and in particular the subiculum, as we have seen, the direct projections from the amygdala to the hippocampus are not the only ones. A large number of these amygdalar projections from the lateral and medial basal nuclei go through the entorhinal cortex (R Insausti et al. 1987), in which the large pyramidal cells in turn project to the hippocampus. So we can concur with R Nieuwenhuys et al.: "The entorhinal cortex is the principal source of cortical input to the hippocampal formation" as "The subiculum is the principal source of hippocampal output".

The other projections are made, either from the same nuclei but with their targets being the CA1 and CA3 fields of the hippocampus or from the baso-lateral nucleus if the target is the subiculum (JG Amaral and WM Cowan 1980).

A Pitkänen et al., in 2000, showed that, in rats, the primary projections from the amygdala to the hippocampus originate in the lateral, baso-lateral, mediobasal nuclei and in the posterior part of the cortical nucleus. These pathways terminate "in the rostral half of the entorhinal cortex, the temporal end of the CA3 and CA1 subfields or the subiculum, the parasubiculum, areas 35 and 36 of the perirhinal cortex, and the postrhinal cortex".

8.2.4.2 Projections to the Neocortex

Given their importance, many teams have studied these projections as a result of the initial work by WJH Nauta in 1961.

The baso-lateral and baso-medial nuclei project towards all the cortical areas of vision, the temporal (located in the lower part of the temporal lobe) and occipital (located on each side of the calcarine sulcus), but there is an occipito-temporal gradient (E Iwai and M Yukie 1987), in the density of projections with a reduction in these projections from the most caudal areas (V1 occipital visual area) to the most rostral area (anterior temporal visual area: TEa).

The baso-lateral and baso-medial nuclei project similarly to the anterior half of the superior temporal gyrus (anterior part of area 22) corresponding to the auditory association cortex (DG Amaral and JL Price 1984).

The baso-lateral and baso-medial nuclei also project to the multisensory association areas: the temporopolar cortex (area 38), the perirhinal cortex (areas 35 and 36) and the cortex of the dorsal bank of the superior temporal sulcus.

The amygdala projects to the insula not only from the lateral basal and medial basal nuclei, like previously, but also from the cortical and medial nuclei (EJ Mufson et al. 1981). The most significant projections go to the agranular cortex. These projections gradually decrease when we examine more caudal areas, dysgranular and granular insular areas (DG Amaral and JL Price 1984). It should also be noted the "paraolfactory" projections from the periamygdaloid cortex to the medial, postero-medial and intermediary agranular insular cortices (R Nieuwenhuys et al. 2008).

The projections towards the frontal lobe have been and are still the most studied. It is now established that the amygdala projects towards the ipsilateral frontal pole, more precisely towards the anterior cingulate cortex, the lateral prefrontal cortex and the posterior orbitofrontal cortex, each of these cortical areas having specific connections (HT Ghashghaei and H Barbas 2002). In 2007 these same authors, with the help of neural tracers surgically injected in vivo into monkeys, in the basal nuclei of one of their amygdalae, were able to demonstrate by microscopic analysis of the density of the axonal endings marked by the tracer that projections predominate (1–6 million points per mm^3) at the caudal orbitofrontal cortex level and the caudal part of the medial prefrontal cortex (anterior cingulate area) but that all the prefrontal cortices contain projections proving their connections with the amygdala. They could also draw from their observations some rules about the laminar specificity of axonal endings at the cortical level: when the target cortex has fewer layers or if the projections are of lesser density, the axonal endings are made in the deeper layers (layers 5–6). When the target cortex has more layers or if the cellular density is greater, the projects are made on the superficial layers (especially layer 1).

But projections to the frontal lobe are not unidirectional and the division of roles is now well known: It is the amygdala that transmits the emotional response and the emotional significance of stimuli to the prefrontal cortex. The orbitofrontal cortex and the cingulate cortex receive projections from all of the limbic cortical and subcortical structures (H Barbas 2000) and will relay the information received to the amygdala. They also send more projections to the amygdala than they receive (HT Gashghaei et al. 2007).

In 2009, S. Yuasa was able to clarify the interconnections between the amygdala, the cortex and the face's emotional expression, which controls a large number of behavioural relationships between individuals, as noted by J.S. Morris et al., in 1996. S. Yuasa was able, in fact, to demonstrate in monkeys that the motor cingulated cortex M3, to which projects the amygdala's baso-lateral nucleus, itself somatotopically projects to the facial nucleus in the brainstem. The author suggests, "The system might be involved in the pathophysiology of temporal lobe epilepsy and the facial expression of patients with Parkinson disease".

It is also interesting to observe the role of the neocortex in the loops in which the amygdala is somehow the alpha and omega. An excellent example is provided by the limbic loop[3] conceived by by R Nieuwenhuys et al. This loop that occurs particularly in the reward circuit is comprised of several brain structures that will be successively implicated and that will ultimately participate in the amygdala's response to an initial stimulus. The baso-lateral and baso-medial nuclei of the amygdala that have just received one or more stimuli send them to the prefrontal cortex in which the projections will be made to the ventral striatum (in particular accumbens nucleus) and then from there to the ventral tegmental area that will directly flood the cortex and the accumbens with dopamine, which initiates the return part of the loop. This ventral striatum projects to the ventral pallidum from which the return to the amygdala occurs, through the magnocellular part of the medial dorsal nucleus of the thalamus and then through the prefrontal cortex. The last projection goes from this cortex to the amygdala's lateral nucleus (receptor nucleus). The amygdala may then induce appropriate behavioural responses.

[3]There are many cortical loops but none of them are as important as the amygdala: This is the case with the punishment pathway, called the PVS, periventricular system, because it involves the thalamus and the hypothalamus (plus the midbrain PAG). It is also the case with the action inhibitor pathway, called the behavioural inhibitory system (BIS), in which the noradrenergic fibres, born from the locus coeruleus, and serotoninergic fibres, born from the median raphe nuclei, are involved.

8.3 Overview of the Amygdala's Physiology

From the previous study of inputs, we can conclude that the amygdala is a strategic, "central" brain structure, receiving all internal and external stimuli successively or concomitantly, directly or indirectly, and permanently supported with information from other brain structures and from those of the brainstem for optimal evaluation of these stimuli (in particular, the distinction between favourable stimuli and unfavourable stress inducers).

From the study of outputs, we can get a glimpse of the physiological areas in which the amygdala intervenes to react and respond to stimuli it receives and for which we provide a brief overview, below.

8.3.1 Actions on Hormonal Secretions

These secretions are produced by the endocrine glands and particularly the adrenal gland (secretion of cortisol and corticosterone) during stressful emotions. They are controlled by the anterior pituitary gland that sends specific pituitary hormones to the other glands (thus, the pituitary hormone adrenocorticotropic or ACTH acts on the cortical part of the adrenal gland, which causes it to secrete cortisol and corticosterone).

The pituitary itself is controlled by the hypothalamus, the actual conductor of the endocrine glands, that sends it, by a capillary system (the hypophyseal portal system), its secretions, the releasing factors that will trigger the secretion of pituitary hormones. Thus, ACTH secretion by the pituitary is triggered by the hypothalamic corticotropin-releasing factor. The amygdala intervenes with the hypothalamus and therefore manages, through it, the hormonal impregnation of the body. As we have seen, the amygdalar neurons can reach the hypothalamus by different routes: They can either take the very direct route through the amygdalo-hypothalamic pathway or take the stria terminalis and relay in the BST. From the BST, a second neuron connects the hypothalamus and, in particular in the chosen example, the paraventricular nucleus. The stimulating effect of the amygdala on the cells producing the corticotropin-releasing factor will trigger the flow of adrenal secretion whose consequences are well known. Note that all the other limbic structures projecting to the hypothalamus play a complimentary role to that of the amygdala, either to strengthen the neural structures and pathways containing the glutamatergic excitatory neurons or to inhibit the neural structures and pathways containing GABAergic neuron inhibitors.

Like that the hippocampus, being very close and very connected to the amygdala, will have opposite effects on the periventricular nucleus of the hypothalamus and would have an inhibitory effect on it (JP Herman and WE Cullinan 1997).

8.3.2 Impacts on the Autonomic Nervous System

It is directly through these axons that reach the brainstem nuclei and indirectly through the BST relay, whose targets are similar, that the amygdala performs these actions. These are manifested by abrupt bodily changes interpreted by individuals as emotion itself, whereas they are only a repercussion. The amygdala is also indirectly involved in the brainstem nuclei through the hypothalamus, which is also connected to them.

During emotions, the amygdaloid control system may then produce sympathetic or parasympathetic effects. We know, for example, the sympathetic manifestations that can occur with great fear: a rapid pulse, an increase in blood pressure, a dry mouth, mydriasis, paleness, sweating, rising hairs, etc.

Similarly, hyperactivity of the amygdala in some subjects suffering from increased sensitivity and who are very emotional may cause gastrointestinal functional disorders which can range from colic crises, bloating and gas to actual chronic diseases such as irritable bowel syndrome (IBS). A wonderful study by B Myers et al., in 2009, perfectly demonstrated the relationship of cause and effect between hyperactivity of the amygdala in anxious people and this pathology: the central nucleus of the amygdala exerts its action on the various structures to which it is directly connected and which act not only on anxiety but also on the implementation of the vegetative nervous system on gastrointestinal motility (hypothalamus, locus coeruleus and Barrington's nucleus, nuclear complex of the vagus nerve, BST) or on the control of the nociception (raphe nucleus, periaqueductal grey, parabrachial nucleus). In order to describe this physiopathology, the authors firstly observed that this affection more frequently occurs in anxious subject, especially patients under stress and in women. They then observed that the stimulation of the central nucleus of the amygdala can inhibit gastric motility (by acting on the vagal complex) and colic motility (by acting on the pelvic parasympathetic) (OA Lyubashina 2004). They then studied the repercussions of colorectal distension experiments and observed, by means of fMRI, that this visceral stimulation induced in IBS patients a more significant amygdalar activity than in the control subjects (C Wilder-Smith et al. 2004), combined with increased growth of the CRF (L Chang et al. 2009) and of corticosterone in the central nucleus of this structure and with a saturation of its glucocorticoid receptors (GR receptors). The authors also showed greater amygdalar activity in women with IBS than in men with IBS, this being due to an influence of ovarian hormones which accentuate the symptomatology of the irritable bowel syndrome, by exacerbing the activity of the amygdala, particularly during the peak hormonal cycle (like that was demonstrated in 2008, by Y Ji et al).

8.3.3 Motor Actions

They are not only due to striatal projections that allow the amygdala to influence the somatomotor system but also to the neuronal connections with the limbic component of the motor system which is comprised of recently discovered specific pathways. This "third motor system of Holstege" (G Holstege 1991) explains the motor reactions which may be related to certain emotional conditions (vomiting or nausea, for instance, when witnessing a scene which generates deep disgust; tachypnea during an emotion triggered by fear; blocked locomotion and immobility (which explains the French expression "cloué sur place" (nailed to the spot), in the event of an aggression); attitudes when preparing for combat (defence reactions) or flight ("flight or fight"); various cries and shouts; and vocal emotional expressions). Holstege also mentions chewing, sucking movements, swallowing movements, head rotations, urinary losses, etc.

We should also recall the existing interactions between the aspect of a face and the faces pulled by another person that can be our emotional primum movens, as well as the aspect that our own faces and eyes can adopt when seeing a scary face (R Adolphs 2008) and a worrying vision or when learning painful or shocking news (the French expression "Je n'en crois pas mes yeux" (I don't believe my eyes) is, from this point of view, perfectly significant). All of these facial reactions are explained by the direct connections between the amygdala and the nucleus of the facial nerve discussed earlier.

8.3.4 Memorising Emotions

Considering the proximity and intimate relationship of the amygdala, the essential pivot of the emotions, with the hippocampus, recognised support and characteristic of the general memory, it is logical that storing emotions involves both structures (GW Van Hoesen 1985). Many authors have studied how this storing occurs, but it is to JE LeDoux and his team that we owe the most holistic approach and most complete description of that process.

In fact, without being a neuroscientist, reflection on the subject very quickly allows anyone to glimpse what returns to memory when we try to recall an emotional episode: on the one hand, the time, the place and the circumstances reappear, placing us in the context which was that of the emotion; on the other hand, we may find that slightly delayed, prints, unrest and the feelings felt are sometimes the same as that of the initial emotion. These two states correspond to two kinds of long-term memory: the first is for the explicit memory (episodic memory, conscious), and the second to implicit memory (unconscious perceptual memory). It is this second kind of memory that characterises the emotions. The popular French language is capable of expressing it in his own way and phrases: "d'y repenser, cela me fait froid dans le dos" (think again, it sends shivers down my spine) or even "c'était horrible…y repenser me donne la chair de poule" (it was horrible … just thinking about, it gives me goose pimples again) and "ne me faites pas repenser à ces moments, cela me donne la nausée" (do not make me think back to those moments, because it makes me sick) sentences that show the impact on the autonomic nervous system are strong to the point that they manifest themselves again with the recall of emotion.

JE LeDoux was able to demonstrate that the two types of memories cited belong respectively to the hippocampus and to the amygdala:

It is in the amygdala that emotional events that have come to it, after undergoing a specific process in the sensory systems, will be stored in the "emotional" memory.

It is in the hippocampus that emotional events that have reached it, after undergoing a specific process in the sensory systems, will be stored in the "event" memory of the emotion.

Additionally, the retrieval of emotional memories takes place through identical pathways, and it is in the amygdala that will resurface the expression of emotional responses, while it is in the hippocampus that the conscious memory will reconstitute the emotional event.

The amygdala is thus the site of implicit memory, and the hippocampus is the site of explicit memory.

But interaction between the two structures exists, as demonstrated by EA Phelps 2004: "The amygdala can modulate the encoding and storage of hippocampal-dependent memories. Similarly the hippocampus, by forming episodic representations of emotional meaning and interpretation of events may influence the amygdala response when emotional stimuli are encountered". Thus, thanks to the hippocampus, the vision of a snake in a zoo causes an amygdalar reaction which is much less than the vision of the snake even seen in the turning of a forest road. As made clear by EA Phelps: "Although these are independent memory systems, they act in concert when emotion meets memory".

But this beautiful harmony can be disrupted and cause post-traumatic stress disorder (B Roozendaal et al. 2009). After intense stress causing an overproduction of stress hormones with the release of plenty of glucocorticoids, it results as a consequence for the subject, an actual reorganisation of the hippocampus–amygdala couple: From now on the encoding of the emotions, even the most mundane, is essentially carried out through the amygdala (without hippocampal weighting context [4]) and thus with the association of an emotion akin to the feeling of fear. The person's memory is disturbed and "the person is no longer able to modify his response to fear in the context or to the predictive elements that deserve it". Fears in all forms, anxieties, become permanent, haunting "even to the extent of preventing a normal life" (PV Piazza and A Desmedt in N Kaouane et al. 2012)!

The way which the amygdala consolidates long-term memorisation of emotional experiences is best known since the work on animals by JL McGaugh in 2004 and CK McIntyre et al. in 2007. The amygdala uses "memory-modulators" effects of the stress-related hormones (epinephrine secreted by adrenal medulla and corticosteroids and cortisol by the adrenal cortex). The modulation is regulated with adrenaline as a neurotransmitter, by the baso-lateral complex of the amygdala, towards which converge noradrenergic and cholinergic muscarinic activations. The baso-lateral complex interacts with other brain regions, including the medial prefrontal cortex, to consolidate the long-term emotional memory and perform storage in other brain regions. Research conducted in humans (AH Van Stegeren et al. 2005) confirmed the role of the amygdala (in fMRI, activation proportional to the degree of emotional intensity of the images presented in the subject examined) but cannot affirm that the phenomena of encoding occur in the baso-lateral complex. On the contrary, the functional MRI (fMRI) performed in patients to whom a beta-blocker was administered (blocking noradrenergic response) has shown that the amygdala activation decreases strongly, confirming the essential nature of the presence of norepinephrine as a mediator of amygdalar activity.

Another interesting example of the amygdala memorisation is that it causes olfactory emotions: speaking of pheromones and amygdala would lead us very far away and probably unnecessarily so and, given the relative impact of these substances on the rhinencephalon rather involutive of man and given the abundant literature about them. On the contrary we cannot ignore the activating emotions that the amygdala will have to encode keeping, for example, the heady scent of perfume and/or the scented woman. Recall that the odours are captured in the olfactory system by specific receptors, which are genetically determined. These odours are transmitted directly to the piriform cortex and the amygdala (bypassing the thalamus). From the latter projections are made to the orbitofrontal cortex and to the entorhinal cortex (the latter projecting towards the hippocampus to ensure the long-term memory of the olfactory episode).

[4]In fMRI, one observes hyperactivation of the right amygdala and hypoactivation of the hippocampus (N Kaouane et al. 2012).

It is in the same spirit that the famous author Marcel Proust[5] could write such wonderful lines about gustative and odour memories of the famous madeleine, memories linked to those of his mother and his aunt Leonie.

In a recent work (2013), M Kadohisa, after having studied the way certain odours can regulate emotion and cognition, raises the possibility of treating psychological disorders with aromatherapy that may reduce the effects of stress.

8.3.5 Interactions with the Prefrontal Brain

They are particularly interesting because they are based on a true dialogue between the neural pathways of emotions and those of reasoning. Long gone are the days when emotions were considered as disruptions in the thinking process as a generator of adapted activities. We now know, in particular from observations made by neuropsychologists, in patients with brain damage that emotions are necessary so that reasoning allows appropriate and effective behaviours.

"La capacité d'exprimer et de ressentir des emotions est indispensable à la mise en oeuvre des comportements rationnels" (The ability to express and feel emotions is indispensable for the functioning of rational behaviour) (AR Damasio 1994; 2004). As H Barbas (2000) recalls, a real synergy exists between the prefrontal cortex and the amygdala and the integrity of the two formations (A Izquierdo and EA Murray 2005) is necessary for creating the "anatomic dialogue" (HT Ghashghaei et al. 2007) and an individual's behaviour results from an adapted decision. More is known about the microscopic anatomy of the previous dialogue through remarkable work in neuroscience. Thus, we now know that there are reciprocal relationships between the amygdala and the prefrontal cortex. "The amygdala appears to extract the affective significance of stimuli, and the prefrontal cortex guides goal-directed behaviour" (AR Damasio 2004; H Barbas 2000) by using the inhibitory properties of the reticular thalamic nucleus to which it projects (H Barbas et al. 2011).

Disturbances in the amygdala-prefrontal cortex coupling explain a number of behavioural disorders.

In this way the problems of adolescents (appetite for risk; various dangerous experiences like binge-drinking; impulsive behaviour; unawareness of the harmful consequences of certain behaviours; aggressiveness; refusing advice from their elders, seniors or parents; eating disorders; depression; etc.) result from hyperactivity of the amygdala induced by hormonal secretions during puberty and not regulated by the insufficient maturity of the frontal cortex that is not fully developed until age 25 (S Whittle et al. 2008).

Apart from the fact that the full maturity of the frontal cortex will be able to regulate the amygdalar activity when adulthood is reached, some authors (University of Michigan) evoke an additional mechanism seen in animals but not yet shown in humans: during adolescence, a group of nerve cells develop in the amygdala enabling it to better communicate with the prefrontal cortex and thereby promoting better social behaviour.

Likewise in adults, some psychiatric disorders are also explained by disruptions in the behavioural and emotional control system involving the amygdala and the cerebral cortex.

8.3.6 Interactions with the Brain Reward Circuit

As we have seen during the study of the connections and their enumeration of incoming and outgoing impulses, the amygdala is connected not only to the nucleus accumbens, the pleasure and reward nucleus,[6] but also to all the structures that secrete dopamine and to the tracts and bundles that interconnect all of these structures (in particular the medial forebrain bundle and the dorsal longitudinal fasciculus) thus to the entire pleasure and reward circuit. Under normal operation, this pathway can play a role by conferring a rewarding signal to all positive emotions.[7]

On the other hand, a malfunction or abnormal stimulation of this circuit may be the origin of the most diverse addictions (alcohol, drugs, gambling, etc).

The stimulus caused by the first experience with the addictive substance will reach the amygdala and be recognised as a pleasant sensory stimulus, a reward announcement. The amygdala will transmit this stimulus as such to the prefrontal cortex, the nucleus accumbens and ventral tegmental area. This area, stimulated by the cortex which takes the complying measure, will generate a veritable flood of reward structures (mainly the nucleus accumbens[8]) but also the prefrontal cortex itself, the amygdala, the septum

[5]Marcel Proust, French author (1871–1922). His primary work is called "La recherche du temps perdu" (In search of lost time) (Prix Goncourt 1919). Volume 1, "Du côté de chez Swann" (Towards at Swan), is the one in which the famous page about the memories of tea and madeleine is written!

[6]Deep brain stimulation (DBS) through bilateral implantation of electrodes in the nuclei accumbens has been successfully tested in cases of major depression, extremely resistant to all treatments (Th E Schlaeper et al. 2008).

[7]Until recently, the amygdala was considered as the receptacle for only emotions of fear or negative emotions in general. Recent work suggests that it also has a role in the receipt of positive emotions (MG Baxter and EA Murray 2002), and this role may be at least as important as the first one.

[8]The intense iterative stimulation of the nucleus accumbens may cause an involution of the striatum such as observed by S Kühn et al. (2014) by fMRI in young men with a habit of watching pornographic videos: The greater the stimulation (assessed in number of hours of video watching/week!), the greater the reduction of the volume of the right caudate nucleus and left putamen.

nuclei, etc., with dopamine (more than ten times the amount associated with the natural reward of positive emotion in everyday life). All of these structures activated by the unusual amount of dopamine will in turn trigger euphoric, motor and sexual effects of the drugs acting on the mesocorticolimbic dopaminergic system (part of the medial forebrain bundle) in which the nucleus accumbens is one of the key relays (see Fig. 8.1) and that continues rostrally, the large bundles of the brainstem.

Habituation and dependence will quickly occur, as the dopamine interacts with another neurotransmitter, the glutamate, to take control of the reward brain pathway. The state of euphoria induced by taking drugs is accompanied by a constant need to have them. Now not only does taking drugs become a vital need (at the same level as the need to eat), but its supply also becomes a vital search (just as having a supply of food is necessary if we do not wish to starve). This is facilitated by the rapid disappearance of the retro-control prefrontal cortex in which the Dopa receptors are saturated. But gradually, the brain responds less to addictive stimuli either by eliminating or by reducing its stimulation of the ventral tegmental area. For the drug user, it results in the need to increase consumption in the hope that stimulation of the reward structures returns to the initial levels.

The amygdala, in synergy with the hippocampus, will also play another major role in the addictive phenomenon. The two structures will in fact memorise not only approval of the addictive stimulus but also the environmental factors that led to the occurrence of this stimulus. This memory is formidable as it persists and may be the cause for a relapse in the subject who had been able to wean himself from his addiction, even after many years of abstinence. This is how such an addict may relapse at the mere sight of a syringe and needle or how a former alcoholic may "fall off the wagon" if he finds himself alone with a bottle of whiskey.

8.3.7 Role in Behaviour: The Example of the Sexual Arousal

The essential role of the amygdala in the behaviour of individuals is now better known. We know that the memorisation of successive emotions, their effects and the consequences of managing them, is a wonderful help to enlightened reasoning and therefore to sound reasoning and good behaviour. Subjects suffering damage to their amygdalae and/or prefrontal brains are unable to make decisions or modify their behaviour to the situations that arise (AR Damasio 2003 and 2004).

A demonstration of the amygdala's role in behaviour is provided in the work of Y Salu, who, in 2013, published an interesting study about how the amygdala contributes to the development of the mental component of sexual arousal, "an essential process for the survival of the species".

This process involves the amygdala from childhood. It is in fact the result of successive conditioning of non-sexual experiences that create the root cues of the sexual arousal that the adult individual will need:

- Conditioning to voice tone that will make him recognise to what genre the speaker belongs.
- The innate emotions of fear and safety, postures, tested during human interactions from childhood (children's games like cops and robbers, cowboys and Indians, playing doctor, etc., but also meeting adults unknown to the child[9]).
- Emotions generated by sexual information provided by other children or sometimes older people (even adults), from reading, pictures, etc.
- The emotions of puberty: the subject can become excited at the sight of a person from the opposite sex or the same sex, according to his/her tendency.

In a parallel to this basis knowledge, the combination of hormonal secretions and the full development of the genitalia will give the pubescent individual the full extent of his/her sexual potential. Additional signals resulting from learning during sexual activities will complete the long process of development of sexual arousal.

As for the amygdala, which was involved in all phases of the process that are memorised, it will react almost automatically (as shown by the functional MRI) to erotic stimuli (e.g., videos) viewed by the subject and with the ensuing sexual arousal (S Karama et al. 2002; S Hamann et al. 2004; T Ball et al. 2009; T Sundaram et al. 2010).

Another study on this subject (S Karama et al. 2002) comparing the sexual arousal of men and women watching erotic films in fMRI showed that this arousal, which is stronger in men than in women, is due not to differences in the activation of the response structures (in particular the amygdala, the ventral striatum, anterior cingulate, orbitofrontal cortex, insula and occipitotemporal cortex) but to the additional activation, in men, of the hypothalamus (the activation degree of this hypothalamus is perfectly correlated to the degree of arousal measured by the BOLD signal blood-oxygen-level-dependent signal).

---

[9]The effect of this emotion related to fear is obvious when you see a child turn his head so as not to look, to huddle with his mother while clutching her mother's skirt, simply because the mother asked him to say hello to the neighbour she had just met. The apprehension to an unknown face, even though female like his mother! If the mother had the misfortune to add: "this woman is going to look after you for a few hours,... while I go...", the child's fear could turn into panic with the fear of being abandoned, and this will involve "I don't want to" and crying manifestations, negative emotions which are part of the amygdala.

9 Vascularization of the Cerebral Amygdala

The vascularisation of the amygdala has been much less studied than the vascularisation of the neighbouring hippocampus (which was the subject of excellent research work conducted by HM Duvernoy 1988; HM Duvernoy et al. 2013). However, due to the importance currently related to epilepsy surgery performed as an open procedure or by stereotaxy (E Mempel 1975) and to the need for preoperative stereotaxic investigations in order to detect or identify epileptogenic areas, renewed research has been undertaken on this topic.

9.1 The State of Our Knowledge

Anatomical studies have obviously focused on arterial vascularisation and certain studies have been mentioned in neurosurgical literature (HO Luders 2008).

This study is difficult and misleading because, given the extreme thinness of the arterioles that irrigate the amygdala, we are led to believe that, following a global vascular injection of an anatomical subject and even after a direct injection into the common carotid system (or even better into the internal carotid artery), the vascularisation of the amygdala is poor. Many of the standard contrast products, such as anatomical coloured resins, do not reach the thin arterioles of the amygdala. Thus, in the study performed by R Marinkovic and L Markovic in 1991, the authors referred to the failure of the injections made into the middle cerebral artery: "By injecting the middle cerebral artery, the amygdala body remained unfilled by the mixture". We encountered the same difficulty! It is therefore necessary to achieve a real assessment of the amount of amygdaloid arteries, not only by performing selective injections in cerebral arteries but also by using very fluid contrasting agents, the most suitable being Indian ink (process used by HM Duvernoy et al. 2013).

Three sources supply the amygdala according to B Goetzen and Sztamska E (1999), a main source and two secondary sources:

- The main source is represented by 1–3 branches from the proximal segment of the middle cerebral artery. A single branch corresponds to the most common arrangement. This arteriole, whose diameter is between 200 and 350 μm, penetrates the amygdala, perpendicularly to its surface and then passes toward the rear section. After having progressed over a few mm in the amygdala, it divides into multiple branches.
- The second source is one of the secondary sources: It is represented by tiny ramifications (80–100 μm in diameter) of the superficial arterioles (cortical) of the middle cerebral artery.
- The third source, which is also secondary, is represented by sparsely branched, thin and straight arterioles, which emerge from the superficial branches (cortical) of the anterior choroidal artery. They penetrate near the uncus and primarily target the basal part of the amygdala. In our experience, we have frequently observed these arterioles and the choroidal source[1] is considered by many authors as the main arterial source of the amygdala (G Salamon 1971; G Lazorthes et al. 1976).

We have also observed, on sections injected with ink, very fine branches from the periamygdaloid cortex and adjacent cortical areas, which are orientated perpendicular to the surface of the cortex and which penetrate, at the level of its deep surface, the amygdaloid nuclei. We must finally point out some thin vascular interconnections between the head of the hippocampus and the amygdala in areas where the two formations join up.

Microscopic studies also provided a good appreciation of the arterial vascularisation of the amygdala as they show the very rich and numerous capillaries of this formation.

[1] The anterior choroidal artery is born from segment C4 (supra-clinoid segment or cerebral segment) of the internal carotid artery. This artery can also develop from the middle cerebral artery or the posterior communicating artery. Through the choroid plexus of the lateral ventricle (whose lower end it supplies), it establishes a remarkable anastomotic system between the carotid system, which it belongs to, and the vertebro-basilar system to which belong the posterior choroidal arteries, branches of the posterior cerebral arteries.

V. Di Marino et al., *The Amygdaloid Nuclear Complex: Anatomic Study of the Human Amygdala*,
DOI 10.1007/978-3-319-23243-0_9

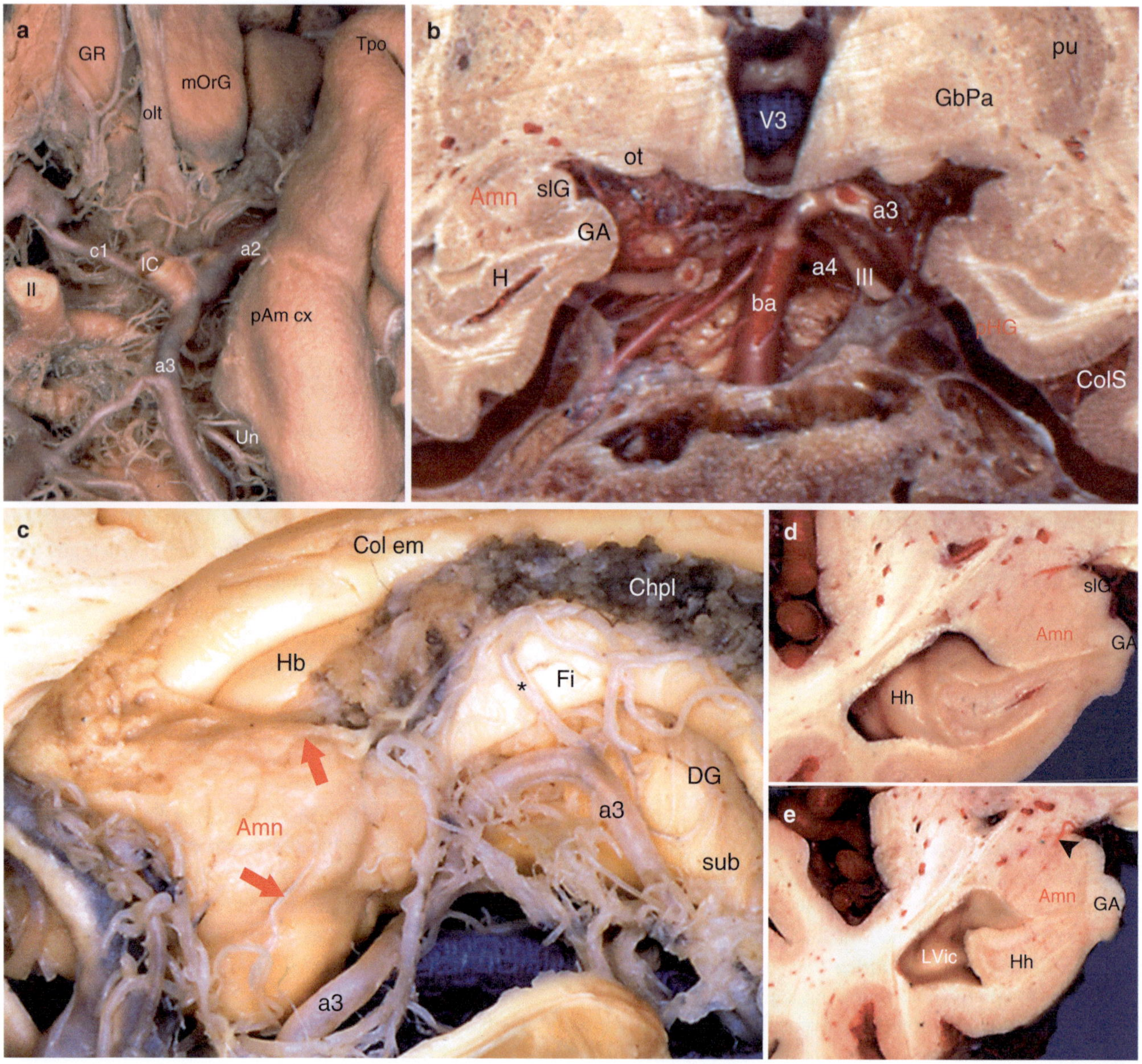

Fig. 9.1 Arterial vascularisation of the amygdaloid nuclear complex. (**a**) inferior view of the arteries edging with the left amygdalo-hippocampal area; (**b**) coronal section of the brain in front of the basilar artery (arteries injected by a red-coloured resin); (**c**) dorsal view of the amygdalo-hippocampal area; (**d**, **e**) coronal sections targeting the amygdala to show the arteries of the neighbourhood. *Amn* amygdaloid nuclear complex, *ba* basilar artery, *Chpl* choroid plexus, *Col em* collateral eminence, *ColS* collateral sulcus, *DG* dentate gyrus, *Fi* fimbria, *GA* gyrus ambiens, *GbPa* globus pallidus, *GR* gyrus rectus, *H* hippocampus, *Hb* hippocampus body, *Hh* hippocampus head, *IC* internal carotid, *LVic* lateral ventricle, inferior cornu, *mOrG* medial orbital gyrus, *olt* olfactory tract, *ot* optic tract, *pAm cx* periamygdaloid cortex, *pu* putamen, *slG* semilunaris gyrus, *sub* subiculum, *Tpo* temporal pole, *Un* uncus, *V3* third ventricle, *a2* middle cerebral artery, *a3* posterior cerebral artery, *a4* superior cerebellar artery, *c1* anterior communicating artery, *II* optic nerve, *III* oculomotor nerve, *red arrows* they show two arterial rami ending in the superior face of the amygdala, *black arrowhead* showing a ramus from a3, *black asterisk* posterolateral choroidal ramus

As for the venous drainage of the amygdala, we know that it essentially takes place into the lower choroidal vein or choroid plexus vein which emerges from the temporal horn of the lateral ventricle and targets the basal vein.

It should also be recalled that the amygdala is largely in contact with the subarachnoid spaces (on the one the hand, at the uncal recess of the temporal horn of the lateral ventricle and, on the other hand, at the level of the lateral part of the transverse fissure) and therefore with the cerebrospinal fluid, whose role of transporting neurotransmitters is not the least, in addition to its nutritional and mechanical functions. The choroid plexus of the lateral ventricle only appears beyond the hippocampal head but immediately provides thin vascular branches for the cortical and baso-medial nuclei of the amygdala (Fig. 9.1).

9.2 Vascular Density Variations of the Amygdala?

This concept may seem impossible to believe if we continue to believe that the vessel density of an anatomical structure is fixed and immutable. But the reality is different, especially concerning the amygdala, as well as the hippocampus, and therefore in the case of two essential structures of the nervous system.

We know, from the work conducted by A Vyas et al. 2002, that chronic stress can disrupt the cyto-architecture of these two structures. However, we did not know that it could cause changes in their internal vascularisation. This shows the interest of the work performed by GN Neigh et al. in 2010, which we wish to mention (although it was conducted in rats), given its extreme importance and the fact that it suggests modulations of nervous system vessels.

The authors, after removing the amygdala and the hippocampus in adult rats exposed to stress during 2 h a day for 2 weeks, counted the areas occupied by the vessels (vascular areas) within both structures by using a very accurate stereological method of area fraction. They then conducted a statistical analysis with the collected data. They were able to see that chronic stress in adults increases the capillary vascular areas of the amygdala (study at the level of the basolateral nucleus) by 30°/° and decreases those of the hippocampus by 20°/°[2]. Similar results have been obtained by NN Gretchen et al. in 2010.

If these great variations in the vascular surface of the nuclei were observed in the amygdala of human subjects exposed to stress, it would be interesting to study the correlations with the cyto-architecture and the function of their amygdaloid nuclei.

9.3 fMRI and Blood Flow in the Amygdala

We know that fMRI is a functional brain imaging technique used to study the changes of blood flow in various brain structures during a specific task or in the presence of a particular feeling such as emotion. The structures employed show an increased blood flow, resulting in particular MRI images (colour scale according to the degree of activation T Iidaka 2007). The amygdala, a very recognisable structure which is very sensitive to emotions, is the perfect structure for an fMRI study which can be conducted to study its activity. Work on this topic has been accumulating since 1990. KL Phan et al., in 2002, completed the first meta-analysis of 55 studies published between 1993 and 2000. We can remember the following conclusions: It is fear, the vision of frightened faces and all aversive stimuli which almost systematically activate the amygdala. But now we know (SB Hamann et al., 1999) that pleasant stimuli also act on the amygdala. It actually reacts to all emotions (JE LeDoux 1998), whether positive or negative. It behaves like a true vigilance and alert system to make us react in order to maintain the good condition and integrity of our body. The meta-analysis of KL Phan et al. 2002 also shows what other parts of the brain are activated at the same time as the amygdala, particularly the prefrontal cortex which is known to constantly interact therewith (see chapter on inputs and outputs). This meta-analysis also shows the limitations of this imaging technique when multiple areas are activated at the same time, thus making it difficult or impossible to discriminate between the areas reacting to the causal stimulus.

fMRI has an additional advantage: It provides a better understanding of the hemispheric lateralisation according to the preferential activation of one of the two amygdalae, in the presence of a particular emotional stimulus. In 1994, the work carried out by ED Ross et al. suggested that "the most primitive forms of emotions preferentially activate the right amygdala" whereas "the most elaborate emotions, referred to as social, concern the left amygdala." In 1998, T Canli et al. performed an fMRI study on a homogeneous group of 14 women. They observed that negative images preferentially activate the right hemisphere and amygdala while positive images concern the left hemisphere and amygdala. Similarly, in his meta-analysis of 385 studies, S Costafreda et al., in 2008, observed an activation of the left amygdala when the stimulus included language while the right amygdala is responsive to masked stimuli.

There is also a gender difference for the lateralisation effect such as demonstrated by L Cahill et al in 2004: The second vision, after 2 weeks during which two equivalent groups of men and women viewed slides, showed that the memorisation of slides considered as exiting, concerned rather the left amygdala in women, whereas in men, the right amygdala was stronger activated during this "emotional memorisation" process.

However, such as pointed out rightly by F Staub et al. (2002), a specific lateralisation of the amygdala is not always necessary, but rather a consequence for the amygdala due to the well-known lateralisation of the hemispheres including prefrontal regions, with 2 circuits of affectivity (RE Wheeler et al., 1993): the left prefrontal circuit "which generates positive emotions and appetitive behaviours" and the right prefrontal circuit, "the avoidance system, which generates negative emotions and flight behaviours when confronted with aversive situations".

[2] J Hellsten et al., in 2005, had already observed that therapeutic treatments, which are effective in treating depressive syndromes, can increased the vascularity of the hippocampus by an extra 30°/°!

10 The Bed Nucleus of the Stria Terminalis

10.1 Generalities

These terms and their abbreviation in English "NBST or BST" are used to refer to a practically seamless continuum of neuronal cell clusters bordering the concave edge of the caudate nucleus, from its head to its tail, the septal region of the amygdaloid body, and within which passes the stria terminalis.

Various names have been used to refer to these cell clusters such as the "support nucleus" (M Volsch 1910) or "secondary nuclei of the stria" (CWA Kappers 1921) or even the "enveloping nucleus" (EW Lauer 1945) and "interstitial nuclei" (R Diepen 1962).

In this bed, in addition to the stria terminalis (see Chap. 6), the large thalamo-striate vein and its tributaries are routed before it drains into the internal cerebral vein. Cell clusters of the BNST are in contact with the two aforementioned elements and are usually interposed there between.

The BST, which we seem to have more or less rediscovered in recent years, is already well known from the work mentioned above and especially since the memorable anatomy works performed by JB Johnston, in 1923. In a long article devoted to the evolution of the forebrain, the author described this structure from the early stages of embryonic life and showed its central position and its role: establishing communication between the amygdaloid complex and the olfactory area. Subsequently, many contributors have clarified the anatomy of the BST, including OJ Andy and H Stephan, in 1968, who studied its paraseptal portion; F Sanides, in 1957, who studied its caudal portion; and, above all, H Strenge et al., in 1977, who carried out an exhaustive and accurate histological study of the entire cell mass using serial sections perpendicular to the arch that forms the BST. We should also recall the work of JS De Olmos (2004) who studied the connections between the BST and the amygdala in particular.

Three portions are usually identified for the BST: the rostro-medial portion, the intermediate portion and the caudal portion.

- The *rostro-medial portion* is special. It forms three areas: the most anterior area, sometimes referred to as "pre-septal"; the second area located behind the previous area, often referred to as the "commissural" considering its proximity to the anterior commissure; and the third area, the "retro-commissural" area.

At the level of pre-septal area, the BST is in the centre of the striatum and positioned medially with respect to the anterior branch of the internal capsule. It insinuates itself between the caudate nucleus, element of the dorsal striatum and the nucleus accumbens (Fig. 10.1a, a'), element of the ventral striatum which has just received one or more branches of the stria terminalis. It is medially covered by subependymal epithelium, which forms, at this level, the inferior recess of the front horn (or anterior cornu) of the lateral ventricle.

At the commissural level, the BST always underlies the caudate nucleus and is positioned against the lateral ventricular wall, inside the crus anterior of the internal capsule. It remains in a dorsal position with respect to the anterior commissure, which forms its median trajectory, slightly arched, with a dorsal convexity (Fig. 10.1b, b' and c, c' and Fig. 10.2).

At the rear of the anterior commissure, the BST remains connected to the head of the caudate nucleus, which it follows up to the posterior third (Fig. 10.1d, d').

For R Walter et al. (1991), who have studied the differential distribution of histochemical markers in the BST of the human brain, this rostro-medial portion is the part which specifically corresponds to the BST due to the fact that beyond this point, the presence of the dorsal part of the magno-prothalamic and parvocellular nuclei, initially described by H Brockaus, makes it difficult to interpret the results. Thus, the marks of A Walter et al. only concern the rostro-medial part of the NBST.

- The *intermediate portion* (see Fig. 7.15a) is the longest of the three portions of the BST. It follows the entire body of the caudate nucleus, i.e. almost half of the caudate arch.

V. Di Marino et al., *The Amygdaloid Nuclear Complex: Anatomic Study of the Human Amygdala*,
DOI 10.1007/978-3-319-23243-0_10

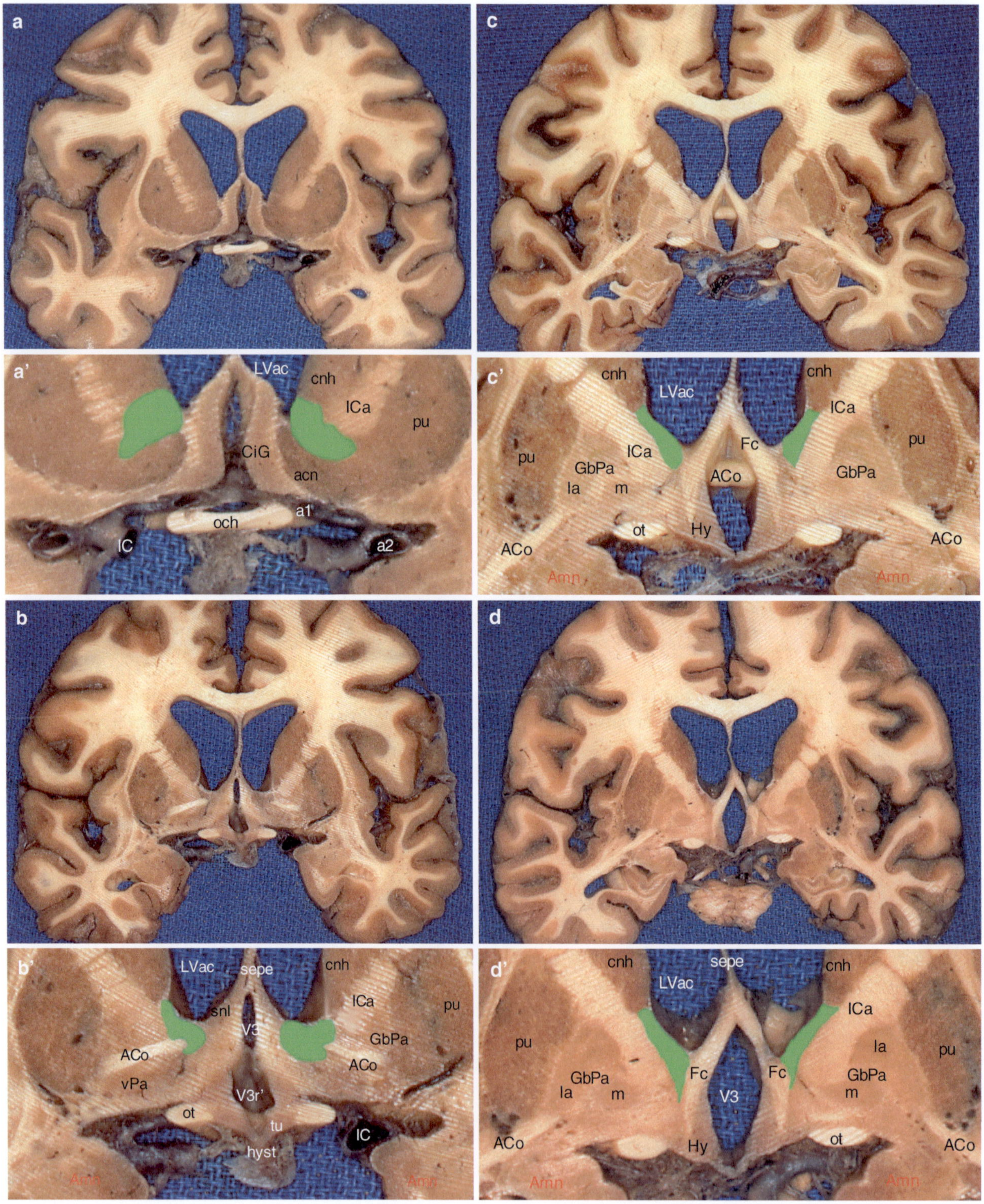

Fig. 10.1 Location of the BST on a series of brain's coronal sections in the area of the anterior commissure. (**a**) Section through the optic chiasm; (**b**) section through the optic recess of V3; (**c**) section behind the median part of ACo; (**d**) section through the hippocampal digitations; (**a′**–**d′**) enlargements of **a**, **b**, **c** and **d**, showing the green-coloured BST. *acn* accumbens nucleus, *ACo* anterior commissure, *Amn* amygdaloid nuclear complex, *CiG* cingulate gyrus, *cnh* caudate nucleus head, *Fc* column of fornix, *GbPa* globus pallidus, *la* pars lateralis, *m* pars medialis, *Hy* hypothalamus, *hyst* hypophyseal stalk, *IC* internal carotid, *ICa* internal capsule, *LVac* lateral ventricle, anterior cornu, *och* optic chiasm, *ot* optic tract, *pu* putamen, *sepe* septum pellucidum, *snl* septal nucleus (lateralis), *tu* tuber, *V3* third ventricle, *V3 r′* optic recess, *vPa* ventral pallidum, *a1* anterior cerebral artery, *a2* middle cerebral artery

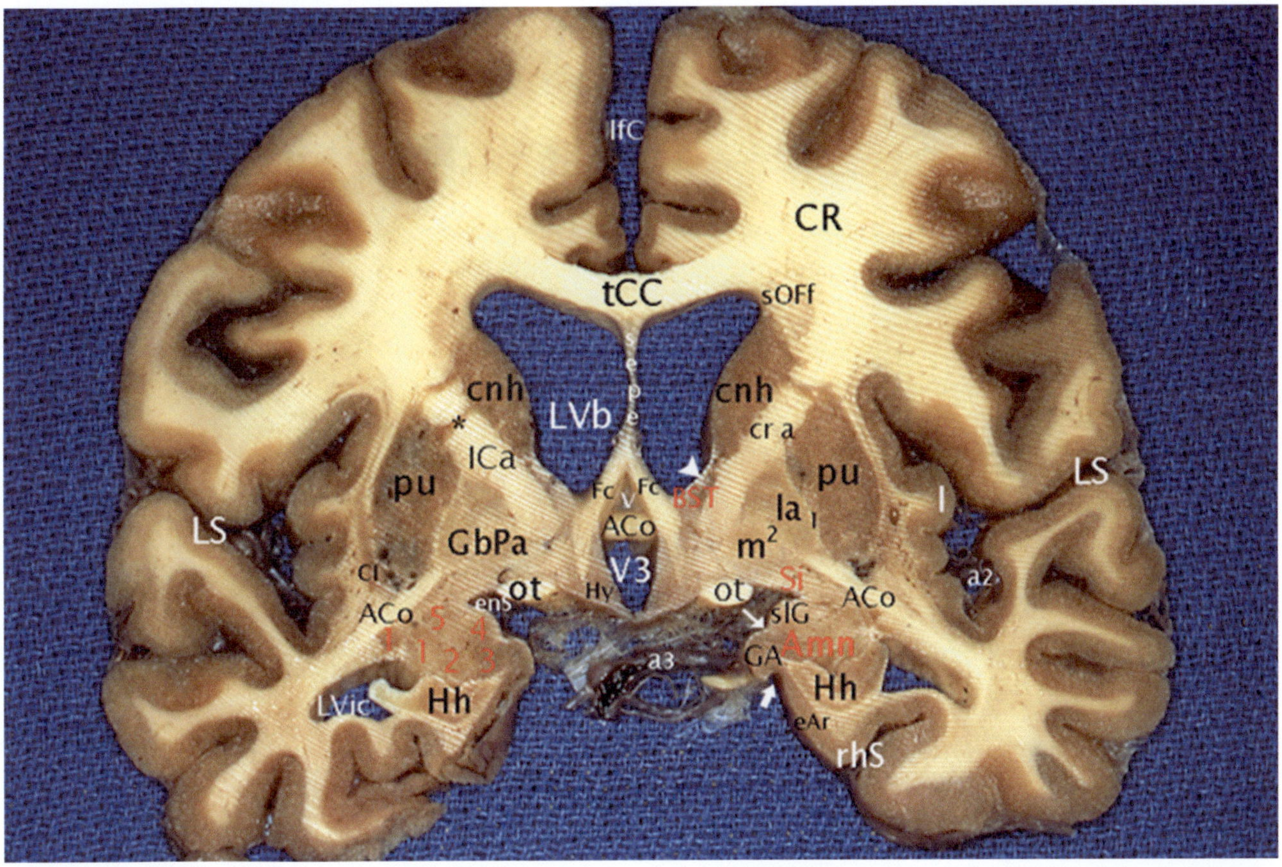

Fig. 10.2 The anterior commissure, the amygdalae and the BST on a coronal section of the brain through the columns of the fornix. *ACo* anterior commissure, *Amn (red keys)* amygdaloid nuclear complex; *1* lateral, *2* baso-lateral, *3* baso-medial, *4* cortical, *5* central; *BST (red key)* bed nucleus of the stria terminalis, *Cl* claustrum, *cnh* caudatus nucleus (head), *CR* corona radiata, *eAr* entorhinal area, *enS* endorhinal sulcus, *Fc* column of the fornix, *GA* gyrus ambiens, *GbPa* globus pallidus, *la* pars lateralis, *m* pars medialis, *1* external medullary lamina, *2* internal medullary lamina, *Hh* hippocampus head, *Hy* hypothalamus, *I* insula, *ICa* internal capsule (*cr a* crus anterior), *lfC* longitudinal fissure of the cerebrum, *LS* lateral sulcus, *LVb* lateral ventricle body, *LVic* lateral ventricle, inferior cornu; *ot* optic tract, *pu* putamen, *rhS* rhinal sulcus, *sepe* septum pellucidum, *Si (red key)* substantia innominata, *slG* semi-lunaris gyrus, *sOFf* superior occipito-frontal fasciculus, *tCC* truncus of the corpus callosum, *v* vulva, *V3* third ventricle, *a2* middle cerebral artery, *a3* posterior cerebral artery, *putamino-caudate bridge, *thin white arrow* semi-annular sulcus, *thick white arrow* uncal notch, *white arrowhead: la* lamina affixa

The front concave trajectory includes three segments, the superior segment with an inferior concavity, the middle segment with an anterior concavity and the inferior segment with a downward, almost straight trajectory.

This section begins beyond the head of the nucleus where the thalamus contacts the body of the caudate nucleus. A thalamo-caudate sinus forms immediately and extends along a groove in a circular arch to the tail of the caudate nucleus. The cells, which form the BST and especially those adjacent to the thalamus, play an important role such as demonstrated by numerous recent studies. This is not surprising if we refer to embryology as it is clear that the thalamo-caudate sinus is a strategic area where the striatum and thalamus become closer to each other so as to allow the diencephalo-telencephalic junction and the incorporation of the diencephalon in the median brain (see Chap. 3). At the level of the dorsal surface of the thalamus, the furrow occupied by the BST is also bordered, on the ventricular side, by a thin white substance lamina, the lamina affixa or lamina cornea, an embryonic residue of the telencephalic white substance separating the striated body from the thalamus. The role played by this lamina affixa is to attach the thalamo-striate vein to the bottom of the furrow. It also transforms this furrow into a triangle, which corresponds to the location of the BST. This triangle will also expand into a square or rectangle according to the areas being observed (Fig. 10.3).

Beyond this point and especially at the level of the posterior surface of the thalamus, the lamina affixa disappears and the ventricular wall of the BST consists only of the ependymal layer.

- The *inferior portion*, although close to the amygdala, is smaller in size. Its volume seems to decrease proportionally to the volumetric reduction of the caudate nucleus which tapers progressively towards its tail section.

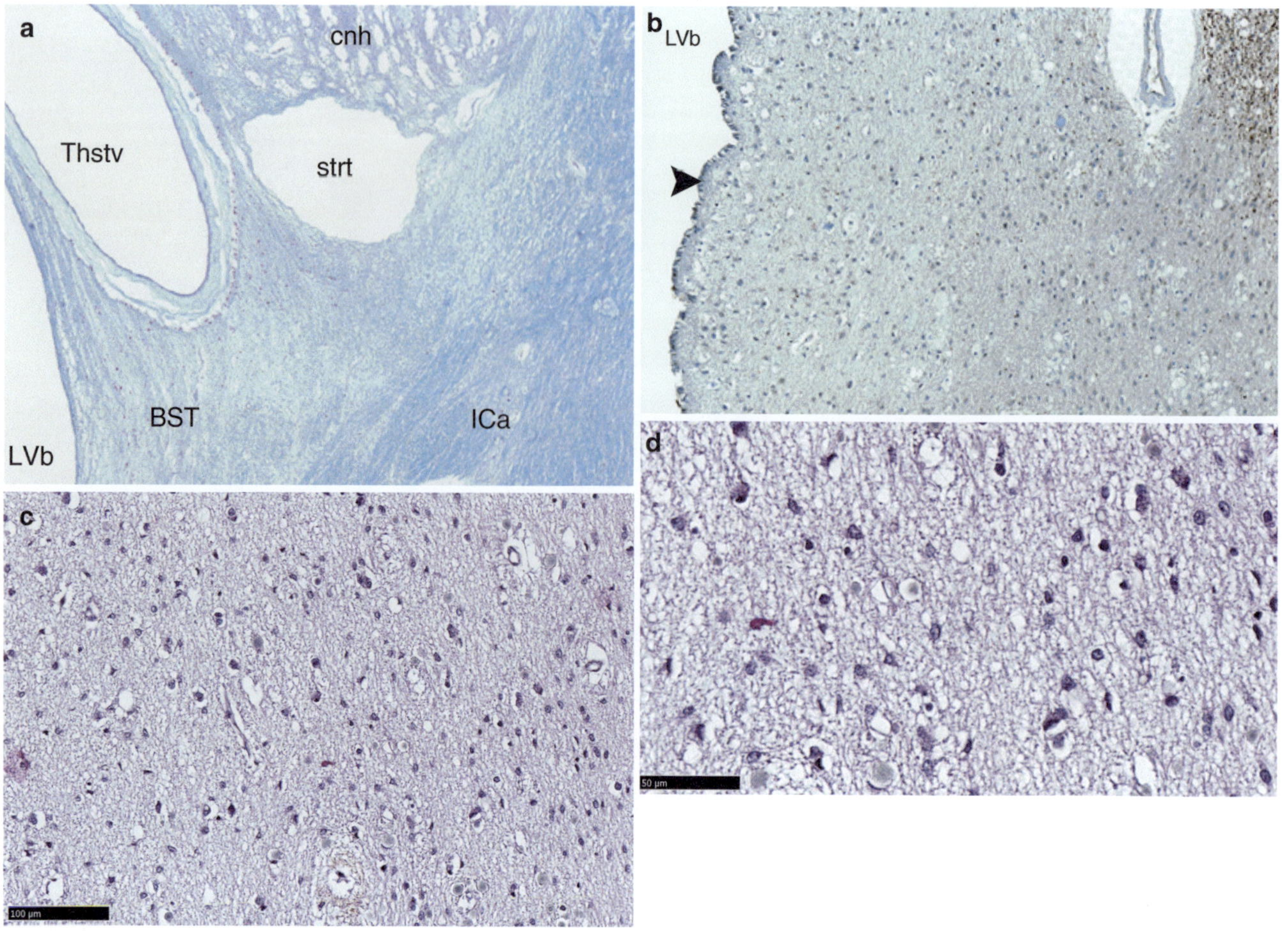

Fig. 10.3 Microscopic aspects of the BST (bed nucleus of the stria terminalis) in front of the anterior commissure. (**a**) Location of BST on a coronal section, (**b**) low-power view (X100), (**c**) intermediate-power view (×200), (**d**) high-power view (×400): *we observe little scattered neurons*. *BST* bed nucleus of the stria terminalis, *cnh* caudate nucleus head, *ICa* internal capsule, *LVb* lateral ventricle body, *strt* stria terminalis, *Thstv* thalamo-striate vein, *black arrowhead* ependymal layer

The NBST cells (Figs. 10.3 and 10.4) have been described by H Strenge et al. (1977):

- On sections passing through the septum, two cell types (slightly pigmented small cells and medium-sized cells with intense staining) are mixed and distributed unevenly.
- On sections passing through the anterior commissure, we can observe small medial cells, along the ventricular wall and medium-sized cells on the lateral part of the NBST.
- On the most posterior sections, the NBST cells can be easily distinguished from the reticular nucleus cells of the neighbouring thalamus, which are larger and more pigmented. Here again, the medial cells, along the ventricular wall (Nsi contingent), are small and the lateral cells (Nse contingent) are larger. Nse is laterally bordered by prothalamic cell clusters.
- On sections concerning the posterior part of the thalamus (level of the lateral geniculate body), the cells are arranged differently, and we can now observe a medial contingent (Nsm) and a paracaudate contingent (Dk), which alone will persist when approaching the amygdala. The Nsm contingent will become increasingly sparse as we approach the end of the caudate nucleus.
- Close to the amygdaloid body, the amount of neurons is reduced and may even be minimal. In addition to Nsp cells, whose quantity is also reduced, we can sometimes observe some Nsm cells, of medium size and with an intense coloration. They are arranged ventrally relative to the tail of the caudate nucleus.

H Strenge et al. also included in their work good anatomical diagrams of the sections that he had performed for the BST study. These sections are still used as references for locating the BST. The observations made by H Strenge et al. (1977) were purely histological, even if they could seem, at that time, of little interest. They actually played a major role in the BST study because they showed, for the first time, that the BST had a complex constitution and that we could observe a real zoning with several different cell areas.

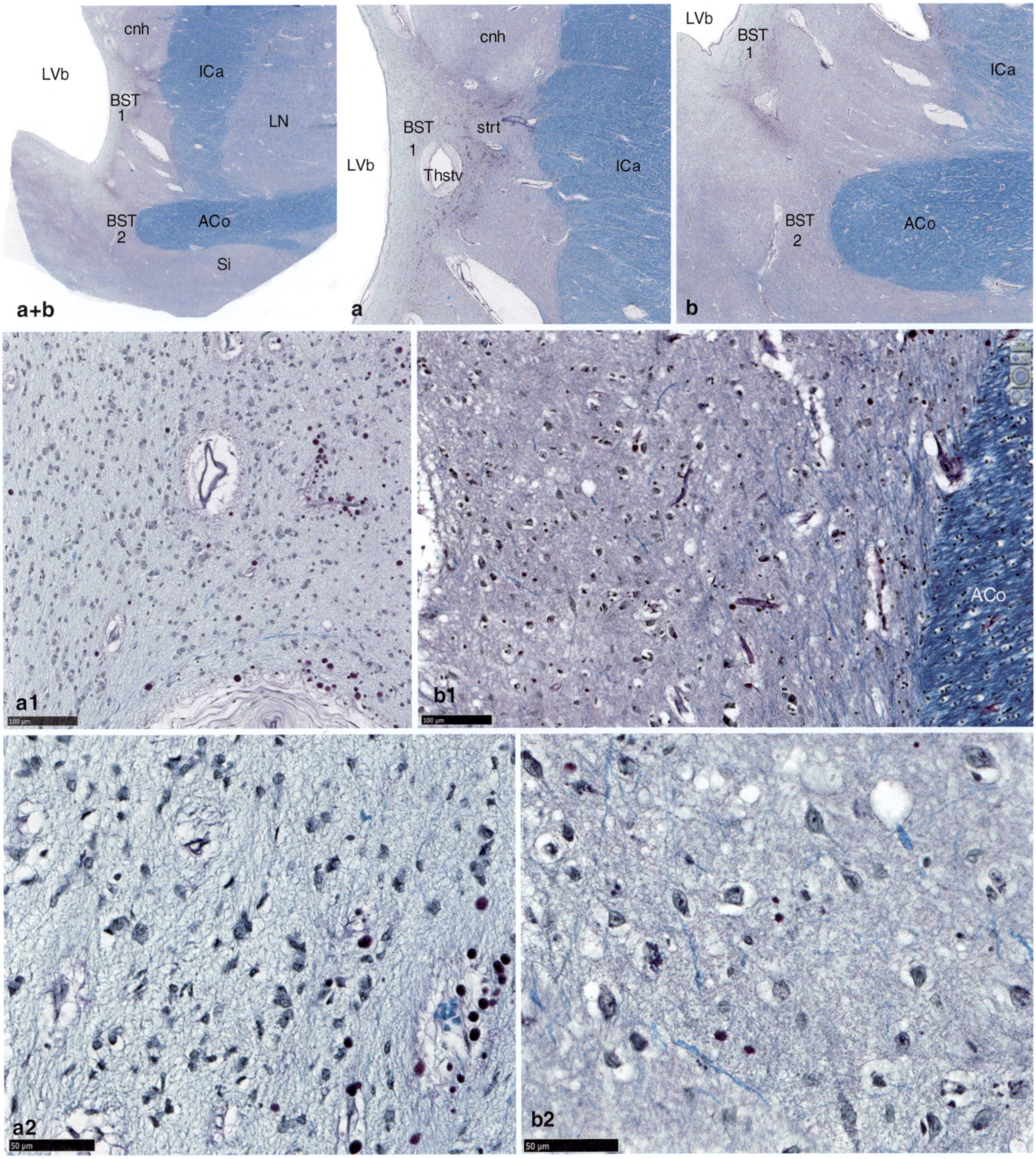

Fig. 10.4 Microscopic aspects of the BST (bed nucleus of the stria terminalis) through the anterior commissure. (**a** + **b**) Global view of the BST; (**a**) BST under the head of the caudate nucleus (BST1); (**b**) BST around the medial end of the left part of ACo (BST2); (**a1**, **a2**, **b1**, **b2**) luxol fast blue–PAS stainings (×200, ×400, ×200, ×400), of BST 1 (**a1**, **a2**) and BST 2 (**b1**, **b2**). *ACo* anterior commissure, *BST* bed nucleus of the stria terminalis, *BST1* pars under caudate nucleus, *BST2* pars around the ACo, *cnh* caudate nucleus head, *ICa* internal capsule, *LN* lentiform nucleus, *LVb* lateral ventricle body, *Si* substantia innominata, *strt* stria terminalis, *Thstv* thalamo-striate vein. *Note that the neurons of BST have small dimensions, an oval-shaped morphology (often fusiform) and big central clear nuclei with conspicuous nucleoli*

Thus, other contributors (JS De Olmos et al. 1985; MM Moga et al. 1989) showed, in mammals, by means of histological and histochemical methods, that there is a cytoarchitecture and a chemo-architectony specific to their NBST.

According to these authors, five main sectors were identified in animals, a lateral sector, a medial sector, a central sector, a latero-ventral sector and a postero-medial sector.

The outcome of this research in animals generated a particular enthusiasm, and parallel studies on the human NBST were rapidly conducted. In 1985, P Gaspar et al. carried out immuno-cytochemical studies on the septal region of the human brain and demonstrated its catecholaminergic innervation. The following year, E. Fliers et al. (1986) discovered the presence of vasopressin cells in the BST. Due to the fact that immunohistochemical methods could now be used to locate cell islands with a specific function, this technique was rapidly used to prepare a chemical map of the NBST. The anatomical subdivisions of this formation in humans were thus observed by using these methods, successively by a French team, A Lesur et al. (1989), and then by a German-Hungarian team, A Walter et al. (1991).

According to Lesur et al., there are only *four identifiable sectors in humans*: the lateral, the central, the medial and the latero-ventral areas. Four markers were used by the authors: synenkephalin, metenkephalin, somatostatin and tyrosine hydroxylase. For the lateral sector, there is no predominance and histochemical markers overlap. The central sector is marked by a "high density of somatostatin neurons". The medial sector is characterised by "a less dense aminergic and peptidergic innervation". The latero-ventral sector shows a predominance of the "peptidergic plexi (somatostatin and enkephalin)".

Walter et al. used a very wide range of histochemical markers (13 in total). In addition to the Lesur markers, they used VIP, substance P, NPH (neurophysins), MBP (myelin basic protein), Cab (calbindin), Syn (synaptophysin), CHR-A (chromogranin), NT (neurotensin), NPY (neuropeptide Y) and the glial fibrillary acid protein. They were able to conclude that there are, in fact, *in humans, only three identifiable sectors*: the lateral sector, the central sector and the medial sector. The lateral sector, located against the internal capsule, is heterogeneous and includes small neurons and other medium-sized neurons. Most of these neurons are NPY or Syn positive. The central sector consists of a homogeneous set of neurons of similar shape and diameter. This sector has the distinction of being divided into two areas, on coronal sections: a supra-commissural area and a peri-ventral and sub-commissural area. The authors rightly consider that these two areas correspond to the same formation as they all have the same immunohistochemical characteristics: VIP innervation density and presence of somatostatin fibres and neurons (DF Swaab 2003). The medial area is located below the tapered end of the lower edge of the lateral ventricle. It is a heterogeneous sector (such as the lateral sector). It is in this sector that the BST neurons with the smallest diameters are observed. This sector has a VIP innervation (JN Zhou et al. 1995) as well as a less dense peptidergic and aminergic innervation.

It is important to consider the fact that A Walter et al. have only focused their work on the pre- and peri-commissural portions of the BST because they believe that beyond this point, the presence of parvo- and magnocellular central prothalamic nuclei, such as observed and described by H Brockaus 1942, shows a different histochemical area which would require a histochemical analysis involving other immune markers than those used in their investigation!

The BST is connected to many areas of the brain by input and output nerve fibres.

10.2 Inputs

These are the nerve fibres which reach the neural clusters of the BST. They have been the subject of many neuroscientific research studies. Although this work has been conducted with mammals, considering the amount of studies and the convergence of their results, they can be extrapolated for humans.

- The BST receives many cortical inputs not only from the limbic lobe (and especially from its rostral and ventral portions) but also, such as shown by the work of AJ McDonald (1998) and AJ McDonald et al. (1999), from the insula, the subiculum and the entorhinal cortex.
- The BST receives amygdaloid inputs of such significance that it has been said that it is the main receptor of amygdaloid projections. In fact, this is especially true for the central nucleus of the amygdaloid complex and especially its lateral portion (GD Petrovich and GW Swanson 1997) and the medial nucleus (see Chap. 11). KL Weller and DA Smith (1982) have observed, on the other hand, that the cortical and baso-medial nuclei badly target the NBST. As for the lateral and baso-lateral nuclei, they are considered, according to HW Dong et al. (2001), as having no significant projection to the NBST.
- The BST receives hypothalamic inputs, particularly from the ventromedial and paraventricular nuclei.
- The BST is also in direct communication with the brainstem centres (as already shown in 1983 by MY Sofroniew, by using the immunochemical detection of marker proteins), which generate many projection fibres. The BST is thus a projection area of the periaqueductal grey substance, the ventral tegmental area, the bulbar nuclei and parabrachial, solitary and dorsal nuclei of the vagus nerve.

10.3 Outputs

These are the numerous structures of the central nervous system targeted by the axons of the NBST nerve cells. Besides the amygdala, these projections occur to the diencephalon (thalamus and hypothalamus) and to the brainstem nuclei.

- The projections of the NBST to the amygdala mainly target the central and medial nuclei, which they reach by passing through the ventral amygdalofugal pathway (amygdalopetal in this case!). KL Weller and DA Smith (1982) also observed projections to the baso-lateral nucleus and amygdalo-hippocampal area.
- The downward projections are complex. They were especially studied by HW Dong et al. (2001). They essentially follow three pathways (Nieuwenhuys et al.): the periventricular tract for the paraventricular nucleus of the hypothalamus, the ventral tract for the medial preoptic, anterior and ventromedial nuclei of the hypothalamus and the lateral tract. The latter follows a long downward pathway, integrates the medial forebrain bundle and distributes its projections to the hypothalamus (dorsomedial, ventromedial, paraventricular, supramammillary nuclei and lateral hypothalamus), the thalamus (thalamic nuclei of the midline) and the various levels of the brainstem. At the level of the midbrain, the projections target the dopaminergic neurons of the pars compacta of the substantia nigra, the periaqueductal grey, the raphe nuclei, the ventral tegmental area and the cuneiform nucleus. At the level of the pons and medulla oblongata, projections reach the nucleus of the solitary tract, the locus coeruleus, the dorsal nucleus of the vagus nerve, the parabrachial nuclei and the rhombencephalic reticular formation.[1]

10.4 The Roles of the BST

In addition to the major role played by the BST in the management of emotions and behaviours, we are starting to consider this nucleus as a real cerebral secondary sexual characteristic enabling to differentiate man from woman.

Thus, in 2003, DF Swaab referred to his own work and that of his team and wrote in chapter 7 of his *Clinical Anatomy* book that the volume of the central nucleus of the BST (BSTc), characterised by the density of its VIP innervation and by its somatostatin fibres and neurons, is smaller in women and only represents 40 % of the volume of this same nucleus in men. In addition, the BSTc in women contains 40 % less somatostatin neurons than that of men (DF Swaab 2007).

Before drawing this amazing conclusion, a number of preliminary observations had contributed to its argumentation:

- In 1990, LS Allen and RA Gorski observed that the volume of BST-dspm is two and a half times greater in men than in women.
- But it was in 1995 that JN Zhou et al. surprised the entire scientific community, especially neuroscience researchers and endocrinologists, by publishing in *Nature* the very interesting results of their research.

These authors compared the central nucleus of the BST, the BSTc, in three groups of individuals, men or women, heterosexuals or homosexuals or even transsexuals, and made the following observations:

Heterosexual women have a smaller BSTc than that of heterosexual men.
Homosexual men have a BSTc which is the same size as that of heterosexual men.
Homosexual women have a BSTc which is the same size as that of heterosexual women.
"Male-to-female" transsexual men have a small BSTc, which is strictly comparable to that of heterosexual women.

These authors also showed that this last observation was independent of any hormonal cause, such as may be believed within a first approach.

- In 2000, FPM Krujiver et al. supplemented the previous study by presenting the case of a "female-to-male" transsexual woman whose BSTc was of a similar size to that of a heterosexual man.
- Also in 2000, A Fernandez-Guasti et al. were studying androgen receptors, AR-ir (androgen receptor immunoreactivity), at the level of the human hypothalamus and also found this type of receptor in the BST.[2]

We still needed to determine at what stage in an individual's life the size difference of the BSTc in men or women becomes obvious. In 2002, a remarkable study conducted by WCJ Chung et al., who were investigating the BSTc from the

[1] "The BST neurons that project to the parabrachial nucleus stain for the same neuropeptides as those in the central nucleus amygdala, thus further demonstrating the anatomical relations between these two structures" (MM Moga et al. 1989, on line 2004).

[2] The androgen and oestrogen receptors had already been highlighted in the BST, in animals, in 1979 by PJ Sheridan and in 1985 by D Commins and D Yahr.

foetal period until adulthood, provided an answer to this question by showing that:

- The volume of BSTc increases significantly with age, but this growth is lower in women than in men so that the BSTc of the adult male is 39 % larger than that of adult women.
- The total number of neurons in the adult BSTc is significantly higher in men than in women (FPM Kruijver had already demonstrated that the adult male BSTc contained the highest number of somatostatin-IR neurons).
- The differential volume for the BSTc can only really be observed during adulthood. This late divergence in volume appears to be a human BST characteristic (LS Allen and RA Gorski in 1990 had already observed that the BST-dpm only appears to be smaller in women, compared to men, as from the age of 14!).
- Given the fact that this late sexual differentiation does not appear to be related to the precocious secretion of gonadal hormones in the foetus, taking also into account the fact that abnormal steroid levels have no impact on the BST during childhood (case of a congenital suprarenal hyperplasia in a girl, referred to by WCJ Chung), gonadal hormone hypothesis does not seem to explain the BNSTc volume difference based on gender. Other hypotheses have been proposed (Y Arai 2004), mostly genetic, including that of a neurochemical secretion of the adult brain in men, induced by SRY, a gene on chromosome Y. We know, such as specified by K Mayer et al. (1998), that the SRY gene is transcribed in the cortex and the hypothalamus of male individuals.

Nevertheless, the existence in the brain of many areas equipped with endocrine receptors suggests that the chapter on sexual differentiation of the brain is far from over!

The multiple experiments conducted by neuroscientists have contributed to confirming the critical role played by the BST in the behaviour of individuals, whether the personal behaviour and moods, the societal behaviour, the sexual behaviour, the parental behaviour (especially maternal) or the eating behaviour. The BST, which is part of the extended amygdala (see Chap. 11), acts as an interface between the amygdala and the hypothalamus (especially at the level of the paraventricular, preoptic and ventromedial nuclei), which explains the endocrine consequences (especially corticotropic) which are added to the somatosensory and vegetative components of the strongest emotions, perceived by the amygdala. Therefore, it is the BST which maintains anxiety disorders, particularly in cases of post-traumatic stress or post-aggression[3]. As suggested by EC Dumont (2009), the BST appears to be "a relay centre with neurocircuits coordinating the activity of autonomic, neuroendocrine and somatic motor systems into fully organised physiological functions and behaviours".

HW Dong and LW Swanson even assigned specific functions to the various sectors of the BST, in a major series of articles. In 2003, the anterior division seems to be involved in the regulation of the eating behaviour; in 2004, the posterior division seems to be involved in the regulation of the reproductive and defensive behaviours; and in 2006a, the antero-medial area seems to be involved in the neuroendocrine, autonomous and behavioural aspects of the energy balance. Finally, in 2006b, the dorsomedial nucleus seems to be involved in the integration of the neuroendocrine, autonomous and beverage absorption responses.

We have only observed in recent years, thanks to two French researchers ((FR George and O Manzoni), the operating process of the NBST and its position in the complexity of brain circuits (M Jalabert et al. 2009).

The NBST targets not only the hypothalamus but also the dopaminergic neurons of the ventral tegmental area, ATV[4] (which is located in the ventromedial part of the mesencephalic tegmentum, in front of the red nucleus (see Fig. 7.19a), and which is one of the "centres of pleasure and reward" such as the septal nuclei and the nucleus accumbens!). These authors have shown that the BST, which receives the terminations of the neurons of the infra limbic prefrontal cortex, does not simply relay the excitatory information of this cortex by sending the axons of its own cells to the dopaminergic cells of the ventral tegmental area. "It behaves like a real amplifier": Without the BST, only 10 % of tegmental dopaminergic cells would be excited. With the BST, 80 % of these cells are excited! Due to the fact that these cells innervate the lateral and ventromedial nuclei of the hypothalamus, we can observe the endocrine effects on the entire body, via the pituitary gland, of such an excitatory gain.

The previous year, the same team (L Massi et al. 2008) identified, in the BST, the CB1 cannabinoid receptors, therefore showing that the dopaminergic cells of the ventral tegmental area are under the inhibitory control of the cannabinoid system (inhibitory control of the CB1 receptors), which suggests possible future therapeutic solutions for controlling dysfunctions of the NBST!

[3]According to M Davis and C Shi (1999), the specific stimulus of fear activates the central nucleus of the amygdala, but when this fear becomes chronic in a state of anxiety, the BST is activated.

[4]ATV projects to other areas of the brain by two major dopaminergic pathways, the mesolimbic pathway (which reaches the structures of the limbic lobe including the nucleus accumbens) and the mesocortical pathway (which reaches the frontal cortex). Both pathways belong to the MFB (medial forebrain bundle of J Olds and P Milner 1954), an axon bundle connecting the reticular formation, the lateral hypothalamus, the amygdala, and the various reward centres, to the prefrontal cortex (See Fig. 8.1).

The other current lines of research concern the neurotransmitters, receptors and synapses that are observed in the BST:

- From recent anatomical knowledge, which showed that the BST receives a major noradrenergic innervation from the nucleus of the solitary tract, FR George and O Manzoni (M Jalabert et al. 2009) were able to demonstrate the influence of adrenaline on the BST neurons and therefore deduct the key role played by the BST in the dopamine–norepinephrine neurotransmitter interactions, which helps to better understand the implications of the BST in opioid dependency and in the opioid withdrawal syndrome.
- We also know that synaptic conduction within the BNST is glutamatergic. ZA McElligott and DG Winder, in 2009, showed that the dysfunctions of these synapses can have serious consequences on behaviours and affect the reward system, by maintaining or triggering addictions, by developing anxiety syndromes and even by causing psychiatric disorders.
- KS Sink et al. (2011) showed that stimulation of the CGRP receptors (calcitonin gene-related peptide), within the BST, generates anxiety phenomena. This process not only affects the BST but also all regions of the brain involved in anxiety phenomena (areas receiving projections from the BST).

These anxiogenic effects are induced (KS Sink et al. 2013a) by the effect of a neuropeptide, CRF (corticotropin-releasing factor), from the BST neurons that act locally on the R1 CRF receptors. The authors suggest that an "enhanced CRF expression within these neurons may contribute to an inappropriate regulation of emotional memories".

It is now officially recorded, such as written by (KS Sink et al. 2013b), that "the BST is a pivotal relay of cortical information to neural structures mediating behavioural and neuroendocrine responses to stress". This concerns the intensity of current neuropsycho-pharmacological research aiming to specifically impact a component of the BST in order to alleviate or cure a disorder caused by BST dysfunctions. It has thus been proposed to treat addiction or anxiety syndromes, stress-related psychopathologies and even certain psychiatric disorders with inhibitors of CGRP receptors or even of R1 CRF receptors.

The Concept of "Extended Amygdala" 11

In 1923, the embryological research and comparative anatomy works performed by JB Johnston revealed the existence, in embryos of lower non-mammalian vertebrates, of grey matter which surrounds a longitudinal association pathway connecting the forebrain and temporal areas. This gave him a glimpse of the existence, in mammals, of a more or less continuous equivalent column of neurons, which follows the stria terminalis, extending from the nucleus accumbens to the amygdala.

JS De Olmos et al. (a team whose major fundamental work concerning the amygdala has already been mentioned) undertook, in 1985, to update this discovery that medical literature had forgotten for over half a century.

Furthermore, the outstanding work of GF Alheid and L Heimer, in 1988, and then of LJ Martin et al. in 1991, detailed and clarified the concept of "the extended amygdala".

It is to Alheid and Heimer that we owe the English terminology of *Extended Amygdala*, which shows that this structure functionally originates from the amygdala, of which it is a true extension, and thereby increases its functional capacity.

We can define the extended amygdala as a more or less complete neuronal continuum of small neuronal masses occupying the base of the forebrain from the BST to the central and medial nuclei of the amygdala.

GF Alheid, 2003 and 2009, more accurately, defines the extended amygdala as "a large basal forebrain macrostructure encompassing a contiguous but heterogeneous collection in excess of 16 interconnected and functionally related neuronal aggregates or subnuclei, extending from the centromedial amygdala to the bed nucleus of the stria terminalis". Such as remarked by GF Alheid, it is important to note that the cortical nucleus, the lateral and the basal complex are excluded from the above definitions of the extended amygdala, despite the fact that it is known that these nuclei project to this structure.

The description of the extended amygdala is complex due to the fact that its content is heterogeneous while it has a relatively homogeneous macroscopic aspect as a long cellular chain connecting the BST and amygdala. It in fact consists of cellular islands and aggregates, which are dispersed rather than clustered, connected by axonal bundles and cellular bridges. Its shape is unique and includes two bends (Figs. 11.1 and 11.2): the first bend occurs between the amygdaloid nuclei and the sub-lenticular basal horizontal portion; the second bend nearly forms a right angle between this horizontal section and the BST, which is positioned sagittally.

The specificity of the extended amygdala is related to the unity of the cellular stream which forms it:

- The neurons of the sub-lenticular horizontal portion and those of the central and medial nuclei of the amygdala have the same immunohistochemical characteristics, especially regarding VIP (GM Abrams et al. 1985), neurotensin and above all, angiotensin II (GF Alheid and L Heimer 1988).
- Similarly, as recalled by GF Alheid and L Heimer, numerous studies have referred to morphological, histological and histochemical similarities between the BST and the central and medial nuclei of the amygdala (AJ McDonald 1983; G Holstege et al. 1985; JL Price et al. 1987).
- We were able, ourselves, to observe the continuity of the extended amygdala components on MRI images of brain sections (see MRI section, etc.).
- Similarly, the unity of the extended amygdala appears on sections stained by our Fe Cl3 staining process (see anatomical sections).
- Furthermore, we used the tractography process to obtain interesting pictures showing the axonal pathways connecting the BST, the amygdala and the temporal region (Figs. 11.1 and 11.3).

It is to be recalled that the basal, horizontal, intermediate, sub-lenticular portion of the extended amygdala is the most complex (JS De Olmos, L Heimer 1999). It is located under the anterior commissure and laterally comprises, near the

V. Di Marino et al., *The Amygdaloid Nuclear Complex: Anatomic Study of the Human Amygdala*,
DOI 10.1007/978-3-319-23243-0_11

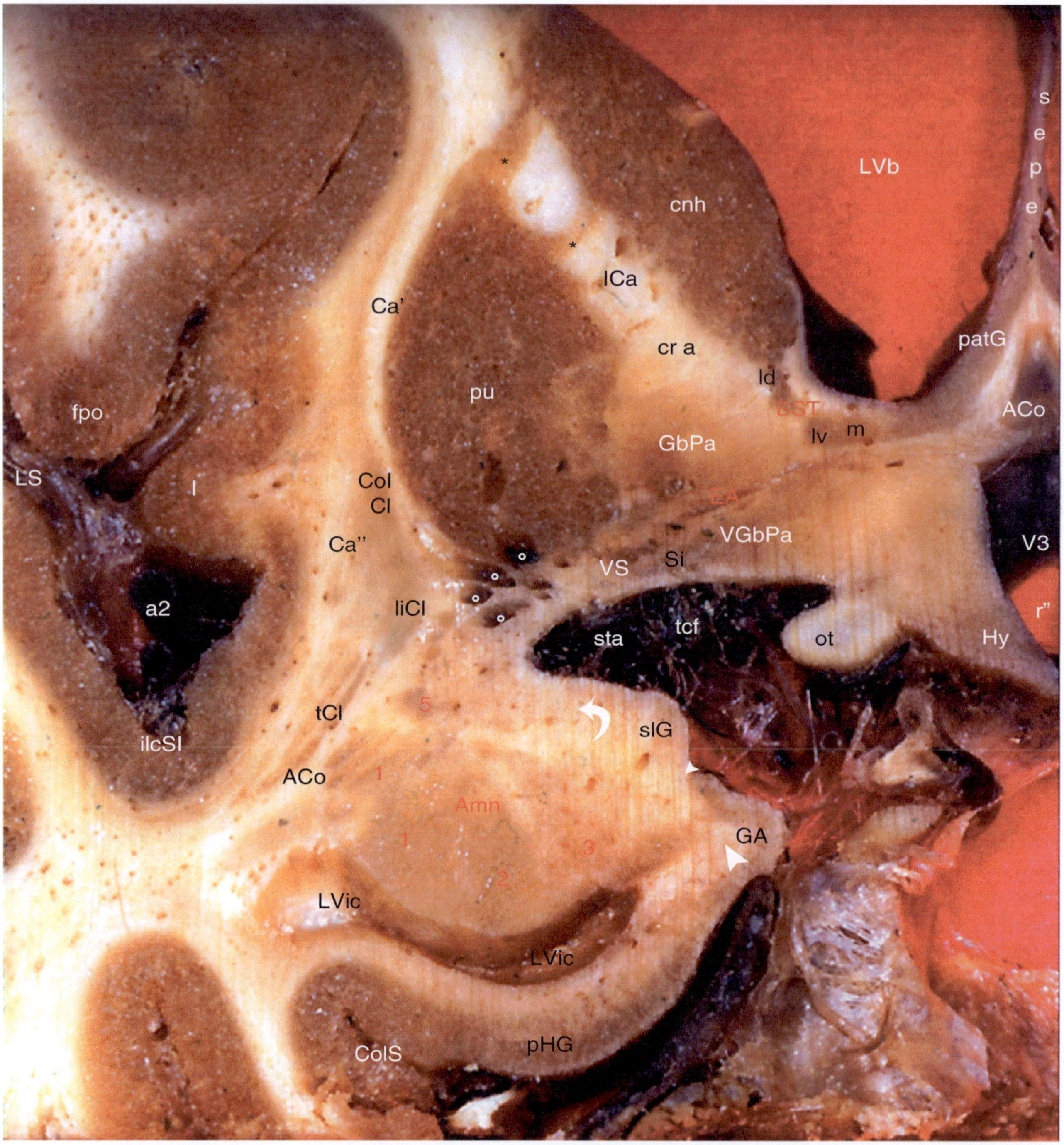

Fig. 11.1 Coronal section of the brain (right half portion) (*Through the amygdaloïd nuclear complex and the BST*). *ACo* anterior commissure, *Amn (red keys)* amygdaloid nuclear complex, *1* lateral, *2* basolateral, *3* baso-medial, *5* central, *BST* bed nucleus of the stria terminalis, *ld* dorsal division of the pars lateralis, *lv* ventral division of the pars lateralis, *m* pars medialis, *Ca'* external capsule, *Ca"* extreme capsule, *cnh* caudatus nucleus head, *ColCl* compact insular claustrum, *ColS* collateral sulcus, *EA* extended amygdala, *fpo* fronto-parietal operculum, *GA* gyrus ambiens, *GbPa* globus pallidus, *Hy* hypothalamus, *I* insula, *ICa (cr a)* internal capsule (crus anterior), *ilcSI* inferior limiting circular sulcus of insula, *liCl* limitans claustrum, *LVb* lateral ventricle (body), *LVic* lateral ventricle (inferior cornu), *LS* lateral sulcus, *ot* optic tract, *patG* paraterminal gyrus, *pHG* para-hippocampal gyrus, *pu* putamen, *r" (inf r)* infundibular recess, *sepe* septum pellucidum, *Si* substantia innominata, *slG* semi-lunaris gyrus, *sta* striate artery, *tcf* transverse cerebral fissure, *tCl* temporal claustrum, *V3* third ventricle, *VGbPa* ventral globus pallidus, *VS* ventral striatum, ** putaminocaudate bridges, °° *white circles* branches of the striate artery, *a2* middle cerebral artery, *curved white arrow* endorhinal sulcus, *little white arrowhead* semi-annularis sulcus, *great white arrowhead sulcus* uncal notch

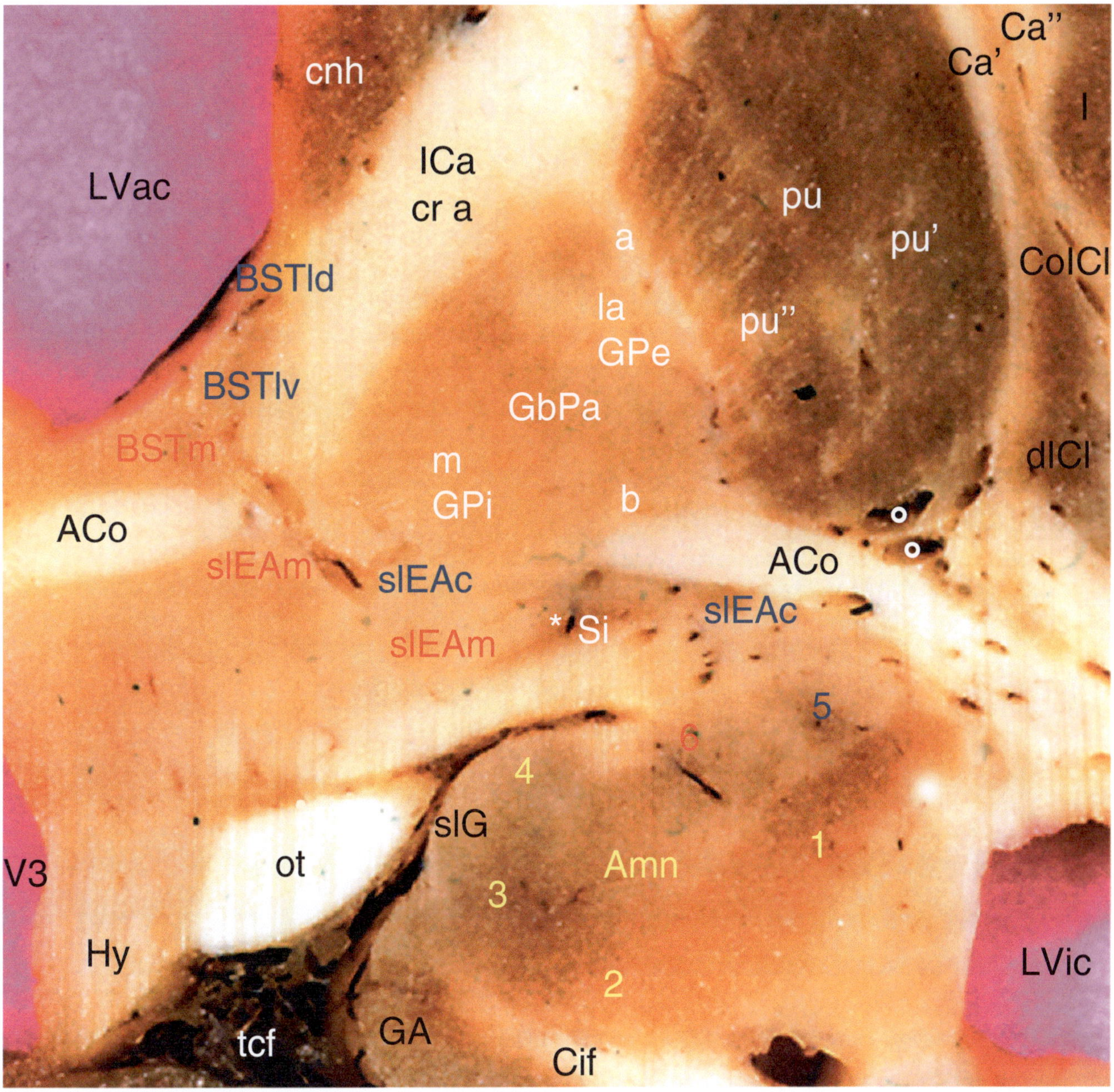

Fig. 11.2 The extended amygdala on a coronal section of the brain through the anterior commissure. *Aco* anterior commissure, *Amn* amygdala, *1* lateral nucleus, *2* baso-lateral nucleus, *3* baso-medial nucleus, *4* cortical nucleus, *5* central nucleus, *6* medial nucleus, *BSTld* dorsal part of the BST's lateral division, *BSTlv* ventral part of the BST's lateral division, *BSTm* medial division of the BST, *Ca'* external capsule, *Ca"* extreme capsule, *Cif* cingulate fasciculus, *cnh* head of the caudate nucleus, *coICl* compact insular claustrum, *dICl* diffuse insular claustrum, *GA* gyrus ambiens (ambient gyrus), *GbPa* globus pallidus of the lentiform nucleus; *la* (*GPe*: globus pallidus, external part), pars lateralis; *m* (*GPi*: globus pallidus, internal part), pars medialis; *a* external medullary lamina; *b* internal medullary lamina, *a* external medullary lamina, *b* internal medullary lamina, *Hy* hypothalamus, *I* insula, *ICa* internal capsule, *cr a* crus anterior, *LVac* lateral ventricle, anterior cornu, *LVic* lateral ventricle, inferior cornu, *ot* optic tract, *pu* putamen, *pu'* lateral part of the putamen, *pu"* medial part of the putamen, *Si* substantia innominata, *slEAc* central division of the sublenticular extended amygdala, *slEAm* medial division of the sublenticular extended amygdala, *slG* semi-lunaris gyrus, *tcf* transverse cerebral fissure, *V3* third ventricle, * basal nucleus of Meynert, ° arteries of the putamen; The elements of the central corridor of the extended amygdala are noticed with blue-coloured letters; the elements of the medial corridor of the extended amygdala are noticed with *red-coloured letters*

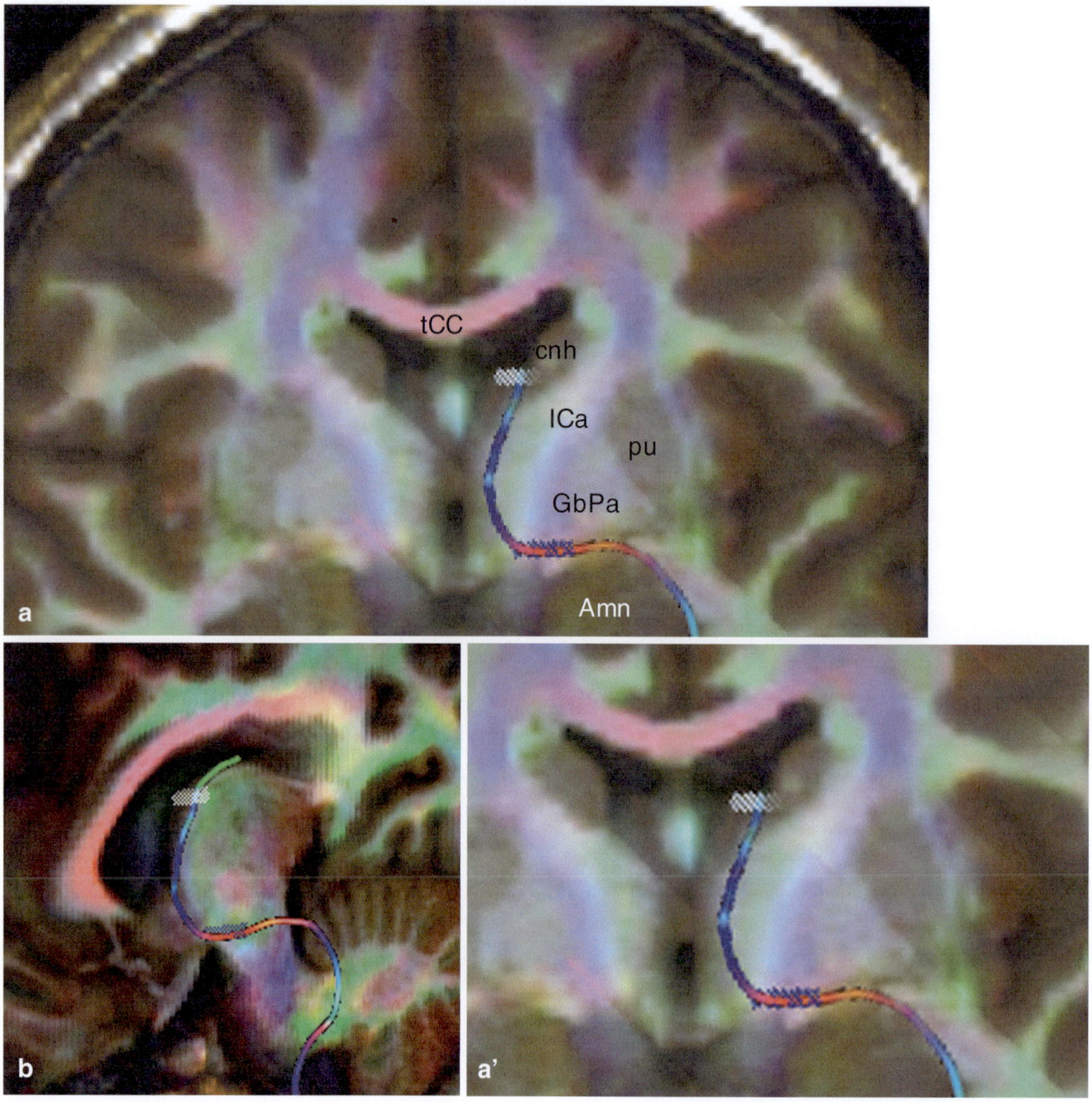

Fig. 11.3 Diffusion Tensor Imaging Fibres Tracking (DTI FT) of the extended amygdala's pathway. (**a**) coronal section; (**a'**) enlargement of **a**. (**b**) three quarter view, in order to visualise the change of plan and the two parts of the pathway. *Amn* amygdaloid nuclear complex, *cnh* caudate nucleus head, *GbPa* globus pallidus, *ICa* internal capsule, *pu* putamen, *tCC* truncus of the corpus callosum, *white-shaded* BST target, *blue-shaded* Si (substantia innominata) target

amygdala, a few elements of the ventral striatum, more medially, the sub-commissural portion of the pallidum or ventral pallidum, and even at the striato-pallidal junction, digitations perpendicular to the axis of the basal portion, which correspond to interpenetrations of the ventral pallidum into the ventral striatum (L Heimer et al. 1991). In addition, at the level of the lower part of the base, elements belonging to the anterior perforated substance mingle with the above-mentioned elements. This complex region, which is crossed by the extended amygdala flow, is what K. Reichert[1] has called, for decades, "the innominate substance"[2] and that JS De Olmos and L Heimer have developed as a concept: "the concept of the ventral striatopallidal system".

It is to be recalled that several systems pass through the innominate substance:

- The striato-pallidal system interposed between the cell columns that connect the cortex to the brainstem via the

[1]Karl Bogislaus Reichert (1811–1883), a German anatomist, known for his embryological work on branchial arches and his anatomical research on the brain and the inner ear.

[2]It is to be recalled that in 1809, Johann Christian Reil, also a German anatomist, had called this area of the base of the brain "die ungennante Mark-Substanz" to express the fact that he was not able to understand its organization!

thalamus, red nucleus, and substantia nigra (see picture b, Fig. 11.4).
- The olfactory system with the olfactory tubercle, the anterior perforated substance, the prepiriform and periamygdaloid cortices.
- The septo-hippocampal system that connects the septal nuclei, the nucleus of the diagonal band, the diagonal band, the amygdala and the hippocampus.
- The extended amygdala, interposed between the cell columns which connect, via the BST, the temporal cortex to the hypothalamus, the thalamus and the brainstem.

The cell columns in the extended amygdala are fairly easy to differentiate from other cells belonging to the striato-pallidal ventral system: "Many of the cells in the sublenticular extended amygdala of the human are medium-sized and elongated, with the larger diameter uniformly diagonally-directed along the orientation of the interconnecting fibre bundles" (L Heimer et al. 1991).

GF Alheid and L Heimer have identified two overlapping sections within the extended amygdala: a medial section and a central section (Fig. 11.2). They are referred to in this manner because each section is a sort of pseudopodium of the corresponding amygdalar nucleus.

We therefore obtain a "medial division of the extended amygdala" whose origin is at the medial nucleus of the amygdala and a "central division of the extended amygdala" whose origin is at the central nucleus of the amygdala. The "medial division" forms a cell column located in the ventral or postero-ventral part of the innominate substance (Siv) and terminates at the medial portion of the BST. The "central division" forms the cell column located in the dorsal or antero-dorsal portion of the innominate substance (Sid) and terminates at the lateral part of the BST. Still according to GF Alheid and L Heimer (1988), due to the connections of each of the two territories, they are perfectly easy to identify from an anatomical viewpoint: the "medial division" is more particularly connected to the nuclear groups of the medial hypothalamic area, while the "central division" is preferably connected to the lateral hypothalamic nuclei. The medial division was involved in certain aspects of social behaviour (psycho-endocrine response to intra- and inter-specific aggressions) and reproductive behaviour (GF Alheid 2009). The central division not only has reciprocal connections with the lateral hypothalamus but also with the midbrain tegmentum, the dorso-lateral pons and the medulla oblongata. It projects (GF Alheid 2009) to the premotor areas but also to more complex sensorimotor regions such as the periaqueductal grey matter, the parabrachial complex and even the nucleus of the solitary tract. "Other theoretically important targets include the dopamine cells of the midbrain ventral tegmentum and the serotoninergic and noradrenergic neurons of the pons" (GF Alheid 2009). All of these inputs explain that the neurons of the central division of the extended amygdala are involved in motor, endocrine and autonomic conditioned responses to fear, anxiety and stress but also in the impact of reward, generated by abused drugs.

However, the anatomy of the innominate substance is even more complex due to the fact that in addition to the striato-pallido ventral system, the anterior perforated substance and the extended amygdala, this "poorly defined cell mass" contains a special nucleus: the *nucleus of Meynert*, which has extremely rich and complex afferent and efferent fibres. Moreover, this nucleus is not the only nucleus of the extended amygdala, and we can also associate with this cellular stream, the nucleus of the diagonal band and nucleus accumbens. In the anatomy of this innominate substance, we must also include the fibrillar structures which border it (especially the anterior commissure) or pass through it along the axis of the extended amygdala, not only the long axons that connect the rostral and caudal portions of the extended amygdala (L Heimer et al. 1991; L Heimer et al. 1997), not only the long intrinsic association fibres of the innominate substance but even the fibres of the ansa lenticularis and the inferior thalamic peduncle (see Chap. 6).

The nucleus accumbens (the complete Latin name is nucleus accumbens septi, i.e. nucleus lying against the septum) is a nucleus adjacent to the septal region with an oval shape and a major axis orientated upwards and inwards, which corresponds to the rostral portion of the extended amygdala, where it occupies a surface adjacent to the pre-commissural BST (see Fig. 4.8, 8.1, 8.3 and 10.1a). This nucleus has two portions: a ventral portion, described as a "shell" (or "coquille" in French), which is considered as part of the extended amygdala, and another dorsal portion, the "core", which is often considered as not being part of the extended amygdala.

What is certain is that, in addition, the nucleus accumbens is also part of the striatum. This nucleus has the same characteristics and location as the striatum. It is positioned under the head of the caudate nucleus and near the inferior edge of the putamen, in the inclined part of the striatal area's bottom, and is more or less flush with the lower edge of the base. It therefore forms (with the olfactory tubercle) the ventral striatum. It is also to be acknowledged that the nucleus accumbens has (S Nery et al. 2002) an embryological origin (caudal ganglionic eminence), which it shares with the BST and the central nucleus of the amygdala (the latter point still being subject to discussion!).

The nucleus accumbens is an important nucleus for the pleasure and reward system. It is particularly involved in the "reward" effects of abused drugs.

Treatment attempts (V Sturm et al. 2002 and 2003) of obsessive compulsive disorders (OCD) by deep brain stimulation (DBS) of the ventral portion of the nucleus accumbens have been carried out with patients not responding to medical and psychotherapeutic treatments. According to thera-

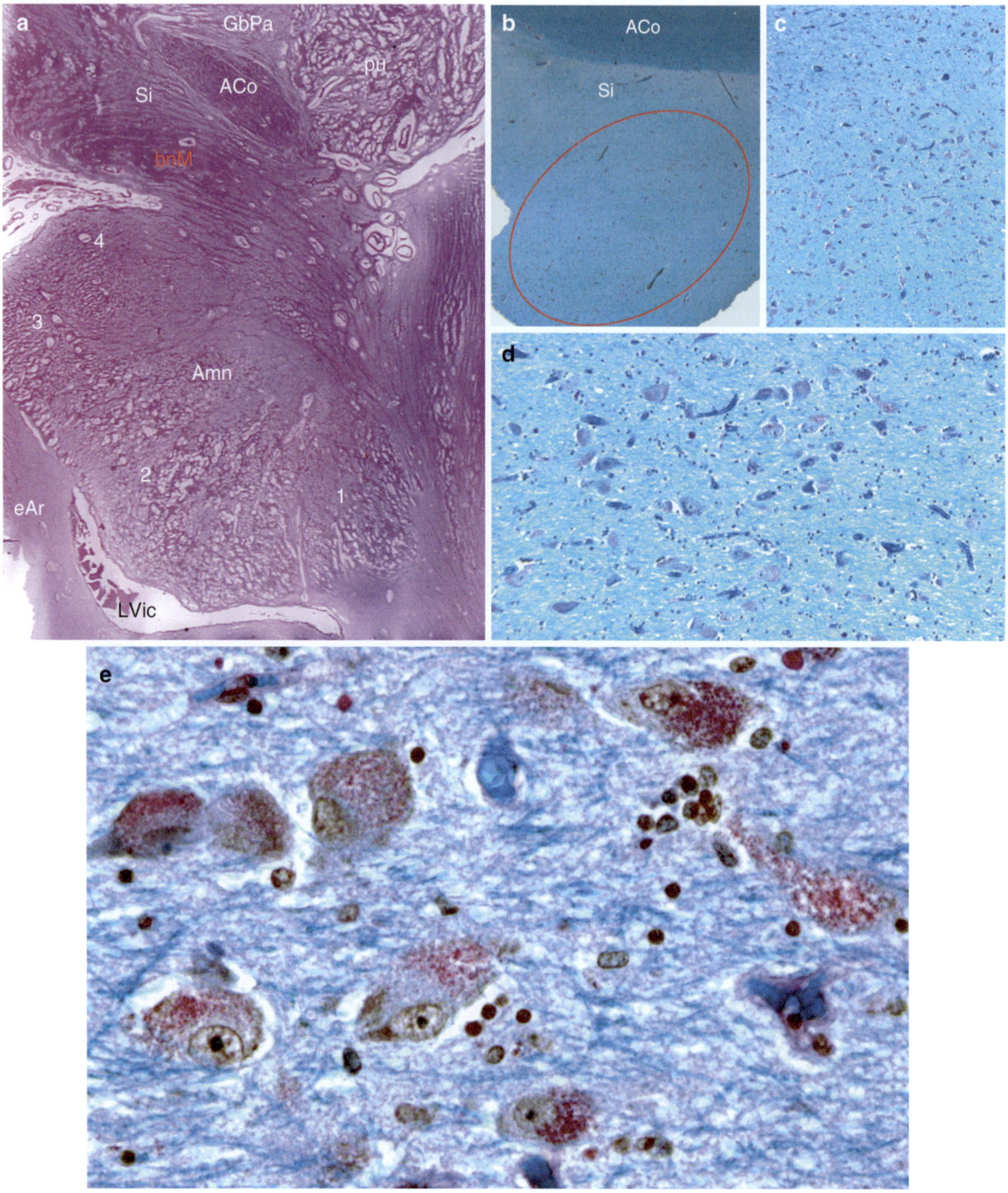

Fig. 11.4 The Basal nucleus or Meynert's nucleus (a very important component on the route of the extended amygdala). (**a**) coronal slice of the brain, stained by hematoxylin–eosin, showing the Meynert's nucleus, situated immediately below the anterior commissure and above the amygdala. (**b**) enlargement of **a**, locating the area of the Meynert's islands in the substantia innominata; (**c–e**) Luxol fast blue–Pas staining (×100, ×200, ×400) reveals globular-shaped neurons; Their cytoplasm have brown and PAS positive granular material due respectively to their neuromelanin and lipofuschine contents. *Aco* anterior commissure, *Amn* amygdaloid nuclear complex, *bnM* basal nucleus of Meynert, *eAr* entorhinal area, *GbPa* globus pallidus, *LVic* lateral ventricle, inferior cornu, *pu* putamen, *Si* Substantia innominata, *1* lateral nucleus of the amygdala, *2* baso-lateral nucleus, *3* baso-medial nucleus, *4* cortical nucleus

pists, the objective was to interrupt the transmission of any pathological information between the amygdala, the basal ganglia and the prefrontal cortex.

Certain results provided by these stimulations were promising but insufficient.

The *nucleus of Meynert*[3] (sometimes assigned to Kolliker[4]). As recalled by F Tagliavini (2009), comparative anatomy has shown that it is a highly evolved structure: "exclusively characteristic of the mammalian brain, it becomes progressively larger and more complex with an increasing cerebralisation and reaches its maximum volume and differentiation in species which have the greatest cortical development such as cetaceans or higher primates" (R Nieuwenhuys et al. 2008). In humans, it appears either as a flattened olive-shaped nucleus, located below the anterior commissure, in the middle portion (approximately in the middle third section) of the ventral pallidum (see picture a, Fig. 11.4), interposed between the dorsal and ventral parts of the innominate substance and therefore between the central and medial divisions of the extended amygdala, or as separate islands, arranged in more or less large groups within the innominate substance. Such as already noted by Ch. Foix and J. Nicolesco, the cells of the nucleus of Meynert are arranged as "cellular islands in a tight but discontinuous configuration and are characterised by their chromophilia" (Fig. 11.4). The shapes of the cells vary and are either multipolar or bipolar, oval, spindle-shaped or ellipsoidal with a major axis which is generally directed obliquely downwards and inwards. The Golgi stain shows multipolar neurons with long dendrites which mingle with those of neighbouring neurons. The size of these neurons is also variable, and medium-sized as well as large cells (the most hyperchromic) can both be observed. The magnocellular contingent in humans at birth has been estimated at around 450,000 (F Tagliavini 1984). The histochemical processes (labelling with choline acetyltransferase or acetyl cholinesterase) show that all the large cells and many medium-sized cells are corticopetal *cholinergic* cells.

The central positioning of the basal nucleus shows that its cells are connected both to the amygdala, the hippocampus, the diagonal band and the anterior perforated substance; to the components of the lenticular nucleus, the BST and the hypothalamus; and to the entire limbic cortex. The role of these cholinergic neurons is essential: they are activators whose action is to modulate the inhibitory neurons of the cingulum by slowing down their action. The activating effect of cholinergic neurons is linked to the production of acetylcholine.[5]

The cells of the basal nucleus not only directly synthesise this major neurotransmitter but they also synthesise choline acetyltransferase, an enzyme which catalyses the following reaction: choline + acetyl coenzyme → acetylcholine.

The work of MM Mesulam et al. (1983), on primates, involving the use of retrograde axonal transport methods, showed that the basal nucleus neurons target "the amygdala as well as various regions of the neo-cortex: the frontoparietal operculum, the prefrontal, insular, parietal posterior, inferior temporal, peristriated areas and the posterior sector of the superior temporal gyrus".

In 1984, MM Mesulam and EJ Mufson also demonstrated, by using autoradiography studies, and still in primates, that the basal nucleus receives inputs only from a limited number of cortical areas: "orbitofrontal cortex, anterior part of the insula, temporal pole, prepiriform olfactory cortex, perirhinal cortex, entorhinal and medial temporal areas". As for the inputs of subcortical structures, they come from the "septal nuclei, of the nucleus accumbens-ventral pallidum area, from the medial dorsal nucleus of the thalamus, from the medial hypothalamus, from the amygdala and from the interpeduncular nucleus".

Overall, such as highlighted by F Tagliavini et al (1984), most afferent nerves at the basal nucleus "originate from limbic and paralimbic cerebral structures. This suggests that the nucleus acts as a cholinergic relay between emotional and motivational areas and the cortex".

11.1 The MBF System

The study of this entity, the "Magnocellular Basal Forebrain System" is intimately linked to the basal nucleus of Meynert, which itself is referred to as the "magnocellular basal nucleus". It has in fact been noted by most authors that cells identical to those of the basal nucleus are sometimes located in aberrant positions, outside the nucleus and even outside the innominate substance, especially in the external medullary lamina of the lenticular nucleus between the putamen and the pallidum (which had already been observed by Ch Foix and J Nicolesco in 1925), in the dorsal pallidum and its lateral edge or in even in the BST! In 1984, JC Hedreen et al. found other locations for these cells: in the white matter, underneath the putamen and nucleus accumbens, on the

[3] Meynert Theodor (1833–1892), a German anatomist and neuro-psychiatrist, author of "Clinic of the forebrain, based on its structure, its functions and its nutrition"), translated in French "Clinique ds maladies du cerveau antérieur, basée sur sa structure, ses fonctions et sa nutrition" by Dr. G Cousot. Manceaux editions, Brussels, 1888

[4] Kölliker Rudolph Albert (1817–1905), a Swiss anatomist and physician, was a professor of Physiology and Comparative Anatomy in Zurich. His works on Microscopic Anatomy and Embryology remain outstanding.

[5] This is the reason why, in the case of neurodegenerative diseases in which alterations are observed in the basal nucleus (rarefaction of its cells in the case of Alzheimer's disease or invasion of the nucleus by protein inclusions in the case of dementia with Lewy bodies), neurologists are attempting to treat induced cognitive impairment by administrating cholinergic drugs or anticholinesterases (for neutralising the cholinesterase, enzyme which destroys acetylcholine).

medial edge of the internal capsule, in the lateral hypothalamus and on the edges of the anterior commissure. There are also scattered neurons within the putamen or nucleus accumbens (which can be recognised by their large size, which contrasts with the small neurons of these two formations).

This MBF, which is a characteristic of primates, appears as a mixture of isolated or clustered cell populations. Some are very specific. The case of the nucleus of Meynert is not unique and other components are included in the MBF: the nucleus of the diagonal band (dorsal and ventral portions, according to JC Hedreen), the largest component of this magnocellular system (L Heimer et al.) and the medial nuclei of the septum. These nuclei, as well as the basal nucleus, consist of large and extremely basophil and cholinergic cells, which project to the amygdala and the hippocampus, but also and especially to the cortex.

However, the MBF (especially in its non-nuclear areas) contains not only cholinergic neurons. Such as has been recalled, it is a mixture of neurons, within which are also observed GABAergic neurons and other types of neurons whose neurotransmitters are neither acetylcholine nor GABA. All of these neurons are corticopetal and have axons that come in contact with cortical pyramidal neurons (especially neurons of the frontal, premotor, motor, somatosensory and cingulate areas and hippocampal pyramidal neurons that they are either to inhibit or activate. This explains that the MBF is involved in various situations including the regulation of sleep, awakening,[6] vigilance and sexual arousal.

The cholinergic neurons of the MBF were labelled in four categories, ch1, ch2, ch3 and ch4 (classification of MM Mesulam et al. 1983). Each of these categories specifically projects to the different parts of the telencephalon but also the diencephalon and brainstem. Therefore, the ch4 group, to which belongs the nucleus of Meynert, projects to the telencephalon (neocortex, amygdala and hippocampus), the thalamus (reticular nucleus), the hypothalamus (anterior, posterior and preoptic nuclei) and the brainstem (interpeduncular nucleus, ventral tegmental area, periaqueductal grey matter, dorsal nucleus of the vagus nerve, nucleus of the solitary tract, locus coeruleus, parabrachial nuclei and septal nuclei).

The MBF includes both ascending and descending fibres. The ascending fibres use, on the one hand, the mesolimbic pathway, which connects the ventral tegmental area to the structures of the limbic system and especially to the nucleus accumbens, and, on the other hand, the mesocortical pathway, which essentially terminates at the frontal cortex.

Cholinergic neurons are impaired in various neurological diseases (their number decreases, they degenerate or they are "smothered" by the local production of an abnormal protein).[7] This is the cause of many serious and invalidating deficiencies.

L Heimer et al. (1991) have the merit to have shown that many degenerative diseases involve structures that belong to the basal brain (in Parkinson's disease, impairment of the dopaminergic system concerning the ventral striatum, such as the dorsal striatum; in Huntington's disease, progressive neuronal loss throughout the entire striatum, including the ventral striatum; in Alzheimer's disease, joint impairment of the hippocampus, entorhinal area, amygdaloid body and extended amygdala). Heimer also showed that such impairments of the basal brain, which are often linked to impairments of the prefrontal areas, characterise psychotic disorders and even schizophrenia. In actual fact, in some patients, related symptoms can be observed: "Alzheimer patients" showing Parkinson's disease symptoms (myoclonias, rigidity, bradykinesis), "Huntington patients" suffering from memory loss and maniac depressive psychotic disorders, a schizophrenic developing motor disorders.

These examples show that the common substratum of all these neurological impairments is located at the level of the innominate substance (L Heimer et al. 1997) of the basal brain and especially of this major extended amygdala, which links structures whose integrity is essential for normal brain operation!

[6]The MBF itself is a relay level of the large reticulo-cortical pathway which has been referred to by two names, which can be superimposed, the AAS (ascending arousal system) or ascending reticular arousal system, ARAS and the RAS (reticular activating system). It manages, from the reticular nuclei of the brainstem and with the help of the thalamus, states of sleep and consciousness (JRL Schwartz et al. 2008).

[7]DL Price et al. have shown, in monkeys, that (AChE)-rich terminals of nbM axons may be involved in the formation of neuritic plaques.

From Prefrontal Lobectomies to Amygdalectomies

12

The first psychosurgery attempts date back to 1891. The alienist, Dr Gottlieb Burckhardt,[1] performed a cortical resection using a curette on either side of the central sulcus of the brain in six patients presenting severe agitation. In one of the patients, agitation was significantly attenuated, but another died and the last one became epileptic. These results provoked strong emotions in the local medical community, forcing Burckhardt to stop his surgeries.

It was not until 1935 that psychosurgery truly entered an active phase. In this period between two wars, E. Moniz,[2] a Portuguese neurologist, performed the first prefrontal lobotomy in restless and anxious subjects. Moniz, whose jurisdiction was unanimously recognised (he developed the cerebral angiography) and who also benefited from political influence (he was a foreign minister), disseminated psychosurgery widely, which had a particularly significant impact in the USA, especially in Stanford University with WJ Freeman[3] who performed 3,500 lobotomies!

The psychological sequelae presented by soldiers returning from World War II represented a land of choice for these prefrontal lobotomies. In 1943, the "Veteran's Administration" also encouraged neurosurgeons to acquire additional training in this area.

Moniz was awarded the Nobel Prize in 1949 for his efforts and his achievements in psychosurgery. It should be noted that at that time, neuroleptics were unknown; they were introduced in psychiatry by J. Delay and P. Deniker only from 1952. These "chemical straitjackets" transformed psychiatry and the image of the hospitals involved. The prefrontal lobotomy (illustrated in 1975 by Milos Forman's film "One Flew Over the Cuckoo's Nest" starring Jack Nicholson) gradually fell into disuse as the use of neuroleptics increased.

Admittedly, surgical results were more beneficial to the patient's entourage than to the patient. Indeed, these patients' aggressive behaviour after prolonged military conflicts or their psychologically trauma caused by the emotions of combat represented an ongoing risk to those around them and to society.

Advances in behavioural neurobiology revived surgical techniques. In 1937, the work of H. Klüver and PC Bucy (see History chapter) on the limbic lobe of the monkey, and more specifically the amygdala, revealed this structure as the regulatory centre of emotions. WH Mark and F Erwin (1970) emitted the idea that violence was the expression of a deficient inhibitory control of aggression called "dyscontrol syndrome" that only the removal of the amygdala could cure.

In this period, new contemporary armed conflicts (Vietnam War from 1965 to 1975), a Japanese surgeon, H Narabayashi (1972), treated hyperkinetic children with localised destruction of the amygdala. Surgical results were satisfactory for aggressiveness but altered considerably the affectivity of operated patients. Again, it must be emphasised that the surgical indication by the admission of Narabayashi was primarily driven by the entourage, perhaps with a view towards academic achievement and upward mobility. The Japanese surgeon confided that it was mostly the father who prompted the consultation and was satisfied with the result (the child becoming more obedient), while the mother reconsulted him privately to ask if he could return her child to the same state as before the surgery.

Despite advances in knowledge of the limbic system, D Carroll and MAJ O'Callaghan reviewed the numerous surgeries and declared in 1981 in a publication that the surgical procedures were not based on solid and indisputable scientific basis. The most expressed reservations about ethics were made immediately. However, as was the case for lobotomy, the discovery of drug treatment, methylphenidate (Ritalin), quietly sounded the death knell of this "rational psychosurgery". Thus, treatment of hyperkinetic children was no longer by surgery.

[1]Gottlieb Burckhardt (1836–1907), Swiss psychiatrist, Medical Director of small mental hospital in Prefargier, Switzerland.

[2]Egas Moniz (1874–1955), Portuguese neurologist and psychosurgeon.

[3]Walter Jackson Freeman (1874–1955), neurologist, controversial father of lobotomy.

V. Di Marino et al., *The Amygdaloid Nuclear Complex: Anatomic Study of the Human Amygdala*,
DOI 10.1007/978-3-319-23243-0_12

We mention, just as a reminder, cingulectomies. In 1980, S Corkin, who had performed the procedure in 85 patients, wrote frankly that he did not understand why the surgery proved effective in some cases and not in others. We also cite hypothalamotomies performed by German neurosurgeons (based on Hess's hypothalamic stimulation experiments) with their adverse metabolic consequences.

Although the era of psychosurgery is now over, epilepsy surgery, especially ATL (anterior temporal lobectomy) and SAH (selective amygdalohippocampectomy), has highlighted the role of the amygdalo-hippocampal block (A Pitkänen et al. 1998).

Indeed, resections (selective or not) on the dominant hemisphere result in impairement of verbal memory and short term cognitive changes (LH Goldstein CE Polkey 1993), while those on the minor hemisphere reach the visuospatial memory (AS Wendling 2012). Interpretation of these problems is difficult because the role of antecedent epilepsy or amygdala damage (such as Kluver–Bucy syndrome, where only lack of emotion and fearlessness are associated with amygdala lesions) cannot be completely excluded.

Another iatrogenic consequence of amygdalohippocampectomy is severe depression (found in 8 % of patients in the Danish series studied in 1994 by AS Naylor et al.).

In rare cases in humans, the effects of bilateral amygdala destruction have led to interesting clinical findings. Total surgical and exclusive amygdala lesions induce a state of emotional indifference and calmness but also behavioural problems. AR Damasio wrote in his book "L'erreur de Descartes [The Error of Descartes]", the case of a woman with this type of bilateral lesion: "extravagant behaviour, deficient emotional reactions, inadequate social relations… with little concern for problematic situations in which she often finds herself".

Various neuropsychological studies (M Brand et al. 2007, D Tranel and BT Hyman 1990), on a rare disorder, Urbach–Wiethe disease (leading to the gradual and complete destruction of both brain amygdalae), allowed a clear view in humans of the consequences of such bilateral deprivation: foremost among them was the disappearance of any feeling of fear but also decision-making (especially ambiguous or risky decision-making) and even executive functioning disorders were severely affected. These patients recognised faces, but were unable to have any emotion to facial expressions whether negative or positive (series of 10 patients studied by M Siebert et al. 2003). These patients, due to the absence of fear (they can handle snakes and tarantulas without any problem!), can often put their lives in danger. However, recent work (JS Feinstein et al. 2013) has highlighted the absolute role of the amygdala in fear. The authors studied three "Urbach–Wiethe" patients, each breathing air containing 35 % of CO_2, and found in all cases that the patients were not only seized by fear due to the impression of asphyxiation but experienced intense panic. The authors therefore concluded the existence of two types of fear: external fear that requires the presence of the amygdala and internal fear with the sense of impending doom that is under the control of another part of the brain…yet to be discovered dorso medial thalamus (WW Roberts, 1962)?

Conclusion

Un peu de science éloigne de Dieu
mais beaucoup de science y ramène – L. Pasteur[1]

After our extraordinary journey through the amygdala and the central nervous system, we are pleased to be able to share with our colleagues and our students our observations and discoveries.

Our anatomist's vision was immediately captivated by the ideal positioning of the amygdala, offering unprecedented ease of connection to all structures of the central nervous system.

Studying the amygdala prompted us to transition from macroscopy to microscopy as we soon realised that the anatomical dissection allowing us to visualise and extract the little almond for ex situ study was actually a unified complex of small nuclei with different functions. Thus, we used ultra-dissection to clear and separate the constituent nuclei, which enabled us to observe them under the microscope, with the help of our microscopy specialist colleagues, and distinguish their respective structures: the parvo- and magnocellular areas, groups of specialised neurons, the different neuronal morphologies, etc.

For networks connecting the amygdala to other brain structures, we used macroscopy, often overlooked by neuroscientists at all levels and especially by beginners that launch headlong into the study of the projections in animals, at the request of their research directors, when they do not know or may have forgotten the neuroanatomy of the human brain. Our work has led us to establish stunning pictures of networks, already observed by our predecessors and described in the literature, but without any images. We also highlighted circuits poorly described or not at all, such as the one that connects the amygdala to the various targets of the diagonal band, hoping to arouse the interest of new researchers.

From all the data we have collected, the most recent raise various issues:

- The expanding the role of the amygdala, which is now known to be activated by both aversive emotional stimuli and hedonic stimuli
- The necessary integration within the holistic limbic system whose boundaries are constantly pushed to the point of soon affecting the whole brain
- The fact that the amygdala is not the only fear centre and two additional candidates for this important role emerge: the thalamus (as already mentioned by Roberts in 1962) and the insula
- The ongoing dialogue established between the orbitofrontal cortex and the amygdala, a wonderful example of modulation, which probably explains in humans the quality of decision-making, motivation and attention at the mercy of emotions experienced
- The involvement of the amygdala, through its links with the septum and the MBF (medial forebrain bundle), to the continuum connecting the cerebral cortex and brainstem structures (or vice versa), connections that we have been able to demonstrate by dissection and photographs

Our knowledge of the amygdala is still its infancy: its targets and its multiple roles that we know help preserve good homeostasis of the body and thus preserve the individual. A wide range of expertise is opening up to us where current resources allow us to expand our knowledge: increasingly powerful MRI 7T, which gives us hope of viewing and exploring the amygdala routinely, fMRI to study how the amygdala reacts to others and to all scenes, situations, joys and sorrows, in short, everything in a moment of life, the study of isolated amygdala neurons and the study of brain circuits using diffusion tensor imaging tracking methods.[2] It also requires progress in fundamental research in the enumeration of all categories of neurons and amygdala interneurons; all types of synapses involved in the nuclei; all neurotransmitters, receptors, and neurohormones; and even

[1] "A little of science takes away from God but a lot of science returns to it" (Louis Pasteur).

[2] The "Human Connectome Project", which collects all the data on intra-connections in the living brain (combining fMRI, tractography, metabolism, infusion, etc.), should help to better understand structural and functional connectivity and thus progress towards making maps of the neural networks underlying the brain's function and the behaviour of individuals, to be able to navigate through the entire brain! This will demonstrate once again the richness and complexity of the amygdala's relationships and of its remarkable functioning.

V. Di Marino et al., *The Amygdaloid Nuclear Complex: Anatomic Study of the Human Amygdala*,
DOI 10.1007/978-3-319-23243-0

all components of the cerebrospinal fluid[3] on the periphery of the amygdala, etc.

Research teams need to collaborate on key targets. The amygdala is often viewed through the distorting image of the single animal model. We must now move on to humans not only for imaging or in vivo electrophysiology (as is done already) but also at the tissue level by collecting all resection specimen post-amygdalectomy and attempting to extract the maximum of information (neuronal cultures, neuronal immunohistochemistry studies, immunofluorescence; optogenetics, electron microscopy, receptor studies; study of the reactions to neurotransmitters, hormones, chemicals including pharmaceuticals, genetic studies, etc.). There is no doubt that such collaboration between clinicians and researchers should help to better determine the characteristics of the human amygdala and its connections and to learn to detect (and treat) anomalies causing the nervous system diseases that we have discussed in several pages of this book. This is our wish.

Marseille, 31 May 2015
Vincent Di Marino, Yves Etienne, Maurice Niddam

[3]The amygdala, like the BST, has structures that bathe in the ventricular cavities and the CSF can move various physiological or even pharmaceutical molecules towards these structures.

References

Abrams GM, Nilaver G, Zimmerman EA (1985) VIP-containing neurons. In: Björklund A, Hökfelt T (eds) Handbook of chemical neuroanatomy, part 1, GABA and neuropeptides in the CNS 4. Elsevier, Amsterdam

Adey WR, Meyer M (1952) Hippocampal and hypothalamic connections of the temporal lobe in the monkey. Brain 75:358–384

Adolphs R (2008) Fear, faces, and the human amygdala. Curr Opin Neurobiol 18(2):166–172

Adolphs R (2010) What does the amygdala contribute to social cognition? Ann N Y Acad Sci 1191:42–61

Adolphs R, Tranel D, Damasio H, Damasio AR (1995) Fear and the human amygdala. J Neurosci 75(9):5879–5981

Aeby Ch. Th (1871) Lehrbuch der Anatomie. F.C.W.Vogel, Leipzig

Aggleton JP (1985) A description of intra-amygdaloid connections in old world monkeys. Exp Brain Res 57:390–399

Aggleton JP (2000) The amygdala. A functional analysis. Oxford University Press, Oxford

Aggleton JP, Burton MJ, Passingham RE (1980) Cortical and subcortical afferents to the amygdala of the rhesus monkey (Macaca mulata). Brain Res 190(2):347–368

Aggleton JP, Mishkin M (1984) Projections of the amygdala to the thalamus in the cynomolgus monkey. J Comp Neurol 222:56–68

Alheid GF (2009) Extended amygdala. In: Binder MD, Hirokawa N, Windhorst U (eds) Encyclopedia of neuroscience. Springer, Berlin/Heidelberg/New-York, pp 1501–1506

Alheid GF, Heimer L (1988) New perspectives in basal forebrain organization of special relevance for psychiatric disorders; The striatopallidal amygdaloid and corticopetal components of substantia innominata. Neuroscience 27(1):1–39

Alheid GF (2003) Extended amygdala and basal forebrain. Ann N Y Acad Sci 985:185–205

Allen LS, Gorski RA (1990) Sex difference in the bed nucleus of the stria terminalis of the human brain. J Comp Neurol 302(4): 697–706

Amaral DG, Cowan WM (1980) Subcortical afferents to the hippocampal formation in the monkey. J Comp Neurol 189(4):573–591

Amaral DG, Price JL (1984) Amygdalo-cortical projections in the monkey (Macaca fascicularis). J Comp Neurol 230:465–496

Andy OJ, Stephan H (1968) The septum in the human brain. J Comp Neurol 133:383–410

Arai Y (2004) Sex differentiation of central nervous system – brain of man and woman. Nippon Rinsho 62(2):281–292

Augustine JR (2007) Human neuroanatomy. An introduction, 1st edn. Academic Press, London; Elsevier, USA

Ball T, Derix J, Wentlandt J, Wieckorst B, Speck O, Schulze-Bonhage A, Mutschler I (2009) Anatomical specificity of functional amygdala imaging of responses to stimuli with positive and negative emotion valence. J Neurosci Methods 180(1):57–70

Ban T, Omukai F (1959) Experimental studies on the fiber connexions of the amygdaloid nuclei in the rabbit. J Comp Neurol 113: 245–279

Barbas H (2000a) Connections underlying the synthesis of cognition memory and emotion in primate prefrontal cortices. Brain Res Bull 52:319–330

Barbas H (2000b) Complementary roles of prefrontal cortical regions in cognition, memory, and emotions in primates. Acta Neurol 84:87–110

Barbas H (2007) Flow of information for emotions through temporal and orbito-frontal pathways. J Anat 211:234–249

Barbas H, Zikopoulos B, Timbie C (2011) Sensory pathways and emotional context for action in primate prefrontal cortex. Biol Psychiatry 69(12):1133–1139

Barger N, Stefanacci L, Schumann CM, Sherwood CC, Annese J, Allman J, Buckwalter JA, Hof PR, Semendeferi K (2012) Neuronal populations in the basolateral nuclei of the amygdala are differentially increased in humans compared with apes: a stereological study. J Comp Neurol 520(13):3035–3054

Bauer RM (1982) Visual hypoemotionality as a symptom of visual-limbic disconnection in man. Arch Neurol 39:702–708

Baxter MG, Murray EA (2002) The amygdala and reward. Nat Rev Neurosci 3:563–573

Bechara A, Damasio H, Damasio AR, Lee G (1999) Different contributions of the human amygdala and ventromédial prefrontal cortex to decision-making. J Neurosci 19(13):5473–5481

Ben-Ari Y (1981) The amygdaloid complex. Proceedings of the international symposium on the amygdaloid complex Senlis; France: 1–4 September 1981. Inserm Symposium n°20. Editor: Yehezkel Ben-Ari. Elsevier/North Holland Biomedical Press

Ben-Ari Y, Tremblay E (1982) Organization of the amygdala with special reference to various pathological syndromes. Scandinavian J Psychol (Suppl 1):26–36

Bonn M, Schmitt A, Lesch KP, Van Bockstaele EJ, Asan E (2013) Serotonergic innervation and serotonic receptor expression of NPY-producing neurons in the rat lateral and basolateral amygdaloid nuclei. Brain Struct Funct 218(2):421–435

Bossy J (1990) Neuro Anatomie. In: Chevrel JP (ed) Anatomie Clinique. Springer, Paris

Brabec J, Rulseh A, Hoyt B, Vizek M, Horinek D, Hort J, Petrovicky P (2010) Volumetry of the human amygdala-an anatomical study. Psychiatry Res 182(1):67–72

Brand M, Grabenhorst F, Starcke K, Vandekerckhove MMP, Markowitsch HJ (2007) Role of the amygdala in decisions under ambiguity and decisions under risk: evidence from patients with Urbach-Wiethe disease. Neuropsychologia 45(6):1305–1317

Brierley B, Shaw P, David AS (2002) The human amygdala: a systematic revue and meta-analysis of volumetric magnetic resonance imaging. Brain Res Rev 39(1):84–105

Broca P (1878) Le grand lobe limbique et la scissure limbique dans la série des mammifères, in Anatomie comparée des circonvolutions cérébrales. Revue d'Anthropologie 1:385–498

Brockhaus H (1938) Zur normalen und pathologischen Anatomie der Mandelkerngebiete. J Psychol Neurol 49:1–136

V. Di Marino et al., *The Amygdaloid Nuclear Complex: Anatomic Study of the Human Amygdala*,
DOI 10.1007/978-3-319-23243-0

Brockhaus H (1942a) Zur feineren Anatomie des septum und des Striatum. J Psychol Neurol, Leipzig 51:1–56
Brockhaus H (1942b) Vergleichend-anatomische Untersuchingen über den Basalkernkomplex. J f Psychol und Neurol 51:57–95
Burdach KF Vom Baue und Leben des Gehirns, vol 3. Dyk'sche Buchhdl, 1819–1826, Leipzig
Burstein R, Potrebic S (1993) Retrograde labelling of neurons in the spinal cord that project directly to the amygdala or the orbital cortex in the rat. J Comp Neurol 235(4):469–485
Buss CL, Davies EP, Shababa B, Pruessner JC, Head K, Sandman CA (2012) Maternal cortisol over the course of pregnancy and subsequent child *amygdala* and hippocampus volumes and affective problems. Proc Natl Acad Sci U S A 109(20):E1312–E1319
Cahill L, Uncapher M, Kilpatrick L, Alkire MT, Turner J (2004) Sex-related hemispheric lateralization of amygdala function in emotionally influenced memory: an FMRI investigation. Learn Mem 11(3):261–266
Canli T, Desmond JE, Zhao Z, Glover G, Gabrieli JD (1998) Hemispheric asymmetry for emotional stimuli detected with fMRI. Neuroreport 9(14):3233–3239
Canteras NS, Swanson LW (1992) Projections of the ventral subiculum to the amygdala, septum and hypothalamus: a PHAL anterograde tract-tracing study in the rat. J Comp Neurol 324(2):180–194
Capogna M (2014) GABAergic cell type diversity in the basolateral amygdala. Curr Opin Neurobiol 26:110–116
Carmichael ST, Clugnet MC, Price JL (1994) Central olfactory connections in the macaque monkey. J Comp Neurol 346(3):403–434
Carroll D, O'Callaghan MAJ (1981) Psychosurgery and the control of aggression. In: Brain PF, Benton D (eds) The biology of aggression. Sijthoff and Noordhoff, Alphen aan den Rijn, p 457–472
Chance SA, Esiri MM, Crow TJ (2002) Amygdala volume in schizophrenia: post-mortem study and review of magnetic resonance imaging. Br J Psychiatry 180:331–338
Chang L, Sundaresh S, Elliott J, Anton PA, Baldi P, Licudine A, Mayer M, Vuong T, Hirano M, Naliboff BD, Ameen VZ, Mayer EA (2009) Dysregulation of the hypothalamic-pituitary-adrenal (HPA) axis in irritable bowel syndrome. Neurogastroenterol Motil 21:149–159
Cho YT, Ernst M, Fudge JL (2013) Cortico-amygdala-striatal circuits are organized as hierarchical subsystems through the primate amygdala. J Neurosci 33(35):14017–14030
Choi CY, Han SR, Yee GT, Lee CH (2010) A understanding of the temporal stem. J Korean Neurosurg Soc 47(5):365–369
Chung WC, De Vries GJ, Swaab DF (2002) Sexual differentiation of the bed nucleus of the stria terminalis in humans may extend into adulthood. J Neurosci 22(3):1027–1033
Ciriello J, Schultz CG, Roder S (1994) Collateral axonal projections from ventrolateral medullary non-catecholaminergic neurons to central nucleus of the amygdala. Brain Res 663(2):346–351
Coenen VA, Panksepp J, Hurwitz T, Urbach H, Mädler B (2012) Human medial forebrain bundle (MFB) and anterior thalamic radiation (ATR): imaging of two major subcortical pathways and the dynamic balance of opposite affects in understanding depression. J Neuropsychiatry Clin Neurosci 24(2):223–236
Concha L, Gross DW, Beaulieu C (2005) Diffusion tensor tractography of the limbic system. AJNR Am J Neuroradiol 26:2267–2274
Corkin S (1980) A prospective study of cingulectomy. In: Valenstein ES (ed) The psychosurgery debate. WH Freeman and Co, San Francisco, pp 164–204
Costafreda SG, Brammer MJ, David AS, Fu CH (2008) Predictors of amygdala activation during the processing of emotional stimuli: a meta-analysis of 385 PET and fMRI studies. Brain Res Rev 58:57–70
Cowan WM, Raisman G, Powell PS (1965) The connexions of the amygdala. J Neurol Neurosurg Psychiatry 28:137–151
Crosby EC, Humphrey T, Lauer EW (1962) Correlative anatomy of the nervous system. The Macmillan company, New-York
Crosby EC, Humphrey T (1941) Studies of the vertebrate telencephalon. II. The nuclear pattern of the anterior olfactory nucleus, tuberculum olfactorium and the amygdaloid complex in adult man. J Comp Neurol 74(2):309–352
Curran E (1909) A new association fiber tract in the cerebrum. J Comp Neurol Psychiat 19:645–657
Damasio AR (1994) Descartes'error, emotion, reason and the human brain. A Grosset/Putman Books
Damasio AR (2003) Spinoza avait raison. Joie et tristesse, le cerveau des émotions. Odile Jacob éd, Paris
Damasio AR (2004) L'erreur de Descartes. La raison des émotions. Odile Jacob éd, Paris, (Traduction de l'édition originale de 1994, en langue anglaise)
Davis M (1992) The role of the amygdala in fear and anxiety. Annu Rev Neurosci 15:353–375
Davis M (1997) Neurobiology of fear responses: the role of the amygdala. J Neuropsychiatry Clin Neurosci 9:382–402
Davis M, Shi C (1999) The extended amygdala: are the central nucleus of the amygdala and the bed of the stria terminalis differentially involved in fear versus anxiety? Ann N Y Acad Sci 877: 281–291
De Olmos JS (2004) Amygdala. In: Praxinos G, Mai JK (eds) The human nervous system. Elsevier, Amsterdam, pp 739–868
De Olmos JS, Alheid GF, Beltramino CA (1985) Amygdala. In: Praxinos G (ed) The rat nervous system, vol 1, Forebrain and midbrain. Academic, New York, pp 223–314
De Olmos JS, Heimer L (1999) The concepts of the ventral striatopallidal system and extended amygdala. Ann N Y Acad Sci 877: 1–32
Déjerine J. Anatomie des centres nerveux. Vol I, Rueff et Cie, 1895. Vol II, Rueff et Cie, 1901
Diepen R (1962) Der Hypothalamus. In: W.v. Möllendorff und W. Bargman (eds) Handbuch der mikroskopischen Anatomie des Menschen, vol IV/7. Springer ed, Berlin
Di Marino V, Etienne Y, Niddam M, Tacher V (2008) Rehaussement des structures du système nerveux central par traitement des coupes anatomiques avec le chlorure ferrique. Morphology 90:144–150
Di Marino V, Etienne Y, Niddam M (2011) Atlas photographique en couleur du système nerveux central. Springer Verlag Ed, France
Dong HW, Petrovich GD, Swanson LW (2001) Topography of projections from amygdala to bed nuclei of the stria terminalis. Brain Res Rev 38:192–246
Dong HW, Swanson LW (2003) Projections from the rhomboid nucleus of the bed nuclei of the stria terminalis: implications for cerebral hemisphere regulation of ingestive behaviors. J Comp Neurol 463:434–472
Dong HW, Swanson LW (2004) Projections from bed nuclei of the stria terminalis, posterior division: implications for cerebral hemisphere regulation of defensive and reproductive behaviors. J Comp Neurol 471:396–433
Dong HW, Swanson LW (2006a) Projections from bed nuclei of the stria terminalis, antero-medial area: cerebral hemisphere integration of neuroendocrine, autonomic and behavorial aspects of energy balance. J Comp Neurol 494:142–178
Dong HW, Swanson LW (2006b) Projections from bed nuclei of the stria terminalis, dorsomedial nucleus: implications for cerebral hemisphere integration of neuroendocrine, autonomic, and drinking responses. J Comp Neurol 494:75–107
Dumont EC (2009) What is the bed nucleus of the stria terminalis? Prog Neuro-Psychopharmacol Biol Psychiatry 33(8):1289–1290
Duvernoy HM (1988) The human hippocampus. An atlas of applied anatomy. J.F. Bergmann Verlag, München
Duvernoy HM (1992) Le cerveau humain. Springer, Paris
Duvernoy HM, Cattin F, Risold PY (2013) The human hippocampus, functional anatomy, vascularization and serial sections with MRI, 4th edn. Springer, Berlin/New York

Ebeling U, Reulen HJ (1988) Neurosurgical topography of the optic radiations in the temporal lobe. Acta Neurochir 92:29–36

Edeline JM, Weinberger NM (1992) Associative retuning in the thalamic source of input to the amygdale and auditory cortex: receptive field plasticity in the medial division of the medial geniculate body. Behav Neurosci 106(1):81–105

Ekman P (1992a) An argument for basic emotions. Cognition and Emotion 6(3/4):169–200

Ekman P (1992b) Are there basic emotions? Psychol Rev 99(3): 550–553

Ekman P, Friesen WV, Ellsworth P (1972) Emotion in the human face: guideline for research ad an integration of findings. NY Pergamon Press, New-York

Ekman P, Davidson RJ, Friesen WV (1990) Emotional expression and brain physiology/II: The Duchenne smile. J Pers Soc Psychol 58(2):342–353

Emre M, Heckers S, Mash DC, Geula C, Mesulam MM (1993) Cholinergic innervation of the amygdaloid complex in the human brain and its alterations in old age and Alzheimer's disease. J Comp Neurol 336(1):117–134

Entis JJ, Doerga P, Feldman Barrett L, Dickerson BC (2012) A reliable protocol for the manual segmentation of the human amygdala and its subregions using ultra-high resolution MRI. Neuroimage 60:1226–1235

Feinstein JS, Adolphs R, Damasio A, Tranel D (2011) The human amygdala and the induction and experience of fear. Curr Biol 21:34–38

Feinstein JS, Buzza C, Hurleman R, Follmer RL, Dahdaleh NS, Coryell WH, Weish MJ, Tranel D, Wemmie JA (2013) Fear and panic in humans with bilateral amygdala damage. Nat Neurosci 16:270–272

Felten DL, Jozefowicz RF (2006) Atlas de neurosciences humaines de Netter. Neuroanatomie, Neurophysiologie. Elseviers Saunders, 2003; Traduction française, Kubis N, Masson Ed, Paris

Fernandez-Guasti A, Kruijver FPM, Fodor M, Swaab DF (2000) Sex differences in the distribution of androgen receptors in the human hypothalamus. J Comp Neurol 425(3):422–435

Fliers E, Guldenaar SE, van de Wal N, Swaab DF (1986) Extrahypothalamic vasopressin and oxytocin in the human brain. Brain Res 375(2):363–367

Foix Ch, Nicolesco J (1925) Les Noyaux gris Centraux et la Région Mésencéphalo-sous-optique. Masson et Cie éds, Paris

Foville LA (1844) Traité complet de l'Anatomie, de la Physiologie et de la Pathologie du Système Nerveux Cérébro-Spinal. Fortin, Masson et Cie éd, Paris

Fox CA (1943) The stria terminalis, longitudinal association bundle and precommissural fornix fibers in the cat. J Comp Neurol 79:277–295

Fudge JL, Kunishio K, Walsh P, Richard C, Haber SN (2002) Amygdaloid projections to ventromedial striatal subterritories in the primate. Neuroscience 110(2):257–275

Gardner E, Gray DJ, O'Rahilly R -Saunders Ed., 1960. Ed. française, Doin, 1979 (Adaptation française J. Bossy)

Gaspar P, Berger B, Alvarez C, Vigny A, Henry JP (1985) Catecholamin innervation of the septal area in man: immunocytochemical study using TH and DBH antibodies. J Comp Neurol 241(1):12–33

Ghashghaei HT, Barbas H (2002) Pathways for emotion: interactions of prefrontal and anterior temporal pathways in the amygdala of the rhesus monkey. Neuroscience 115(4):1261–1270

Ghashghaei HT, Hilgetag CC, Barbas H (2007) Sequence of information processing for emotions based on the anatomic dialogue between prefrontal cortex and amygdala. Neuroimage 34(3): 905–923

Giacomini CH (1884) Fascia dentata du grand hippocampe dans le cerveau de l'homme. Arch Ital Biol 5:1–16, 205–219, 396–417

Goetzen B, Sztamska E (1999) Vascularisation artérielle du noyau amygdaloïdien chez l'homme et le mouton. Morphologie 83(260): 29–30

Goldstein LH, Polkey CE (1993) Short term cognitive changes after unilateral temporal lobectomy or unilateral amygdalo-hippocampectomy for the relief of temporal lobe epilepsy. J Neurol Neurosurg Psychiatry 56:135–140

Gouazé A, Laffont J (1994) Neuroanatomie clinique, 4th edn. Expansion Scientifique Française, Paris

Gray TS, Carney ME, Magnuson DJ (1989) Direct projections from the central amygdala nucleus to the hypothalamic paraventricular nucleus: possible role in stress-induced adrenocorticotropin release. Neuroendocrinology 50(4):433–446

Gretchen NN, Owens MJ, Taylor WR, Nemeroff C (2010) Changes in the vascular area fraction of the hippocampus and amygdala are induced by prenatal dexamethasone and/or adult stress. J Cerebral Blood Flow Metabol 30:1100–1104

Hall E (1960) Efferent pathways of the lateral and basal nuclei of the amygdala in the cat. Anat Rec 136:205–215

Hamilton JP, Siemer M, Gotlib IH (2008) Amygdala volume in major depressive disorder: a meta-analysis of magnetic resonance imaging studies. Mol Psychiatry 13(11):993–1000

Hamann S, Herman RA, Nolan CL, Wallen K (2004) Men and women differ in amygdala response to visual sexual stimuli. Nat Neurosci 7(4):411–416

Hamann SB, Ely TD, Grafton ST, Kilts CD (1999) Amygdala activity related to enhanced memory for pleasant and aversive stimuli. Nat Neurosci 2(3):289–293

Hanaway J, Woolsey Th, Gado MH, Roberts MP (2001) Atlas du Cerveau. De Boeck Université, traduction française par Dhem A

Hanson JL, Suh JW, Nacewicz BM, Sutterer MJ, Cayo AA, Stodola DE, Burghy CA, Wang H, Avants BB, Yushkevich PA, Essex MJ, Pollak SD, Davidson RJ (2012) Robust automated amygdala segmentation via multi-atlas diffeomorphic registration. Front Neurosci 6:166, 1–8

Hariri AR, Tessitore A, Mattay VS, Fera F, Weinberger DR (2002) The amygdala response to emotional stimuli: a comparison of faces and scenes. Neuroimage 17(1):317–323

Hedreen JC, Struble RG, Whitehouse PJ, Price DL (1984) Topography of the magnocellular basal forebrain system in human brain. J Neuropathol Exp Neurol 43(1):1–21

Heimer L (1978) The olfactory cortex and the ventral striatum. In: Livingston KE, Hornykiewicz O (eds) Limbic mechanisms: the continuing evolution of the limbic system concept. Plenum, New York, p 95–187

Heimer L (1995) The human brain and spinal cord. Functional neuroanatomy and dissection guide, 2nd edn. Springer Verlag, New York

Heimer L (2008) Anatomy of neuropsychiatry: the anatomy of the basal forebrain and its implication for neuropsychiatric illness. Academic, Amsterdam/Boston/Heidelberg/London

Heimer L, de Olmos J, Alheid GF, Zaborszky L (1991) Chapter 7. "Perestroika" in the basal forebrain: opening the border between neurology and psychiatry. In: Progress in brain research, vol 87, p 109–165

Heimer L, Harlan RE, Alheid GF, Garcia MM, de Olmos J (1997) Substantia inominata: a notion which impedes clinical-anatomical correlations in neuropsychiatric disorders. Neuroscience 76(4): 957–1006

Hellsten J, West MJ, Arvidsson A, Ekstrand J, Jansson L, Wennstrom M, Tingstrom A (2005) Electroconvulsive seizures induce angiogenesis in adult rat hippocampus. Biol Psychiatry 58: 871–878

Herman JP, Cullinan WE (1997) Neurocircuitry of stress: central control of the hypothalamo-pituitary-adrenocortical axis. Trends Neurosci 20(2):78–83

Herzog AG, Van Hoesen GW (1976) Temporal neocortical afferent connections to the amygdala in the rhesus monkey. Brain Res 115(1):57–69

Hill SY, Wang S, Carter H, McDermott MD, Zezza N, Stiffler S (2013) Amygdala volume in offspring from multiplex for alcohol dependence families: the moderating influence of childhood environment and 5-HTTLPR variation. J Alcohol Drug Depend 12(Suppl 1) pib: 001. doi: 10.4172/2329-6488.S1-001

Hilmazer-Hancke DM (2012) Chapter 22. Amygdala. In: Mai JK, Paxinos G (eds) The human nervous system, 3rd edn. Elsevier/ Academic Press, London, pp 759–784

Höistad M, Barbas H (2008) Sequence of information processing for emotions through pathways linking temporal and insular cortices with the amygdala. Neuroimage 40(3):1016–1033

Holland PC, Gallagher M (1999) Amygdala circuitry in attentional and representational processes. Trends Cogn Sci 3(2):65–73

Holstege G (1991) Descending motor pathways and the spinal motor system: limbic and non limbic components. Prog Brain Res 87:307, 412, ch. 4

Holstege G, Meiners L, Tan K (1985) Projections of the bed nucleus of the stria terminalis to the mesencephalon, pons and medulla oblongata in the cat. Exp Brain Res 58:379–391

Huber D, Veinante P, Stoop R (2005) Vasopressin and oxytocin excite distinct neuronal populations in the central amygdala. Science 308(5719):245–248

Humphrey T (1968) The development of the human amygdala during early embryonic life. J Comp Neurol 132:135–165

Iidaka T (2007) Neuroimaging study of the human amygdala. Toward an understanding of emotional and stress responses. Jpn Soc Plasma Fusion Res 2:S1 1006–S1 1007

Insausti R, Amaral DG, Cowan WM (1987) The entorhinal cortex of the monkey: III. Subcortical afferents. J Comp Neurol 264(3):396–408

Iwai E, Yukie M (1987) Amygdalofugal and amygdalopetal connections with modality-specific visual cortical areas in macaques (Macaca fuscata, M. Mulatta, and M. fascicularis). J Comp Neurol 261(3):362–387

Izquierdo A, Murray EA (2005) Opposing effects of amygdala and orbital prefrontal cortex lesions on the extinction of instrumental responding in macaque monkeys. Eur J Neurosci 22(9):2341–2346

Jalabert M, Aston-Jones G, Herzog E, Manzoni O, Georges F (2009) Role of the bed nucleus of the stria terminalis in the control of ventral tegmental area dopamine neurons. Prog Neuropsychopharmacol Biol Psychiatry 33(8):1336–1346

Ji Y, Tang B, Traub RJ (2008) The visceromotor response to colorectal distention fluctuates with the estrous cycle in rats. Neuroscience 154:1562–1567

Johnston JB (1923) Further contributions to the study of the evolution of the forebrain. J Comp Neurol 35(5):337–481

Jones MP, Dilley JB, Drossman D, Crowell MD (2006) Brain-gut connections in functional GI disorders: anatomic and physiologic relationships. Neurogastroenterol Motil 18:91–103

Jongen-Rêlo AL, Amaral DG (1998) Evidence for a GABAergic projection from the central nucleus of the amygdala to the brainstem of the macaque monkey: a combined retrograde tracing and in situ hybridization study. Eur J Neurosci 10(9):2924–2933

Kadohisa M (2013) Effect of odor on emotion, with implications. Front Syst Neurosci 7:66

Kahle W, Frotscher M (2005) Anatomie, 3, Système nerveux et organes des sens, 3è ed, Trad. Française E Vitte, D Hasboun. Médecine-Sciences Flammarion, Paris

Kaouane N, Porte Y, Vallée M, Brayda-Bruno L, Mons N, Calandreau L, Marighetto A, Piazza PV, Desmedt A (2012) Glucocorticoids can induce PTSD-like memory impairments in mice. Science 335:1510–1513

Karama S, Lecours AR, Leroux JM, Bourgouin P, Beaudoin G, Joubert S, Beauregard M (2002) Areas of brain activation in males and females, during viewing of erotic film excerpts. Hum Brain Mapp 16(1):1–13

Kelley AE, Domesick VB, Nauta WJH (1982) The amygdalostriatal projection in the rat- an anatomical study by anterograde and retrograde tracing methods. Neuroscience 7(3):615–630

Kemppainen S, Jolkkonen E, Pitkänen A (2002) Projections from the posterior cortical nucleus of the amygdala to the hippocampal formation an parahippocampal region in rat. Hippocampus 12(6):735–755

Kier EL, Staib LH, Davis LM, Bronen RA (2004) MR imaging of the temporal stem: anatomic dissection tractography of the uncinate fasciculus, inferior occipito-frontal fasciculus, and Meyer's loop of the optic radiation. AJNR Am J Neuroradiol 25(5): 677–691

Kim JE, Lyoo IK, Estes AM, Renshaw PF, Shaw DW, Friedman SD, Kim DJ, Yoon SJ, Hwang J, Dager SR (2010) Laterobasal amygdalar enlargement in 6-to7-year-old children with autism spectrum disorder. Arch Gen Psychiatry 67(11):1187–1197

Klemm WR, Vertes RP (1990) Brainstem mechanisms in behavior. Wiley Ed, New York

Klinger J, Gloor P (1960) The connections of the amygdale and the anterior temporal cortex in the human brain. J Comp Neurol 115:333–369

Klüver H, Bucy PC (1937) Psychic blindness and other symptoms following bilateral temporal lobectomy in rhesus monkeys. Am J Physiol 119:352–353

Koelliker A (1896) Handbuch der gewebelehre des Menschen, vol 2. W. Engelmann, Leipzig

Krettek JE, Price JL (1977) Projections from the amygdaloid complex and adjacent olfactory structures to the entorhinal cortex and to the subiculum in the rat and cat. J Comp Neurol 172(4): 723–752

Krettek JE, Price JL (1978) A description of the amygdaloid complex in the rat and cat with observations on intra-amygdaloid axonal connections. J Comp Neurol 178:255–280

Krujiver FP, Zhou JN, Pool CW et al (2000) Male-to-female transsexuals have female neuron numbers in a limbic nucleus. J Clin Endocrinol Metab 85:2034–2041

Kühn S, Gallinat J (2014) Brain structure and functional connectivity associated with pornography consumption: the brain on porn. JAMA Psychiatry 71(7):827–834

Kwon HG, Byun WM, Ahn SH, Son SM, Jang SH (2011) The anatomical characteristics of the stria terminalis in the human brain: a diffusion tensor tractography study. Neurosci Lett 500:99–102

Lange C, Irle E (2004) Enlarged amygdala volume and reduced hippocampal volume in young women with major depression. Psychol Med (Cambridge Journals) 34(06):1059–1064

Laricchiuta D, Petrosini L (2014) Individual differences in response to positive and negative stimuli: endocannabinoid-based insight on approach and avoidance behaviors. Front Syst Neurosci 8(238)

Lauer EW (1945) The nuclear pattern and fiber connections of certain basal telencephalic centers in the macaque. J Comp Neurol 82:215–254

Lauterbur PC (1973) Image formation by induced local interactions. Examples employing nuclear magnetic resonance. Nature 242:190–191

Lautin A (2001) The limbic brain. Kluwer Academic/Plenum publisher, New-York

Lazorthes G, Gouazé A, Salamon G (1976) Vascularisation et circulation de l'encéphale. Tome premier: Anatomie descriptive et fonctionnelle. Masson, Paris

LeDoux JE (1987) Emotion. In: Handbook of physiology. 1: The nervous system, p 419–460. F Plum ed, Bethesda, MD: Am Physiol Soc

LeDoux JE (1998) The emotional brain. Weidenfeld & Nicolson, London

LeDoux JE (2000) Emotion circuits in the brain. Annu Rev Neurosci 23:155–184

LeDoux JE, Iwata J, Cicchetti P, Reis DG (1988) Different projections of the central amygdaloid nucleus mediate autonomic and behavioural correlates of conditioned fear. J Neurosci 7:2517–2529

LeDoux JE, Cicchetti P, Xagoraris A, Romanski L-M (1990a) The lateral amygdaloid nucleus: sensory interface of the amygdala in fear conditioning. J Neurosci 10:1062–1069

LeDoux JE, Farb C, Ruggiero DA (1990b) Topographic organization of neurons in the acoustic thalamus that project to the amygdala. J Neurosci 10(4):1043–1054

LeDoux JE, Phelps E (2008) Emotional networks in the brain. In: Lewis M, Haviland-Jones J, Barrett L (eds) Handbook of emotions. Guilford Press, New York, pp 159–179

Leichnetz GR, Astruc J (1977) The course of some prefrontal corticofugals to the pallidum, substantia innominata, and amygdaloid complex in monkeys. Exp Neurol 54(1):104–109

Lenhardt ML, Shulman A, Goldstein BA (2007) The role of the parabrachial nucleus in the natural history of tinnitus and its implications. Int Tinnitus J 13(2):87–89

Lesur A, Gaspar P, Berger B, Alvarez C (1989) Chemoanatomic compartments in the human bed nucleus of the stria terminalis. Neuroscience 32:181–194

Luders HO (2008) Textbook of epilepsy surgery. Informa Healthcare, London

Lyubashina OA (2004) Possible mechanisms of involvement of the amygdaloid complex in the control of gastric motor function. Neurosci Behav Physiol 34(4):379–388

McDonald AJ (1983) Neurons of the bed nucleus of the stria terminalis: a Golgi study in the rat. Brain Res Bull 10(1):111–120

McDonald AJ (1987) Organization of amygdaloid projections to the mediodorsal thalamus and prefrontal cortex: a fluorescence retrograde transport study in the rat. J Comp Neurol 262(1):46–58

McDonald AJ (1998) Cortical pathways to the mammalian amygdala. Prog Neurobiol 55:257–332

McDonald AJ (1992) Cell types and intrinsic connections of the amygdala. In: Aggleton JP (ed) The amygdala; neurobiological aspects of emotion, memory and mental dysfunction. Wiley, New York, pp 67–96

McDonald AJ, Shammah-Lagnado SJ, Shi C, Davis M (1999) Cortical afferents to the extended amygdala. Ann N Y Acad Sci 877:309–338

McElligott ZA, Winder DG (2009) Modulation of glutamatergic synaptic transmission in the bed nucleus of the stria terminalis. Prog Neuropsychopharmacol Biol Psychiatry 33(8):1329–1335

McGaugh JL (2004) The amygdala modulates the consolidation of memories of emotionally arousing experiences. Annu Rev Neurosci 27(1):1–28

McIntyre CK, Roozendaal B. Adrenal stress hormones and enhanced memory for emotionally arousing experiences. In: Neural plasticity and memory: Fom genes to brain imaging. Chapter 13, Frontiers in Neuroscience 2007, F Bermùdez-Rattoni ed, CRC Press, Boca-Raton

Machado CJ, Emery NJ, Capitiano JP, Mason WA, Mendoza SP, Amaral DJ (2008) Bilateral neurotoxic amygdala lesions in rhesus monkeys (Macaca mulatta): consistent pattern of behavior across different social contexts. Behav Neurosci 122(2):251–256

Macchi G (1951) The ontogenic development of the olfactory telencephalon in man. J Comp Neurol 95(2):245–305

MacLean PD (1958a) Contrasting functions of limbic and neocortical systems of the brain and their relevance to psychophysiological aspects of medicine. Am J Med 25(4):611–626

MacLean PD (1958b) The limbic system with respect to self-preservation and the preservation of the species. J Nerv Ment Dis 127(1):1–11

MacLean PD (1970) The triune brain, emotion and scientific bias. In: Schmitt FO (ed) The neurosciences second study program, Band 2. Rockefeller University Press, New York, pp 336–349

Makris N, Gasic GP, Seidman LJ, Goldstein JM, Gastfriend DR, Elman I, Albaugh MD, Hodge SM, Ziegler DA, Sheahan FS, Caviness VS Jr, Tsuang MT, Kennedy DN, Hyman SE, Rosen BR, Breiter HC (2004) Decreased absolute amygdala volume in cocains addicts. Neuron 44(4):729–740

Mai JK, Paxinos G, Voss T (2008) Atlas of the human brain, 3rd edn. Elsevier Academic Press, London

Mansfield P, Maudsley AA (1976) Line scan proton spin imaging in biological structures by NMR. Phys Med Biol 21(5):847–852

Marinkovic R, Markovic L (1991) Arterial vascularization of the amygdaloid body in man. Med Pregl 44(5–6):229–230

Mark VH, Ervin FR (1970) Violence and the brain. Harper and Brown, New York

Martin LJ, Powers RE, Dellovade TL, Price DL (1991) The bed nucleus-amygdala continuum in human and monkey. J Comp Neurol 309:445–485

Massi L, Elezgarai I, Puente N, Reguero L, Grandes P, Manzoni OJ, Georges F (2008) Cannabinoïd receptors in the bed nucleus of the stria terminalis control cortical excitation of midbrain dopamine cells in vivo. J Neurosci 28(42):10496–10508

Mayer A, Lahr G, Swaab DF, Pilgrim C, Reisert C (1998) The Y-chromosomal genes SRY and ZFY are transcribed in adult human brain? Neurogenetics 1(4):281–288

Medalla M, Lera P, Feinberg M, Barbas H (2007) Specificity in inhibitory systems associated with prefrontal pathways to temporal cortex in primates. Cereb Cortex 17(Suppl 1):136–150

Mehler WR (1980) Subcortical afferent connections of the amygdala in the monkey. J Comp Neurol 190(4):733–762

Mempel E (1975) Leczenie amygdaloktomia napadow padaczkowych. PZWL, Warszawa

Mesulam MM, Geula C (1992) Overlap between acetylcholinesterase-rich and choline acetyltransferase-positive (cholinergic) axons in the human cerebral cortex. Brain Res 577:112–120

Mesulam MM, Mufson EJ (1984) Neural inputs into the nucleus basalis of the substantia innominata (Ch 4) in the rhesus monkey. Brain 107:253–274

Mesulam MM, Mufson EJ (1985) The insula of Reil in man and monkey. In: Peters A, Jones EG (eds) Association and auditory cortices, vol 4, Cerebral cortex. Plenum, New York, pp 179–226

Mesulam MM, Mufson EJ, Levey AI, Wainer BH (1983) Cholinergic innervation of cortex by the basal forebrain: cytochemistry and cortical connections of the septal area, diagonal band nuclei, nucleus basalis (substantia inominata) and hypothalamus in the rhesus monkey. J Comp Neurol 214:170–197

Meyer A (1970) Karl Friedrich Burdach and his place in the history of neuroanatomy. J Neurol Neurosurg Psychiatry 33:553–561

Meyer G, Gonzalez-Hernandez T, Carrillo-Padilla F, Ferres-Torres R (1989) Aggregation of granule cells in the basal forebrain(Islands of Calleja): Golgi and cytoarchitectonic study in different mammals, including man. J Comp Neurol 284(3):405–428

Meynert Th (1888) Clinic of the forebrain based on its structure, its functions and its nutrition. Translated in French by: G Cousot, Manceaux Ed, Brussels

Millhouse OE, De Olmos J (1983) Neuronal configurations in lateral and basolateral amygdala. Neuroscience 10:1269–1300

Mizuno N, Takahashi O, Satoda T, Matsushima R (1985) Amygdalospinal projections in the macaque monkey. Neurosci Lett 53(3):327–330

Moga MM, Saper CB, Gray TS (1989) Bed nucleus of the stria terminalis: cytoarchitecture, immunohistochemistry, and projection to the parabrachial nucleus in the rat. J Comp Neurol 283(3):315–332, Publié en ligne en 2004!

Moga MM, Gray TS (1985) Evidence for corticotropin-releasing factor, neurotensin, and somatostatin in the neural pathway from the central nucleus of the amygdala to the parabrachial nucleus. J Comp Neurol 241(3):275–284

Morey RA, Gold AL, LaBar KS, Beall SK, Brown VM, Haswell CC, Nasser JD, Wagner HR, McCarthy G (2012) Amygdala volume changes in posttraumatic stress disorder in a large case-controlled veterans group. Arch Gen Psychiatry 69(11):1169–1178

Morley BJ (1986) The interpeduncular nucleus. Int Rev Neurobiol 28:157–182

Mormann F, Dubois J, Kornblith S, Milosavlejic M, Cerf M, Ison M, Tsuchiva N, Kraskov A, Quiroga RQ, Adolphs R, Fried I, Koch C (2011) A category-specific response to animals in the right human amygdala. Nat Neurosci 14(19):1247–1249

Morris JS, Frith CD, Perrett DI, Rowland D, Young AW, Calder AJ, Dolan RJ (1996) A differential neural response in the human amygdale to fearful and happy facial expressions. Nature 383: 812–815

Morris JS, Ohman A, Dolan RJ (1998) Conscious and unconscious emotional learning in the human amygdala. Nature 393(6684): 467–470

Morris JS, Ohman A, Dolan RJ (1999) A subcortical pathway to the right amygdala mediating "unseen" fear. Proc Natl Acad Sci U S A 96(1):1680–1685

Mufson EJ, Mesulam MM, Pandya DN (1981) Insular interconnections with the amygdale in rhesus monkey. Neuroscience 7:1231–1248

Müller F, O'Rahilly R (1989) The human brain at stage 17, including the appearance of the future olfactory bulb and the first amygdaloid nuclei. Anat Embryol (Berl) 180(4):353–369

Müller F, O'Rahilly R (1990a) The human brain at stages 18–20, including the choroid plexuses and the amygdaloid and septal nuclei. Anat Embryol (Berl) 182(3):285–306

Müller F, O'Rahilly R (1990b) The human brain at stage 21–23, with particular reference to the cerebral cortical plate and the development of the cerebellum. Anat Embryol (Berl) 182(4): 375–400

Müller F, O'Rahilly R (2006) The amygdaloid complex and the medial and lateral ventricular eminences in staged human embryos. J Anat 208(5):547–564

Munson J, Dawson G, Abbott R, Faja S, Webb SJ, Friedman SD, Shaw D, Artru A, Dager SR (2006) Amygdalar volume and behavioural development in autism. Arch Gen Psychiatry 63(6):686–693

Murat Yucel I, Solowij N, Respondek C, Whittle S, Fornito A, Pantels C, Lubman DI (2008) Regional brain abnormalities associated with long-term heavy cannabis use. Arch Gen Psychiatry 65(6): 694–701

Myers B, Greenwood-VanMeerveld B (2009) Role of anxiety in the pathophysiology of irritable bowel syndrome: importance of the amygdala. Front Ent Neurosci 1:2. doi:10.3389/neuro.21.002.2009

Nacewicz BM, Dalton KM, Johnstone T, Long MT, McAuliff EM, Oakes TR, Alexander AL, Davidson RJ (2006) Amygdala volume and non verbal social impairment in adolescent and adult males with autism. Arch Gen Psychiatry 63(12):1417–1428

Narabayashi H. Stereotaxic amygdalectomy. In: Eleftheriou BE (ed) The neurobiology of the amygdala. Plenum Press, New-York, 1972, p 459–483

Nauta WJH (1961) Fibre degeneration following lesions of the amygdaloid complex in the monkey. J Anat 95:515–531

Nauta WJH, Haymaker W (1969) Hypothalamic nuclei and fiber connections. In: Haymaker W, Anderson E, Nauta WJH (eds) The hypothalamus. Thomas, Springfield, pp 136–209

Naylor AS, Rogvi-Hansen BA, Kessing L, Kruse-Larsen C (1994) Psychiatric morbidity after surgery for epilepsy: short term follow up of patients undergoing amygdalohippocampectomy. J Neurol Neurosurg Psychiatry 57(1):1375–1381

Neigh GN, Owens MJ, Taylor WR, Nemeroff CB (2010) Changes in the vascular area fraction of the hippocampus and amygdala are induced by prenatal dexamethasone and/or adult stress. J Cereb Blood Flow Metab 30(6):1100–1104

Nery S, Fishell G, Corbin JG (2002) The caudal ganglionic eminence is a source of distinct cortical and subcortical cell populations. Nat Neurosci 5:1279–1287

Nieuwenhuys R, Voogd J, van Huijzen C (2008) The human central nervous system, 4th edn. Springer Ed

Norita M, Kawamura K (1980) Subcortical afferents to the monkey amygdala: an HRP study. Brain Res 190(1):225–230

Olds J, Milner P (1954) Positive reinforcement produced by electrical stimulation of septal area and others regions of the brain. J Comp Physiol Psychol 47:419–427

Olry R, Nicolay X (1994) From Paul Broca to the long term potentiation: the difficulties in confirming a limbic identity. Hist Sci Med 28(3):199–203

O'Rahilly R, Müller F (1999) The embryonic human brain: an atlas of developmental stages, 2nd edn. Wiley-Liss, New-York

Ottersen OP (1980) Afferent connections to the amygdaloid complex of the rat and cat. II. Afferents from the hypothalamus and the basal telencephalon. J Comp Neurol 194:267–289

Ottersen OP (1981) Afferent connections to the amygdaloid complex of the rat with some observations in the cat. III. Afferents from the lower brain stem. J Comp Neurol 202:335–356

Ottersen OP, Ben-Ari Y (1979) Afferent connections of the amygdaloid complex of the rat and cat. I; Afferents from the thalamus. J Comp Neurol 187:401–424

Pandya DN, Van Hoesen GW, Domesick VB (1973) A cingulo-amygdaloid projection in the rhesus monkey. Brain Res 61:369–373

Pape H-C, Pare D (2010) Plastic synaptic networks of the amygdala for the acquisition expression and extinction of conditioned fear. Physiol Rev 90(2):419–463

Pardini DA, Raine A, Erickson K, Loeber R (2014) Lower amygdala volume in men is associated with childhood aggression, early psychopathic traits, and future violence. Biol Psychiatry 75: 73–80

Paturet G (1964) Traité d'Anatomie Humaine. Tome IV. Système nerveux. Masson et Cie Ed.

Peltier J, Verclytte S, Delmaire C, Pruvo JP, Godefroy O, Le Gars D (2010) Neurosurgical anatomy of the temporal stem: clinical relevance and correlations with diffusion tensor imaging fiber tracking laboratory investigation. J Neurosurg 112(5):1–5

Petrovich GD (2001) Amygdala (amygdaloid complex). In: International encyclopedia of the social and behavorial sciences. Elsevier Ed, p 471–473

Petrovich GD, Swanson LW (1997) Projections from the lateral part of the central amygdalar nucleus to the postulated fear conditioning circuit. Brain Res 763:247–254

Peuskens D, van Loon J, Van Calenbergh F, Van den Bergh R, Goffin J, Plets C (2004) Anatomy of the anterior temporal lobe and the frontotemporal region demonstrated by fiber dissection. Neurosurgery 55(5):1174–1184

Pezawas L, Meyer-Lindenberg A, Drabant EM, Verchinski BA, Munoz KE, Kolachana BS, Egan MF, Mattay VS, Hariri AR, Weinberger D (2005) 5HTTLPR polymorphism impacts, human cingulated-amygdala interactions: a genetic susceptibility mechanism for depression. Nat Neurosci 8:828–834

Phan KL, Wager T, Taylor SF, Liberzon I (2002) Functional neuroanatomy of emotion: a meta-analysis of emotion activation studies in PET and fMRI. Neuroimage 16(2):331–348

Phelps EA (2004) Human emotion and memory: interactions of the amygdala and hippocampal complex. Curr Opin Neurobiol 14(2):198–202

Phelps EA, LeDoux JE (2005) Contributions of the amygdala to emotion processing: from animal models to human behavior. Neuron 48:175–187

Pirot S (1999) Structure et fonction de l'amygdale. Neuropsychiatrie: Tendances et Débats 5:35–39

Pitkänen A, Amaral DG (1991) Demonstration of projections from the lateral nucleus to the basal nucleus of the amygdala: a PHA-L study in the monkey. Exp Brain Res 83:465–470

Pitkänen A, Amaral DG (1998) Organization of the intrinsic connections of the monkey amygdaloid complex: projections originating in the lateral nucleus. J Comp Neurol 398:431–458

Pitkänen A, Tuunanen J, Kälviäinen R, Partanen K, Salmenperä T (1998) Amygdala damage in experimental and human temporal lobe epilepsy. Epilepsy Res 32:233–253

Pitkänen A, Pikkarainen M, Nurminen N, Ylinen A (2000) Reciprocal connections between the amygdala and the hippocampal formation, perirhinal cortex, and posterorhinal cortex in rat. A review. Ann N Y Acad Sci 911:369–391

Powell TPS, Cowan WM, Raisman G (1965) The central olfactory connexions. J Anat 99:791–813

Price JL, Amaral DG (1981) An autoradiographic study of the projections of the central nucleus of the monkey amygdala. J Neurosci 1(11):1242–1259

Price JL, Russchen FT, Amaral DG (1987) The limbic region. II: the amygdaloid complex. In: Björklund A, Hökfelt T, LW Swanson (eds) Handbook of chemical neuroanatomy, integrated system of the CNS, part I. Elsevier, Amsterdam. 5, p 279–388

Pruessner JC, Li LM, Serles W, Pruessner M, Collins DL, Kabani N, Lupien S, Evans AC (2000) Volumetry of hippocampus and amygdala with high resolution MRI and three-dimensional analysis software: minimizing the discrepancies between laboratories. Cereb Cortex 10(4):433–442

Quin S (2014) Amygdala volume predicted childhood anxiety. Biol Psychiatry 75:892–900

Quirk GJ, Repa C, LeDoux JE (1995) Fear conditioning enhances short-latency auditory responses of lateral amygdala neurons: parallel recordings in the freely behaving rat. Neuron 1(5):1029–1039

Quirk GJ, Armony JL, LeDoux JE (1997) Fear conditioning enhances different temporal components of tone evoked spike trains in auditory cortex and lateral amygdala. Neuron 19(3):613–624

Rempel-Clower NL (2007) Role of orbitofrontal cortex connections in emotion. Ann N Y Acad Sci 1121:72–86

Retzius G (1896) Das Menschenhirn. Nordstedt and Söner, Stockholm

Roberts WW (1962) Fear-like behavior elicited from dorsomedial thalamus of cat. J Comp Physiol Psychol 55:191–197

Rodrigo J, Alonso D, Bentura ML, Castro-Blanco S, Encinas JM, Fernàndez AP, Fernàndez-Vizzara P, Richart A, Santacana M, Serrano J, Martinez A (2002) Physiology and pathophysiology of nitric oxide in the nervous system, with special mention of islands of Calleja and the circunventricular organs. Histol Histopathol 17:973–1003

Roozendaal B, McEwen BS, Chattarji S (2009) Stress, memory and the amygdala. Nat Rev Neurosci 10:423–433

Rosene DL, Van Hoesen GW (1977) Hippocampal efferents reach widespread areas of the cerebral cortex and amygdala in the rhesus monkey. Science 198(4314):315–317

Ross ED, Homan RW, Buck R (1994) Differential hemispheric lateralization of primary and social emotions. Neuropsychiatry Neuropsychol Behav Neurol 7:1–19

Rosso IM, Killgore WDS, Cintron CM, Gruber SA, Tohen M, Yurgelun-Todd DA (2007) Reduced amygdala volumes in first-episode bipolar disorder and correlation with cerebral white matter. Biol Psychiatry 61:743–749

Russchen FT, Amaral DG, Price JL (1985a) The afferent connections of the substantia innominata in the monkey, Macaca fascicularis. J Comp Neurol 242(1):1–27

Russchen FT, Bakst J, Amaral DG, Price JL (1985b) The amygdalostriatal projections in the monkey. An anterograde tracing study. Brain Res 329(1–2):241–257

Rutishauser U, Tudusciuc O, Wang S, Mamelak AN, Ross IB, Adolphs R (2013) Single-neuron correlates of atypical face processing in autism. Neuron 80:887–899

Sah P, Faber ESL, Lopez de Armentia M, Power J (2003) The amygdaloid complex: anatomy and physiology. Physiol Rev 83(3):803–834

Salamon G (1971) Atlas de la vascularisation artérielle du cerveau chez l'homme. Sandoz Ed, Paris

Salu Y (2013) The role of the amygdala in the development of sexual arousal. Electron J Hum Sex 16:1–6

Sandman CA, Davis EP (2012) Neurobehavioral risk is associated with gestational exposure to stress hormones. Expert Rev Endocrinol Metab 7:445–459

Sanides F (1957) Die Insulae terminales des Erwachsenengehirns des Menschen. J Hirnforsch 3(2–3):243–273

Saunders RC, Rosene DL, Van Hoesen GW (1988) Comparison of the efferents of the amygdala and the hippocampal formation in the rhesus monkey: II. Reciprocal and non-reciprocal connections. J Comp Neurol 271(2):185–207

Schiltz K, Witzel J, Northoff G, Zierhut K, Gubka U, Fellman H, Kaufmann J, Tempelmann C, Wiebking C, Bogerts B (2007) Brain pathology in pedophilic offenders: evidence of volume reduction in the right amygdala and related diencephalic structures. Arch Gen Psychiatry 64(6):737–746

Schlaeper Th E, Cohen MX, Frick C, Kosel M, Brodesser D, Axmacher N, Joe AY, Kreft M, Lenartz D, Sturm V (2008) Deep brain stimulation to reward circuitry alleviates anhedonia in refractory major depression. Neuropsychopharmacology 33:368–377

Schumann CM, Hamstra J, Goodlin-Jones BL, Lotspeich LJ, Kwon H, Buonocore MH, Lammers CR, Reiss AL, Amaral DG (2004) The amygdala is enlarged in children but not adolescents with autism; the hippocampus is enlarged at all ages. J Neurosci 24(28):6392–6401

Schumann CM, Amaral DG (2006) Stereological analysis of amygdala neuron number in autism. Neurosci 26(29):7674–7679

Schwaber JS, Kapp BS, Higgins G (1980) The origin and extent of direct amygdala projections to the region of the dorsal motor nucleus of the vagus and the nucleus of the solitary tract. Neurosci Lett 20(1):15–20

Schwartz JRL, Roth T (2008) Neurophysiology of sleep and wakefulness: basic science and clinical implications. Curr Neuropharmacol 6(4):367–378

Sedat J, Duvernoy H (1990) Anatomical study of the temporal lobe. Correlations with nuclear magnetic resonance. J Neuroradiol 17(1):26–40

Semba K (2000) Multiple output pathways of the basal forebrain: organization, chemical heterogeneity and role in vigilance. Behav Brain Res 115:117–141

Semendeferi K, Barger N, Schenker N (2010) Brain reorganization in humans and apes. In: Broadfield D, Schick KD, Toth N, Yuan M (eds) The human brain evolving: paleoneurological studies in honor of Ralph L. Holloway. Stone Age Institute Press, Gosport, pp 119–155

Senatorov VV, Damadzic R, Mann CL, Schwandt M, George DT, Hommer DW, Heilig M, Momenan R. Reduced anterior insula, enlarged amygdala in alcoholism and associated depleted von Economo neurons. Brain, first published on line 3 Nov 2014

Sheridan PJ (1979) The nucleus interstitialis striae terminalis and the nucleus aygdaloideus medial: prime targets for androgen in the rat forebrain. Endocrinology 104(1):130–136
Shibata H, Suzuki T, Matsushita M (1986) Afferent projections to the interpeduncular nucleus in the rat, as studied by retrograde and anterograde transport of wheat germ agglutinin conjugated to horseradissh peroxidase. J Comp Neurol 248:272–284
Siebert M, Markowitsch HJ, Bartel P (2003) Amygdala, affect and cognition: evidence from 10 patients with Urbach-Wiethe disease. Brain 126:1–11
Sink KS, Chung A, Ressler KJ, Davis M, Walker DL (2013a) Anxiogenic effects of CGRP within the BNST may be mediated by CRF acting at BNST CRFR1 receptors. Behav Brain Res 243:286–293
Sink KS, Walker DL, Freeman SM, Flandreau EL, Ressler K, Davis M (2013b) Effects of continuously enhaced corticotropin releasing factor expression within the bed nucleus of the stria terminalis on conditioned and unconditioned anxiety. Mol Psychiatry 18(3): 308–319
Sink KS, Walker DL, Yang Y, Davis M (2011) Calcitonin gene-related peptide in the bed nucleus of the stria terminalis produces an anxiety-like pattern of behavior and increases neural activation in anxiety-related structures. J Neurosci 31(5):1802–1810
Snell RS (2010) Clinical Neuroanatomy, 7ème edn. Lippincott, Williams and Wilkins Ed
Sofroniew MY (1983) Direct reciprocal connections between the bed nucleus of the stria terminalis and dorsal medulla oblongata: evidence from immunohistochemical detection of tracer proteins. J Comp Neurol 213:399–405
Solano Castiella E, Anwander A, Lohman G, Weiss M, Docherty C, Geyer S, Reimer E, Friederici AD, Turner R (2010) Diffusion tensor imaging segments the human amygdala in vivo. Neuroimage 49(4):2958–2965
Soudry Y, Lemogne C, Malinvaud D, Consoli SM, Bonfils P (2011) Olfactory system and emotions: common substrates. Eur Ann Otorhinolaryngol Head Neck Dis 128(1):18–23
Spampanato J, Polepalli J, Sah P (2011) Interneurons in the baso-lateral amygdala. Neuropharmacology 60(5):765–773
Staub F, Bruggimann L, Magistretti P, Bogousslavsky J (2002) Anatomie des émotions. Schweizer Archiv Für Neurologie und Psychiatrie 153(8):344–353
Stefanacci L, Amaral DG (2000) Topographic organization of cortical inputs to the lateral nucleus of the macaque monkey amygdala: a retrograde tracing study. J Comp Neurol 421(1):52–79
Stefanacci L, Amaral DG (2002) Some observations on cortical inputs to the macaque monkey amygdala: an anterograde tracing study. J Comp Neurol 451(4):301–323
Strenge H, Braak E und Braak H (1977) Uber den Nucleus striae terminalis im Gehirn des erwachsenen Menschen. Z. mikrosk-anat Forsch, Leipzig, 91, 1, p 105–118
Sturm V, Lenartz D, Koulousakis A, Treuer H, Herholz K, Klein JC, Klosterkötter J (2003) The nucleus accumbens: a target for deep brain stimulation in obsessive-compulsive and anxiety disorders. J Chem Neuroanat 26:293–299
Sundaram T, Jeong GW, Kim TH, Kim GW, Baek HS, Kang HK (2010) Time-course analysis of the neuroanatomical correlates of sexual arousal evoked by erotic video stimuli in healthy males. Korean J Radiol 11(3):278–285
Swaab DF (2003) Chapter 7. Bed nucleus of the stria terminalis (BST) and the septum. In: Handbook of clinical neurology, vol 79 (3nd Series Vol.1): The human hypothalamus: basic and clinical aspects. Part I: Nuclei of the human hypothalamus. Elsevier, Amsterdam, p 149–162
Swaab DF (2007) Sexual differentiation of the brain and behavior. Best Pract Res Clin Endocrinol Metab 21(3):431–444
Swanson LW, Petrovich GD (1998) What is the amygdala? Trends Neurosci 21(8):323–331
Tagliavini F (2009) The basal nucleus of Meynert. In: Binder MD, Hirokawa N, Windhorst U (eds) Encyclopedia of neuroscience. Springer, Berlin, pp 115–116
Tagliavini F, Pilleri G, Bouras C, Constantinidis J (1984) The basal nucleus of Meynert in idiopathic Parkinson's disease. Acta Neurol Scand 70(1):20–28
Tarin P (1750) Anthropotomie, ou l'Art de disséquer. AC Briasson, Paris
Tessitore A, Hariri AR, Fera F, Smith WG, Chase TN, Hyde TM (2002) Dopamine modulates the response of the human amygdala: a study in Parkinson's disease. J Neurosci 22(20):9099–9103
Testut L (1911) Traité d'Anatomie Humaine. Tome troisième: Angeiologie; Système nerveux central. O Doin et fils Ed, Paris
Todorov A (2012) The role of the amygdala in face perception and evaluation. Motiv Emot 36(1):16–26
Tranel D, Hyman BT (1990) Neuropsychological correlates of bilateral amygdala damage. Arch Neurol 47(3):349–355
Travers N (2008) L'étude micro-anatomique des fibres du faisceau unciné et ses implications dans la chirurgie fronto-temporo-insulaire. Microsurgical anatomical study of the uncinate fasciculus. e-memoires de l'Académie Nationale de. Chirurgie 7(4):31–41
Van Hoesen GW (1981) The differential distribution, diversity, and sprouting of cortical projections to amygdala in the rhesus monkey. In: Ben-Ari Y (ed) The amygdaloid complex. Elsevier North Holland, Amsterdam, pp 77–90
Van Hoesen GW (1985) Neural systems of the non-human primate forebrain implicated in memory. Ann N Y Acad Sci 444:97–112
Van Stegeren AH, Goekoop R, Everaerd W, Scheltens P, Barkhof F, Kuijer JP, Rombouts SA (2005) Noradrenaline mediates amygdala activation in men and women during encoding of emotional material. Neuroimage 24(3):898–909
Veening JG, Swanson LW, Sawchenko PE (1984) The organization of projections from the central nucleus of the amygdale to brainstem sites involved in central autonomic regulation: a combined retrograde transport-immunohistochemical study. Brain Res 303(2): 337–357
Volsch M (1910) Zur vergleichenden Anatomie des Mandelkerns und seiner Nachbargebilde. II. Teil. Arch mikrosk Anat 76: 373–523
Vyas A, Mitra R, Shankaranarayana RB, Chattarji S (2002) Chronic stress induces contrasting patterns of dendritic remodeling in hippocampal and amygdaloid neurons. J Neurosci 22:6810–6818
Walter A, Mai JK, Lanta L, Görcs T (1991) Differential distribution of immunohistochemical markers in the bed nucleus of the stria terminalis in the human brain. J Chem Neuroanat 4:281–289
Wang RY, Aghajanian GK (1977) Inhibition of neurons in the amygdala by dorsal raphe stimulation. Mediation through a direct serotonergic pathway. Brain Res 120(1):85–102
Warwick R, Williams PL (1973) Gray's anatomy. Longman Group, 35th edn
Watson C, Andermann F, Gloor P, Jones-Gotman M, Peters T, Evans A, Olivier A, Melanson D, Leroux G (1992) Anatomic basis of amygdaloid and hippocampal volume measurement by magnetic resonance imaging. Neurology 42:1743–1750
Webster MJ, Lingerleider LG, Bachevalier J (1991) Connections of inferior temporal areas TE and TEO with medial temporal-lobes structures in infant and adult-monkeys. J Neurosci 11(4):1095–1116
Weller KL, Smith DA (1982) Afferent connections to the bed nucleus of the stria terminalis. Brain Res 232:255–270

Wendling AS (2012) Les déficits mnésiques et émotionnels de l'épilepsie temporomésiale avec sclérose hippocampique sont-ils liés à l'étendue de la résection chirurgicale ? Thèse Strasbourg, Ecole Doctorale des Sciences de la Vie et de la Santé

Whalen PJ, Phelps EA (eds) (2009) The human amygdala. Guilford Publications, New York

Wheeler RE, Davidson RJ, Tomarken AJ (1993) Frontal brain asymmetry and emotional reactivity: a biological substrate of affective style. Psychophysiology 30:82–89

Whittle S, Yap MBH, Yücel M, Fornito A, Simmons JG, Barrett A, Sheeber L, Allen NB (2008) Prefrontal and amygdala volumes are related to adolescents affective behaviours during parent-adolescent interactions. Proc Natl Acad Sci U S A 105(9):3652–3657

Wilder-Smith C, Schindler D, Lovblad K, Redmond S, Nirkko A (2004) Brain functional magnetic resonance imaging of rectal pain and activation of endogenous inhibitory mechanisms in irritable bowel syndrome patient subgroups and healthy controls. Gut 53:1595–1601

Woon FL, Hedges DW (2009) Amygdala volume in adults with posttraumatic stress disorder: a meta-analysis. J Neuropsychiatry Clin Neurosci 21(1):5–12

Woon FL, Hedges DW (2008) Hippocampal and amygdala volumes in children and adults with childhood maltreatment-related posttraumatic stress disorder: a meta-analysis. Hippocampus 18(8): 729–736

Yuasa S (2009) Somatotopy in the emotional expression by the amygdala. Brain Nerve 61(12):1395–1404

Zaborszky L, Alheid GF, Beinfeld MC et al (1985) Cholecystokinin innervation of the ventral striatum: a morphological and radioimmunological study. Neuroscience 14:427–453

Zhang DX, Bertram EH (2002) Midline thalamic region: widespread excitatory input to the entorhinal cortex and amygdala. J Neurosci 22(8):3277–3284

Zhao-Shea R, Liu L, Pang X, Gardner PD, Tapper AR (2013) Activation of GABAergic neurons in the interpeduncular nucleus triggers physical nicotine withdrawal symptoms. Curr Biol 23(23): 2327–2335

Zhou J-N, Hofman MA, Gooren LJ, Swaab DF (1995) A sex difference in the human brain and its relation to transsexuality. Nature 378:68–70

Zuckerkandl E (1887) Uber das Riechcentrum, eine vergleichend – anatomische Studie. Verlag Von Ferdinand Enke, Stuttgart, pp 32–41

Index of Abbreviations

a1	Anterior cerebral artery
a2	Middle cerebral artery
a3	Posterior cerebral artery
a4	Superior cerebellar artery
aCalS	Antecalcarinus sulcus
aclp	Anterior clinoid process
acn	Accumbens nucleus
ACo	Anterior commissure
Afp	Amygdalofugal pathway
agI	Agranular insula
Alv	Alveus
Amn	Amygdaloid nuclear complex
	1 Lateral nucleus
	2 Basolateral nucleus
	3 Basomedial nucleus
	4 Cortical nucleus
	5 Central nucleus
	6 Medial nucleus
	7 Anterior nucleus
	Abbreviations according to the "nomenclature of De Olmos":
	AHi Amygdalo-hippocampal area
	BL Basolateral nucleus (DBL, VBL, IBL)
	BM Basomedial nucleus (DLBM, VLBM, DMBM, VMBM)
	Ce Central nucleus (CeL, CeM)
	Co Cortical nucleus ACo (DACo, VACo) PCo (DPCo, VPCo)
	La Lateral nucleus (MLa, LLa, LiLa)
	Me Medial nucleus (Mea, Mep)
anG	Angular gyrus
aOrG	Anterior orbital gyrus
aOS	Anterior occipital sulcus
ar	Arachnoid
apSb	Anterior perforated substance
ATR	Anterior thalamic radiations (or anterior thalamic peduncle)
ba	Basilar artery
bacr	Basicranium
bc	Brachium conjunctivum
bccli	Brachium of caudal colliculus
BG	Basal ganglia
bmFa	Basomedial frontal artery
bnM	Basal nucleus of Meynert
bpl	Basal plate (subpallium)
Br	Brainstem
brcli	Brachium of rostral colliculus
BST	Bed nucleus of the stria terminalis
	BSTlv Ventral division of pars lateralis
	BSTld Dorsal division of pars lateralis
	BSTm Pars medialis
bte	Basal telencephalon
c1	Anterior communicating artery
CIII	Oculomotor nerve's nuclear complex
ca	Caudal
CA	Cornu ammonis (Ammon's horn)
cA	Cuneiform area
CalS	Calcarine sulcus
CAq	Cerebral aqueduct
caSepe	Cavity of the septum pellucidum
cb	Cerebellum
cb r	Cerebellar rudiment
CC	Corpus callosum
ccli	Caudal colliculus
Ceh	Cerebral hemisphere
CeP	Cerebral peduncle
CeS	Central sulcus
chci	Chiasmatic cisterna
Chf	Choroid fold
Chpl	Choroid plexus
Cif	Cingulate fasciculus
	Cif′ Juxta-hippocampic part of the cingulate fasciculus
	Cif″ Medial part of cingulate fasciculus
	Cif‴ Rostral expansion of cingulate fasciculus
CiG	Cingulate gyrus

V. Di Marino et al., *The Amygdaloid Nuclear Complex: Anatomic Study of the Human Amygdala*,
DOI 10.1007/978-3-319-23243-0

cigcv Cisterna of the great cerebral vein
CiS Cingulate sulcus
CiS′ Marginal part of CiS
Cl Claustrum
cn Caudate nucleus
cnb Caudate nucleus body
cnh Caudate nucleus head
cnt Caudate nucleus tail
cnt′ Para-amygdalar part of caudate nucleus tail
CoICl Compact insular claustrum
Col em Collateral eminence
ColS Collateral sulcus
ColS′ Transverse anterior collateral sulcus
ColS″ Transverse posterior collateral sulcus
Col tri Collateral trigone
cpl Cortical plate
CR Corona radiata
cr Cranial
crCe Crus cerebri
f Parieto-temporo-occipito pontine tract (2)
g Corticonuclear (1′) and corticospinal tracts (1) (CST)
h Fronto-pontine tract (3)
cTf Central tegmental fasciculus
cx Cortex
db Diagonal band (r) rostral part (d) dorsal part
dCl Dorsal claustrum
dG Descendens gyrus
DG Dentate gyrus
DG′ Dentate gyrus in the hippocampal tail
dgI Dysgranular insula
dICl Diffuse insular claustrum
dlf Dorsal longitudinal fasciculus
ds Dorsum sellae
EA Extended amygdala
eAr Entorhinal area
ed External digitations (hippocampus)
enS Endorhinal sulcus
epl Ependymal layer (and mantle layer)
eS Entorhinal sulcus
F Fornix
faG Fasciolaris gyrus
Fb Body of fornix
fc Fasciola cinerea
Fc Column of fornix
Fc (pr) Pre-commissural (Fc′)
Fc (po) Post-commissural (Fc″)
Fcr Crus of fornix
Fi Fimbria
fl Fasciculus liminis (limen)
FIG Frontal inferior gyrus
FIG pop Frontal inferior gyrus, pars opercularis
FIG ptr Frontal inferior gyrus, pars transversalis
FIG por Frontal inferior gyrus, pars orbitalis
FLO Frontal lobe
flV3 Floor of the third ventricle
FmG Fronto-marginal gyrus
FpG Fronto-polaris gyrus
Fpo Frontal pole
fs Fundus striati
FSG Frontal superior gyrus
fSI Falciform sulcus of insula
FuG Fusiform gyrus (MOTG: medial occipito-temporalis gyrus)
G Band of Giacomini
GA Gyrus ambiens (or ambient gyrus)
GbPa Globus pallidus of lentiform nucleus (la: pars lateralis or GPe ; m: pars medialis or GPi; a: external medullary lamina ; b: medial medullary lamina (internal medullary lamina); VGbPa: ventral GbPa)
gCC Genu of corpus callosum
gcv Great cerebral vein
gI Granular insula
GR Gyrus rectus
H Hippocampus
h Habenula
Hb Hippocampus body
Hc Hippocampus cortex
Hh Hippocampus head
hit Habenulo-interpeduncularis tract
HOrS H orbital sulcus
HS Hippocampal sulcus
Ht Hippocampus tail
Hy Hypothalamus
hy Hypophysis
HyS Hypothalamic sulcus
hyst Hypophyseal stalk
I Insula
IaG Insular anterior gyri
IC Internal carotid
ICa Internal capsule (cr a: crus anterior: anterior limb – cr p: crus posterior: posterior limb)
Ca′ External capsule
Ca″ Extreme capsule
ICeS Insular central sulcus
IFS Inferior frontal sulcus
ig Indusium griseum
ilcSI Inferior limiting circular sulcus of insula
ilf Inferior longitudinal fasciculus
ilG Intralimbic gyrus
iOFf Inferior occipito-frontal fasciculus
ipc Interpeduncular cisterna
iPFo Interpeduncular fossa
IpG Insular posterior gyri
ipn Interpeduncular nucleus
IPS Intraparietal sulcus
irclco Intercollicular commissure
Ist Isthmus

ITG Inferior temporal gyrus
iThad Interthalamic adhesion
ITR Inferior thalamic radiations (or inferior thalamic peduncle)
ITS Inferior temporal sulcus
in Infundibular nucleus
inr Infundibular recess (r″)
IPS Interparietal sulcus
ivf Interventricular foramen
la Lateral
laa Lamina affixa
laS Lateral cerebral sulcus
- **mlaS** Medial lateral cerebral sulcus
- **plaS** Posterior lateral cerebral sulcus

le Lenticular fasciculus (field H2 of Forel)
lE Lateral eminence
lfC Longitudinal fissure of the cerebrum
lGb Lateral geniculate body
li Limen insulae
liCl Limitans claustrum
linG Lingual gyrus
linS Lingual sulcus
llst Lateral longitudinal stria
LN Lentiform nucleus
LNP Lentiform nucleus peduncle
lolst Lateral olfactory stria
lOrG Lateral orbital gyrus
lOTG Lateral occipito-temporalis gyrus
lpthn Lateral posterior thalamic nucleus
LS Lateral sulcus (b: basal part, l: lateral part)
lt Lamina terminalis
LV Lateral ventricle
LVa Lateral ventricle (atrium)
LVac Lateral ventricle (anterior cornu) frontal horn
LVb Lateral ventricle (body)
LVic Lateral ventricle (inferior cornu) temporal horn
LVpc Lateral ventricle (posterior cornu) occipital horn
MB Mamillary body
mcp Mesocortical pathway (Mc Lean)
MCo Meynert commissure
mD (mgD) Margo denticulatus
me Medial
mE Medial eminence
Mes Mesencephalon (midbrain)
MFB Medial forebrain bundle
MFG Middle frontal gyrus
mGb Medial geniculate body
ml Marginal layer
mL Medial lemniscus
mlp Mesolimbic pathway
mlst Medial longitudinal stria
Mo Medulla oblongata
molst Medial olfactory stria
mOrG Medial orbital gyrus
MSDB Medial septum-diagonal band
MTG Middle temporal gyrus
NIII or N3 Oculomotor nucleus (NIIIp or N3p: pupillar nucleus of Edinger–Westphal)
O1 Superior occipital gyrus
O2 Middle occipital gyrus
O3 Inferior occipital gyrus
O4′ Posterior part of fusiform gyrus
O5′ Posterior part of lingual gyrus
oc Orbital cavity
och Optic chiasma (chiasm)
ocI Olfactory cortex of insula (primary olfactory cortex)
olb (ob) Olfactory bulb
olt Olfactory tract
oltu Olfactory tubercle
olS Olfactory sulcus
OLO Occipital lobe
Opo Occipital pole
or Optic recess
ORa Optic radiation
OrarS Orbital arcuatus sulcus
OrG Orbital gyrus
OrFf Orbito-frontal fibres
ot Optic tract
OTS Occipito-temporal sulcus
paCalS Paracalcarine sulcus
paCeG Paracentral gyrus
paCeS Paracentral sulcus
pACl Pre-amygdalar claustrum
PAG Periaqueductal grey matter
pal Pallium
pAm cx Peri-amygdalar cortex
patG Paraterminal gyrus
PB Pineal body
PCo Posterior commissure
pclp Posterior clinoid process
perica Pericallosal artery
pericS Pericallosum sulcus
pHG Parahippocampal gyrus
- **pHG′** Anterior part of the pHG (pL: piriform lobe)
- **pHG″** Posterior part of the pHG (pL: piriform lobe)

pHy Peduncle of hypophysis (or hyst: hypophyseal stalk)
plp Planum polare
Po Pons
poCeG Post-central gyrus
pOrG Posterior orbital gyrus
pOS Parieto-occipital sulcus
ppc Prepiriform cortex

pPt (or pPT) Pedunculopontine tract
prCeG Precentral gyrus
prhC (or prhc) Perirhinal cortex
prHr Prehippoccampal rudiment
PTR Posterior thalamic radiations (or posterior thalamic peduncle)
PTS Parieto-temporal sulcus
Pu Putamen
 pu′ Lateral part
 pu″ Medial part
pul Pulvinar
puP Putamen peduncle
R Reticular formation
 r′ Optic recess (or)
 r″ Infundibular recess (inr)
rCalS Retro-calcarine sulcus
rCC Rostrum of corpus callosum
rcli Rostral colliculus
 1 Stratum zonale
 2 Stratum opticum
 3 Stratum lemnisci
 4 Deep white layer
rhS Rhinal sulcus
RN Red nucleus
 a Magnocellular part (paleo rubrum)
 b Parvocellular part (neo rubrum)
ro Rostral
rV3 Roof of V3
sas Semi-annularis sulcus
Sbn substantia nigra
 Sbn′ Substantia nigra pars reticulata (or reticular part)
 Sbn″ Substantia nigra pars compacta (or compact part)
sCG Subcallosus (or subcallosum) gyrus
scba Superior cerebellar artery
scbp Superior cerebellar peduncle (brachium conjunctivum)
sCeG Subcentral gyrus
sCS Subcallosus (or subcallosum) sulcus
se Septum
Sepe ou sepe Septum pellucidum
SFS Superior frontal sulcus
Si Substantia innominata
Sip Sulcus intermedius primus of Jensen
sL Spinal lemniscus
SLEAc Central division of sublenticular extended amygdala
SLEAm Medial division of sublenticular extended amygdala
Slf Superior longitudinal fasciculus
 brF Brachium frontalis
 brp Brachium posterius
 bri Brachium inferius
slG Semilunaris gyrus
SMA Supplementary motor area
smaG Supra marginal gyrus
sn Septal nuclei (m: medial; l: lateral)
sOFf Superior occipito-frontal fasciculus
SPG Superior parietal gyrus (prCu)
Spl Subpallium
splCC Splenium of corpus callosum
SPS Superior parietal sulcus
sPS Subparietal sulcus
ssplG Subsplenial gyrus
sss Superior sagittal sinus
sstrg Substriatal grey (substriatal grey matter)
STG Superior temporal gyrus
sTh Subthalamus
sThn Subthalamic nucleus
stm Stria medullaris thalami
STR Superior thalamic radiations (or superior thalamic peduncle)
strt Stria terminalis
STS Superior temporal sulcus
 STS′ Posterior part of STS (horizontal segment)
 STS″ Posterior part of STS (ascending segment)
Sub Subiculum (or sub)
SuOrS Sus orbital sulcus
tTectum of midbrain
tcb Tentorium cerebelli (cerebellar tentorium)
tCC Truncus of corpus callosum
tcf Transverse cerebral fissure
tCl Temporal claustrum
Tg Tegmentum mesencephali
Th Thalamus
 an Anterior nucleus
 van Ventral anterior nucleus
 mvn Medio-ventral nucleus
Thstv Thalamo-striate vein
TLO Temporal lobe
To Temporal opercule
TOI Temporo-occipital incisure
Tpo Temporal pole
Tste Temporal stem
TtAG Temporal transverse anterior gyrus
TtPG Temporal transverse posterior gyrus
TtiS Temporal transverse intermedium sulcus
TtS Temporal transverse sulcus
tu Tuber
tuf Tuber fasciculus
UG Uncinate gyrus
Un Uncus
Una or Un a Uncus apex

UnF Uncinate fasciculus
- **Ip** Isthmic part
- **Fp** Frontal part
- **Tp** Temporal part

Unf Uncal fissure
Unn Uncal notch
Unr Uncal recess
UnS Uncal sulcus
v Vulva (of V3)
V3 Third ventricle
- **V3r′** Optic recess (enter)
- **V3r″** Infundibular recess (enter)

V4 Fourth ventricle
vCl Ventral claustrum
vd Vertical digitation (hippocampus)
ve Ventral
vPa Ventral pallidum
vpw Ventral pathway
VS Ventral striatum
VTA Ventral tegmental area
II Optic nerve
III Oculomotor nerve

MIX
Papier aus verantwortungsvollen Quellen
Paper from responsible sources
FSC® C105338

If you have any concerns about our products,
you can contact us on
ProductSafety@springernature.com

In case Publisher is established outside the EU,
the EU authorized representative is:
Springer Nature Customer Service Center GmbH
Europaplatz 3, 69115 Heidelberg, Germany

Printed by Libri Plureos GmbH
in Hamburg, Germany